SCIENCE AND ART SYMPOSIUM 2000

SCIENCE AND ART SYMPOSIUM 2000

3rd International Conference on Flow Interaction of Science and Art with Exhibition/Lectures on Interaction of Science & Art, 28.2 – 3.3 2000 at the ETH Zurich

Edited by

ALBERT GYR
IHW-ETH Zurich, Switserland

PETROS D. KOUMOUTSAKOS
IFD-ETH Zurich, Switserland and CTR-NASA Ames, U.S.A.

and

U. BURR
IHW-ETH Zurich, Switserland

KLUWER ACADEMIC PUBLISHERS
DORDRECHT / BOSTON / LONDON

A C.I.P. Catalogue record for this book is available from the Library of Congress.

ISBN 0-7923-6358-2

Published by Kluwer Academic Publishers,
P.O. Box 17, 3300 AA Dordrecht, The Netherlands.

Sold and distributed in North, Central and South America
by Kluwer Academic Publishers,
101 Philip Drive, Norwell, MA 02061, U.S.A.

In all other countries, sold and distributed
by Kluwer Academic Publishers,
P.O. Box 322, 3300 AH Dordrecht, The Netherlands.

Printed on acid-free paper

Printed in the Netherlands.

CONTENTS

INTRODUCTION ix

SCART-LECTURE I: Scientific computing visualization – a new venue in the arts
N. Zabuski 1

COLOR PLATE SERIE I 11

VORTICES

Vortex flows and the perception of movement in still images
L. Schouweiler, T. Leweke, M. Lansak and M. Provansal 21

Pairing and acoustics of azimuthally perturbed vortex rings
C.C.K. Tang, R.C.K. Leung and N.W.M. Ko 29

Air entrainment by a wavy plate entering a poole of stagnant liquid
R. Delfos and J. Kockx 35

Numerical simulation and manipulation of the Vortex dynamics In a confined flow
I. Mortazavi 41

ART IS DOMINANT

Double-diffusive Christmas tree
A. Tsinober 51

Power of beauty: Streamwise vortices
N. Yurchenko 59

SCART-LECTURE II: Interaction between the "two cultures": Art & Science
A. Tamir 65

TURBULENCE: COHERENT STRUCTURES

Vortex packets and the structure of wall turbulence – Extended abstract
R.S. Adrian 77

The hidden structure in a round jet
S. McIlwain, A. Pollard and S. Amalfi 79

The interaction between longitudinal vortices in a channel flow visualized through seeding and hydrogen bubble techniques
H. Kocherscheidt, M. Schmidts and V.I. Vasanti Ram 85

TURBULENCE: STATISTIC ASPECTS

Choice of representation modes and color scales for visualization in computational fluid dynamics
M. Farge 91

COLOR PLATE SERIE II 101

Extracting singularities in turbulent flow with real and complex wavelets
M. Haase 111

ART

SCART-LECTURE III: Vortism a basic art
G.E. Koppenwallner and L. Plotter 117

VISUALIZATION AND REPRESENTATION

SCART-LECTURE IV: Gallery of physical phenomena
S. Alekseenko and D. Markovich 127

The role of color in the optical visualization of flows with variable fluid density
W. Merzkirch and N. Ramesh 139

STABILITIES

Direct Navier-Stokes simulations of unsteady vortex breakdown in a swirling jet
P. Chen and E. Meiburg 143

The transition within the subcritical regime of flow over a circular cylinder
K.W. Lo and N.W.M. Ko 149

CONTROLING, BIOFLUIDMECHANICS AND LAMINAR FLOWS

SCART-LECTURE V: Microfluidic system for biological agent detection
Y.-F. Chen, J.M. Yang, J.-J. Gau, C.-M. Ho and Y.-C. Tai 159

Artificial shark skin on its way to technical application
W. Hage, D.W. Bechert and M. Bruse 169

SCART-LECTURE VI: Medusoid vortex rings
P. Koumoutsakos, J. Walther and J. Sagredo 177

Computational anatomy of human lung based on the fluid dynamics
R. Takaki and H. Kitaoka 179

Paradoxes of fish-like propulsion
L. Koryenna 185

Two-phase flow processes in porous media producing geometric patterns
I. Neuweiler and W. Kinzelbach 191

WAVES

SCART-LECTURE VII: Flow patterns in ballet and fluid dynamics
S. Krasnapolskaya and V. Melechko 201

Postcomputational visualization of baroclinic wave drift
V. Goriachev, N. Ivanov and E. Smirnov 211

Art, history, self-organization and coherent structures in fluid mechanics
S.P. Bardakhanov and S.A. Kozlov 217

ELASTICITY AND RHEOLOGY

An experimental and numerical investigation of the free vibration of an elastic cylinder in a cross-flow
R.M.C. So, Y. Liu and Y. Zhou 227

Elongational behaviour of dilute polymer solutions
A: Gampert, T. Eich and C. Wilkes 233

MHD, ASTRO-PHYSICS AND TOPOLOGY

Vision of numerical relativity 1999
W. Benger, H.C. Hege and S. Heusler 239

COLOR PLATE SERIE III 247

The ordering effect of a horizontal magnetic field on liquid metal natural convective flow and its influence on heat transfer
U. Burr 257

Knots and braids on the sun
R.L. Ricca 263

Reconnections of elliptical vortex rings
A. Gyr and A. Müller 269

Vorticity induced by electrostatic forces
W. Egli and E.A. Gerteisen 275

MISCELLANEOUS

Simulation of flow over two unequal square cylinders
A.Y.W. Chiu and N.W.M. Ko 279

List of addresses of the participants 285

COLOR PLATE SERIE IV 291

Catalogue of the art exhibition 301

Walking on air
N. Gyr 303

T. Wildner 305
Y. Netzhammer 306
G. Kousbroek 307
E.N.J. Hobijn 308
Y. Fontijne 310
Z. Leutenegger 311
J. Werner 313

COLOR PLATE SERIE V 315

T. Jansen 325
M. Frei 326
D. Renggli 327
R. Signer 328
A. Ehrat 329
B. Lademann 330
N. Gyr 331
H. Schmid 332

COLOR PLATE SERIE VI 333

Introduction

Some words about SCART 2000.

SCART stands for science and art. SCART meetings are organized in a loose time sequence by an international group of scientists, most of them fluid-dynamicists. The first meeting was held in Hong-Kong, the second one in Berlin, and the third, and latest, one in Zurich. SCART meetings include a scientific conference and a number of art events. The intention is to restart a dialogue between scientists and artists which was so productive in the past. To achieve this goal several lectures given by scientists at the conference are intended for a broader public. In the proceedings they are denoted as SCART lectures. The artists in turn address the main theme of the conference with their contributions. The lectures at SCART 2000 covered the entire field of fluiddynamics, from laminar flows in biological systems to astrophysical events, such as the explosion of a neutron star. The main exhibition by Dutch and Swiss artists showed video and related art under the title 'Walking on Air'. Experimental music was performed in two concerts.
Video and related art was chosen because this medium of expression is very technical and closely related to science and technology. In addition, a common aspect of this kind of art and science is the dimension of time. New ways of possible creative processes were shown with music. The ear decomposes sound by a Fourier analysis, the brain therefore interprets sound in a phase space, whereas images are seen in an analogous form in a physical space. These two impressions are analyzed in two different processing centers in the brain, and it is more than a philosophical question whether we need both types of information to get a reliable grasp of reality. Such ideas were also discussed in two seminars organized by the Collegium Helveticum during the conference. In the proceedings all those interactions cannot be shown, but we hope to transmit a flavor of the interdisciplinary effort, which could stimulate the creativity of the reader as well.

Some words about these proceedings.

These proceedings contain the scientific contributions, ordered by themes, as they were presented at the conference. To show the close relation between science and art, the proceedings also contain the art part which is reproduced as a catalogue of the exhibition. This part is edited by Nadia Gyr, the curator of the exhibition.
Due to this concept and due to the esthetic aspects of many scientific contributions, this book contains a rather extensive representation in colors, 32 pages. It is the hope of the editors, that this book helps in stimulating the dialog between science and art, and that it may also contribute to the continuation of such common efforts.

Zurich, January 2000 Albert Gyr

List of the members of the committees and of our sponsors

International Conference Committee

E. Meiburg Los Angeles (Chairman)

R. J. Adrian	Urbana-Champaign
G. Buresti	Pisa
R. Delfos	Delft
H. Eckelmann	Göttingen
M. Farge	Paris
C.M. Ho	Los Angeles
N. W.M. Ko	Hong Kong
M.Kiya	Sapporo
B.H.K. Lee	Ottawa
R.E. Luxton	Adelaide
A. E. Perry	Melbourne
A. Pollard	Kingston
P. Orlandi	Rom
A. Tamir	Beer Sheba
A. Tsinober	Tel Aviv
I. Wygnansky	Tel Aviv/Tucson

Local Conference Committee

A. Gyr (Chairman)

U. Burr
W. Kinzelbach
P. Koumoutsakos
A. Müller
F.-S. Rys

Patronage Committee

Dr. H.H. Coninx
Publisher TA Media A.G.
Dr. J. Estermann
Lord Mayor of Zurich
Prof. Dr. A. Waldvogel
Vice-president of research & industrial relations of the ETH Zurich

Ackowledgement:

We would like to express our thanks to our sponsors who made this unique event possible.

The ETH Zurich, the Mondriaan Stichting, the State of The Netherlands, the Swiss National Science Foundation, the City of Zurich, the TA Media AG, the MIGROS Kulturprozent and the US Air Force (EOARD).

Scientific Computing Visualization - a new venue in the arts[§]

Norman J. Zabusky
Laboratory for Visiometrics and Modeling
Mechanical & Aerospace Engineering and CAIP
Rutgers, The State University of New Jersey
Piscataway, NJ 08854-8058
nzabusky@caip.rutgers.edu

Abstract: An overview of nearly half a century of involvement with numerical simulations of wave and fluid phenomena has led me to the "visiometric" or visualization and quantification approach that is illustrated in this paper. I believe that the visiometric approach to simulated scientific data sets will provide opportunities for artistic innovation and expression.
Keywords: Visualization, quantification, computer simulation, fluid dynamics, artistic expression, art, computer visualization practitioners.

1. COMPUTER VISUALIZATION PRACTITIONERS, BEAUTY AND ARTISTIC EXPRESSION

This paper focuses on my personal experiences in the visualization and quantification of images in the fluid sciences. I believe that this arena will provide *computer visualization practitioners (CVP)*, that is scientists, programmers and electronic media specialists, opportunities for artistic innovation and expression. As in other artistic areas, many people may participate in producing the final product, but it is the artist who perceives how the data will finally appear and who directs its production. Let me review some general ideas and images that have led me to make this statement.

The great physicist Werner Heisenberg [Heisenberg, 1974] presented a talk to the Bavarian Academy of Fine Arts in 1970 on "The meaning of beauty in the exact sciences"(chapter XIII). Among other stimulating ideas, Heisenberg repeated one definition of beauty: "Beauty is the proper conformity of the parts to one another and to the whole" and affirmed the Latin motto "Beauty is the splendor of the truth."

[§]This paper is dedicated to MARTIN DAVID KRUSKAL on his 75th birthday. Our work together in the 1960's on nonlinear wave phenomena and solitons first opened for me the opportunity to explore the visualization and quantification of simulation phenomena in the sciences.

A. Gyr et al. (eds.), Science and Art Symposium 2000, 1-10.

Subramanian Chandrasekhar (1910-1998), one of the giants of 20 century astrophysics and fluid dynamics discussed aesthetics, creativity and beauty in his book "Truth and beauty in science" [Chandrasekhar, 1987]. Chandrasekhar discussed Heisenberg's paper and commented on the first quotation above, saying , "... it does appear that this definition touches the essence of what we may describe as 'beautiful': it applies equally to 'King Lear', the Missa Solemnis, and the Principia". Chandrasekhar juxtaposed the creations of mathematicians and physicists with composers and writers. However, he did not discuss beauty and art in the works of sketchers, painters, sculptors, photographers, animators and others who deal with visual images.

In his chapter, Heisenberg quotes (on p.179) physicist Pauli (1900-1958):

> The process of understanding in nature, together with the joy that man feels in understanding , i.e., in becoming acquainted with new knowledge, seems therefore to rest upon a correspondence, a coming into congruence of preexistent internal images of the human psyche with external objects and their behavior. ... the place of clear concepts is taken by images of strongly emotional content, which are not thought but are seen pictorially, as it were, before the minds eye... .

Pauli's words, briefly quoted here (see [Jung et al, 1955] for more), move me deeply. I hope some of the examples that I give below will illustrate Pauli's stimulating words on the role of imagery in understanding and joy.

2. A BRIEF HISTORY OF FLUID IMAGERY

The visualization of fluid motion by fluid scientists has been aimed at a physical and mathematical understanding of the processes observed [Zabusky, 1981,1984]. Understanding requires abstract representations and quantifications of the visualizations, which I call "visiometrics", as I discuss below.

To my knowledge the first person to sketch and comment systematically on fluid images was Leonardo da Vinci (1452-1519), the renaissance engineer-scientist and artist. His artistic renderings include "Rock avalanche and water turbulence", a series of ten drawings illustrating an imagined "deluge" caused by rocks tumbling down a mountainside into a lake and producing violent (large amplitude) surface turbulence including many spiral-like features. Leonardo was enormously excited by the diversity and surprise in fluid motions . He showed great visual acuity, including an ability to "freeze-frame" the rapidly varying images, thereby achieving a sense of cogency as discussed below.

The Japanese Ukiyoe (wood block print) artists were also captivated by fluid motion, in particular, Katsushika Hokusai (1760-1849) well-known shown "Great Wave" and Hiroshige Anders' juxtaposition of waves and a large counterclockwise surface vortex in "Naruto". See Fig. 1 on the color plate for a reproduction of the original print (left) and photographs from overhead by M. Nanjo (right upper) and from a zoom on a whirlpool (right lower- from a video as described in [Zabusky et al., 1996]). There is a museum dedicated to Hokusai in Obuse in Nagano Prefecture, Japan which is worth a visit. Besides numerous prints, it contains two wonderful festival wagons whose inner ceilings are adorned with his rendering of what he calls a male vortex and a female vortex.

If one is looking for the "father" of modern flow visualization and quantification, it surely must be the photographer and artist Etienne-Jules Marey, whose pictures from the 1880-1900 period showed "chronophotographic" (stroboscopic) multiple exposures of the motion of the human form (gymnastics, pole vaulter, nude bike rider, ..) and of the animal form (dogs , horses and birds in flight), as well as 1900 smoke visualizations of flow around a sphere and around a curved surface, as well as particle motion in a water wave. A collection of his work was published in 1984 [Frizot, 1984]. (Consult this volume for a complete list of publications by Marey).

Fluid scientists too have been active in visualization. H. Werle contributed an important early modern review of flow visualization [Werle, 1983] and Milton van Dyke published the first systematic compilation of fluid visualizations in 1982 [van Dyke, 1982]. Van Dyke's adequately sized and well rendered black and white photographs illustrate a diversity of fluid processes, mostly from laboratory experiments. Another beautiful collection by the Visualization Society of Japan is the "Fantasy of Flow (The World of Fluid flow captured on Photographs)." It also contains an appendix on laboratory flow visualization methods [Visualization Society, 1993]. In 1983, Hans. J Lugt published "Vortex flow in nature & technology" [Lugt, 1983], which contained a historical and scientific journey through vortex flows with reproductions from classical artists including da Vinci and photographs and sketches of natural phenomena like tornadoes and planetary nebula as well as laboratory experiments. The recent volume by Lex Smits and Tony Lim is devoted to modern techniques of flow visualization in laboratory experiments and numerical simulations [Lim et al., 1999].

Appearing regularly in the September volume of the journal Physics of Fluids are selections from the "Gallery of Fluid Motion." This Gallery is a competition occurring each year at the annual November meeting of the Division of Fluid Dynamics of the American Physical Society. For example, in Vol 11, September 1999, ten pages are devoted to color reproductions of the winning entries of the 1998 competition. These were selected from 39 poster and 22 video entries.

3. COLOR, COLOR MAPS, THRESHOLDS AND ENVIRONMENTS

In this section, I will review briefly some of the things we have done at the Rutgers University, Laboratory for Visiometrics and Modeling. To view the work of the last decade in which we emphasize *feature extraction and tracking for reduced modeling*, see the Vizlab URL: http://www.caip.rutgers.edu/vizlab.html. In particular, click on "Gallery" to find an array of artistic and scientific images, some of which are mentioned but not shown in this paper.

3.1 Color

Artists know that placing regions of complementary colors (red/green, yellow/purple, orange/blue) maximally enhances contrast. See, for example, the English pre-Raphaelite artist W.H. Hunt's picture , "On English Coasts (Strayed sheep)" on p. 39 of [Cole, 1993], where the coats of sheep are enhanced by such contrasts.

Most CVP's, however, use the familiar "rainbow" or "ROYGBV" color map. It is possible that a color map that is modulated in a manner to highlight the data being presented, e.g. using the complementarity of colors, would be more cogent. I use the word cogent rather than beautiful to indicate that when the viewer of a still or of an animation compares the text and mathematical equations with visualizations of the rendered data, he will be able to "entrain" to memory a harmonious relationship for intuition building and future recall.

We have modulated the the color map in an abstraction below to emphasize features in the data. Consider the numerical simulation of the Euler equations in 2d for the classical problem of a shock wave striking a planar inclined interface between materials of different densities, eg air at left and a heavier R22 (refrigerant) at right (density ratio of 3.0). We have a choice of viewing many variables, the primitive (denisty, pressure, velocity and internal energy) or the derived (e.g. vorticity). In Fig. 2 in the color plates, four images at different times are shown, the first where the shock has just completed traversing the inclined planar interface, which here is the blue vortex layer with rolled-up lower end. The interface is unstable (Kelvin-Helmholtz) and begins to roll up locally, as shown in the second and following frames. We see the localized vortices, which formed early, begin to merge. Note the red-yellow (positive) vortex layer near the lower boundary emanates from the triple point of the transmitted shock.

A useful abstraction is a space-time plot of the integrated positive and negative vorticity, as shown in Fig. 3 in the color plates for the [left] slow-fast (density ratio 0.14) and [right] fast-slow interaction (density ratio 3.0) and both have $M = 1.2$ [Hawley et al., 1989]. The vertical time axis is labeled by frame number. The color map was chosen so that low (high) values are blue (red), and the logarithmic scale was obtained by rendering the function with the filter $Asinh^{-1}(y)$. At left, the sharp white-black transition highlights the wall vortex at 6 . At right, the dominant yellow domain is separating into blue regions, after passing through black/white modulation, which emphasizes the localized (blue) vortices.

We emphasize that the dominant wall vortices are "dipolar" because of the antisymmetry of the inviscid lower boundary. These are examples of the ever present "vortex projectiles" present in the Richtmyer-Meshkov environment, as discussed below.

Two other examples of modulation, one in art and the other in science, are the painting by artist Frank Stella "Concentric Squares" and rendering of data by C. A Pickover [Pickover, 1988].

3.2 Threshold

By threshold, I mean a number which is a fraction of the maximum of a function in a particular data set. The choice of threshold determines the area or volume of a function that appears in a frame. In our work, we have used fixed thresholds of the maximum (e.g. of vorticity) in the field. It may be appropriate to change the color map in space or time as the important areas undergo decay or if some other unusual dynamical event occurs. In three dimensions, color may be used to render the isosurfaces of the extracted object according to another variable (e.g. isosurfaces of vorticity colored by the rate of strain function, or direction of the vorticity vector, etc).

4. ANIMATIONS

Animations enhance the perception of unusual phenomena. When properly thresholded, colored and textured some of these animations are artistically exciting and inspiring and they may direct the viewers attention to important regions of space-time.

4.1 Solitons

The first influential animation of a computer simulation of a partial differential equation was the numerical solution of the Korteweg-deVries equation, eg representing the height of waves on a shallow fluid. These simulations led to the discovery of the soliton [Zabusky et al., 1965]. The film was produced by Gary Deem, myself and Martin Kruskal, at Bell Laboratories in Whippany NJ in 1965 and showed the formation and evolution of solitons and the near recurrence to the initial condition, a cosine function. In the 60's and 70's, the film was available on free loan through the Bell Telephone Labs film library [Deem et al., 1965].

The summary of that motion was rendered by a space-time diagram of the dependent variable. In 1965, the data was hand-rendered to Fig. 3 in [Zabusky, 1984] and now computer rendered in Fig. 4 in the color plates. The near-straight reddish streaks of negative slope are the peaks of the strongest solitons as they traverse the periodic domain and interact with each other and undergo phase shifts. The merger of these streaks at about $x = .75(2\pi)$ is the near-recurrence to the initial state. Viewing the original 1965 cinema always delights me for its languid and recurrent motions and its historic significance. Now it is possible to produce this result on a PC or MAC in a matter of seconds!

4.2 Winkler and Kuwahara

Over the years, I have seen several other artistic and inspiring animations of 2d and 3d processes. I was most strongly impressed by the work of Karl-Heinz Winkler in 1987 and Kunio Kuwahara in 1989. In 1986 at Los Alamos National Laboratory, Winkler, an astrophysicist by training, gathered a team about him and did path breaking simulations of 2d axisymmetric high-Mach number jets (motivated by the newly discovered astrophysical jets) and Richtmyer-Meshkov environments (motivated by laboratory shock-bubble experiments of Haas and Sturtevant). Some of these results were shown as animations at many meetings in the period '86-'90, including the Gallery of Fluid Motion in 1986 [Winkler et al., 1987c]. The results were summarized in papers which contained many color images [Winkler et al., 1987a and 1987b].

Kuwahara stimulated the development of CFD internationally in the 80's and 90's with meetings in Japan and by forming a group, iCFD in Tokyo. They studied a multitude of flows in 2d and 3d and made numerous animations. Some were collected on a large disc with accompanying music in [Kuwahara, 1989], which I viewed in his home on a large screen HDTV. Mainly, it contains simulations of flow around a cylinder in 2d up to a Reynolds number of one million! In the animations we see streamfunction, vorticity and motion of incoming arrays of particles. The disc is accompanied by a brochure, with descriptions in Japanese, of numerous colored pictures. I believe these animations will become a classic of the genre because of their impact! Some of these results and more recent ones may be seen at http://www2.icfd.co.jp/.

4.3 Vizlab at Rutgers

In the work of my Vizlab group, we focussed on the collapse and intensification of vortex tubes in 3d, both incompressibly and compressibly. With Richard Pelz, Deborah Silver, Olus Boratav, Victor Fernandez and Thomas Scheidegger, we used pseudospectral codes to study the interaction of originally orthogonally-offset tubes of like and unlike strength in incompressible and compressible environments and Biot-Savart codes to study the collapse of a Lissajous-elliptic vortex ring.

Fig. 5 in the color plates shows a roll up of a weaker vortex tube by a stronger (initially orthogonally-offset), a figure which made the cover of Computers in Physcics [Bitz et al., 1990] and was for a time used as our logo. Slices (as one moves from left to right) through the region where the tubes are almost antiparallel are shown above.

Fernandez studied vortex collapse and reconnection with a Biot-Savart code and single and multifilament rings. Fig. 6 from [Fernandez et al., 1994, 1995a, 1995b and 1996] shows (upper left) three times (time increases from bottom to top) during the collapse of a single filament "Lissajous-ellipse" ring. (The ring is colored by a scalar vortex straining variable, as shown in the color map). In his study of a 512^3 decaying turbulence pseudospectral simulation data base (see [Chen et al., 1992],

Fernandez produced (right upper and lower) isosurfaces and volume rendering of enstrophy and ellipsodal fits as well as vortex lines (red and blue) launched from these regions.

For the Richtmyer-Meshkov environment of a 3d planar-inclined and sinusoidally perturbed interface, Ravi Samtaney has provided the simulation data for the feature extraction of vorticity in Fig. 6 (lower left) in the color plates [Samtaney et al., 1994]. The dominant feature is the wall vortex (red) which has taken on a "hairpin" form. He has also produced numerous animations showing the effects of vortex deposition following shock passage in 2d and 3d.

4.4 Environments: David and Visualization Mariner

In 1987 Francois Bitz and I began to develop "DAVID", a 2d environment for the visualization scientist to assist him in the creative process [Bitz et al., 1990]. Further additions were made by Akos Feher, Ravi Samtaney and Simon Cooper [Feher et al., 1996]. DAVID facilitates extraction, tracking, quantification and mathematization of evolving and interacting amorphous objects from time-dependent multidimensional simulations. DAVID can also provide an interactive way to compare different variables with a two variable distribution function. Although forward looking, it has not been developed since it was posted on the internet in 1996.[†].

Recently, C-W Feng, D. Silver , P. Rona and K. Bemis developed an environment for exploring simulation data sets of underwater coastal regions. The movie I show (see http://www.caip.rutgers.edu/~cwvon/IMCS.html) has two windows: the left shows an Autonomous Underwater Vehicle (AUV) swimming in a coastal area of New Jersey and the right shows the forward view of temperature, velocity, etc variables of the simulation as the vehicle traverses its trajectory, etc.

5. VORTEX PROJECTILES

"Vortex projectiles" (VPs) are something everyone is familiar with, whether it be in the form of a smoke ring from a cigarette, the mushroom cloud following a bomb explosion, or the cascading drops seen after one has carefully released a drop of colored liquid into a glass of still water, and, possibly, the objects seen in the vicinity of supernovae and planetary nebulae that astrophysicists call "proplyds".

The idea that proplyds are VPs occurred to me as an experience of enlightenment when I viewed, in a close interval of time, three images: the laboratory experiment of the collision of two differently colored coaxial and countermoving vortex rings and their break up into a myriad of smaller two-colored rings [Lim et al., 1992]; the colored image [Dohm-Palmer et al., 1996] showing 2D simulations of the break-up of an interface, following a blast wave, into a myriad of dipole-like mush-

[†]See http://www.caip.rutgers.edu/vizlab_group_files/RESEARCH/VISIOMETRICS/DAVID/index.html

room objects; and the colored image from the Hubble Space Telescope by O'Dell and Handron [O'Dell et al., 1996] showing many "knots" (that are now called "proplyds") in the vicinity of the Helix Nebulae.

The Dohm-Palmer & Jones objects are vortex rings, as are the objects that arise when a $M = 10$ shock strikes a heavy spherical bubble, as shown in Fig. 7 in the color plates [Zabusky et al., 1998, Ray et al., 1999]. Here we render the lower half of the vorticity in an axisymmetric computation where (upper frame) the incoming shock has passed the bubble and is moving to the right closely followed and perturbed by a strong "supersonic" coaxial vortex. The positive (blue) vorticity deposited on the lower-half bubble boundary when the shock first passes the bubble is rolling up near the bubble with a negative layer (red) caused by the shock refracted around the backside.This array of rolled up vorticity (lower frame) moves into complex rings of opposite (red and blue) vorticity. These rings will be unstable and decompose into ringlets by mechanisms related to the Crow instability. At least two laboratories are engaged in experiments and simulations where shock waves, formed from powerful laser beams, interact with very small spherical metal spheres. They should observe these phenomena.

In essence, whenever you have an accelerated flow in the presence of a density inhomogeneity or wall, you produce nearby regions of positive and negative vorticity which "bind" into projectiles. That is, the positive and negative vorticity come together, to form dipoles in two dimensions and rings in three dimensions. Once they form, they travel for a while like weakly bound particles.

6. DISCUSSION

How artistic are the scientific images that I have presented? Will an artistic style emerge that captures the excitement and joy that can lead to innovation and discovery? Since antiquity, artistic practitioners have made use of new technological developments which allowed new forms of expression, for example: colored pigments, oil paints, cameras and video recorders. Now the power of the computer can be harnessed for this purpose by ever inventive CVPs. Who knows what art forms will derive from miniature embedded nanochips, high-speed data transmission, or holographic projectors?

ACKNOWLEDGMENTS

My work in the last five years has been supported mainly by Dr. Fred Howes and Dr. Daniel Hitchcock of Department of Energy (Grant DE-FG02-98ER25364) and also by grants from SROA and the CAIP Center at Rutgers university. I acknowledge many years of fruitful collaboration in visiometrics and fluid dynamics with Deborah

Silver and Ravi Samtaney. Also over the last decade many Vizlab graduate students and posdocs have made contributions, particularly V. Fernandez, J. Ray, X. Wang, H-B Yao, C-W Feng, M. Gao, N. Gagvani, and T. Scheidegger. In a fruitful dialogue, my wife Charlotte Fox Zabusky helped me make the presentation more coherent and A. Kotelnikov and Y. Gulak assisted in producing and posting the submitted version of this paper.

REFERENCES

Chandrasekhar, S; 1987, " Truth and beauty in science"; Univ. of Chigago Press.

Cole, A.; "Colour"; 1993;Harper-Collins.

Chen, S.-Y.; Shan, X; 1992; "High resolution turbulent simulations using the connection-machine";*Comput. Physics* **6**, 643-646.

Deem, G. S.; Zabusky, N. J.; Kruskal, M. D.; 1965; Formation, propagation, and interaction of solitons in nonlinear dispersive media"; This 16 mm film had three parts: (1) Solution of the KdV equation with a sinusoidal initial condition; (2) Solution of the modified-KdV equation with a compressive and rerefractive soliton initial condition ; (3) Solution of the nonlinear lattice with a high frequency ("optical") initial condition.

O'Dell, C. R.; Handron, K.; 1996; "Comentary knots in the Helix nebulae"; *Astronomical J.* **111**, 1630- 1639.

Dohm-Palmer; Jones, R.; Jones, T. W.; 1996; "Young supernova remnants in non-uniform media"; *Astrophys. J.* **471**, 279.

Van Dyke, Milton; 1982; "A gallery of fluid motion"; Parabolic Press.

Feher, A.; Zabusky, N. J.; 1996; An interactive imaging environment for scientific visualization an quantification"; *Int. J. of Imaging Systems and Technology* **7**, 121- 130.

Fernandez, V. M.; Zabusky, N. J.; Gryanik, V.M.; 1994; "Near singular collapse and local intensification of a Lissajous- elliptical vortex ring: Nonmonotonic behavior and zero approaching local energy densities"; *Phys. Fluids* **6**, 2242-2244.

Fernandez, V. M.; Zabusky, N. J.; Gryanik, V.M.; 1995 a; "Vortex intensification and collapse of the Lissajous- elliptical ring: Single and multi-filament Biot- Savart simulations and visiometrics"; *J. Fluid Mech.* **299**, 289- 331.

Fernandez, V. M.; Zabusky, N. J.; Bhat, S.; Silver, D.; Chen, S.-Y.; 1995b, "Visualization and feature extraction in isotropic Navier- Stokes turbulence"; In: *Proc. Of AVS'95 Conf.*; Boston, Ma.

Fernandez, V. M.; .;Silver, D.; Zabusky, N. J.; 1996; "Visiometrics of complex physical Processes: Diagnosing vortex dominated flows"; *Computer in Physics* **10**, 463- 470; Also cover illustration (5).

Frizot, M.; 1984; "Etienne-Jules Marey. Introduction, documentation etnoticeses; le Centre National de la Photographie avec le concours du Ministere de la Culture; 60 plates; ISBN 2-86754-013-5.

Havley, J. F.; Zabusky, N. J.; 1989; "Vortex paradigm for shock accelerated density-stratified Interfaces"; *Phys. Rev. Letter* **63**, 1241- 1244.

Heisenberg, W.; 1974; "Across the frontiers"; Harper and Row.

Jung, C. G.; Pauli, W.; 1955; " The influence of archetypal ideas on scientific theories of Kepler"; In: *The interpretation of nature and the psyche*; pp. 151-212; Bollingen Series LI, Pantheon, N.Y.; See pp. 151-153.

Kuvahara, K.; 1989; " Dynamics of fluid motion [Cylinder]";Music: Part I, by Harold Budd and Haraaji; Part II, by Henry Wolff & Nancy Hennings; Part III, by Sohmei Sato. Also the accompanying pamphlet; ISBN 4-98706-008-4C3842.

Lim, T. T.; Nickels, T. B.; 1992; "Instability and reconnection in the head on collision of two vortex rings"; *Nature* **357**, 225-227.

Lim, T. T.; Smits, A.; 1999; "Flow visualization: Techniques and examples; Imperial College Press; To be published in 1999.

Lugt, H. J.; 1983; "Vortex flow in nature & technology" John Wiley & Sons.

Pickover, C.A.; 1988; "A note on rendering chaotic 'repeller distance towers'"; Computers in Physics; May/June pp. 75.

Ray, J.; Samtaney, R; Zabusky, N. J.; 1999; "Circulation deposition and emergence of vortex clusters from shock-bubble interactions"; In: *Proc. 2nd Intern. Workshop of Laser-Astrophysics*; to be published.

Samtaney, R.; Silver, D.; Zabusky, N. J.; Cao, J.; 1994; "Visualizing features and tracking their revolution"; *Computer*; (a publication of IEEE); pp. 20-27, July

Visualization Society; 1993; "Fantasy of flow (The world of fluid flow captured on photographs)"; ISBN 90-5199-138-x.

Werle, H. ; 1973; "Hydrodynamic flow and visualization"; In: *Ann. Rev. Fluid Mech.* **5**, 361-382.

Winkler, K. H.; Chalmers, J.; Woodward, P. L.; Zabusky, N. J.; 1987 a; "A numerical laboratory"; *Phys. Today*, 28-39.

Winkler, K. H.; Hodson, S.; Chalmers, J.; McGowen, M.; Tolmie, D. E.; Woodward, P. L.; Zabusky, N. J.; 1987 b; "Ultra speed graphic for computational science"; *Cray Channels*; summer pp 4-9.

Winkler, K. H.; Chalmers, J.; Hodson, S.; Woodward, P. L.; Zabusky, N. J.; 1987 c; *Phys. Fluids 30*, pp. 2605; VCR cinema winner at the Gallery of Fluid Motion; APS DFD meeting in Colombus Ohio, 11/86.

Zabusky, N. J.; Kruskal, M.D.; 1965; "Interaction of 'solitons' in a collisionless plasma and the recurrence of initial states"; *Phys. Rev. Lett.* **15**, 140-143.

Zabusky, N. J.; 1981;"Computational synergetics and mathematical innovation"; *J. Computational Physics* **43**, 195-249.

Zabusky, N. J.; 1984; "Computational synergetics"; *Phys Today*, 34-45 July.

Zabusky, N. J.; Boratav, O. N.; Pelz, R. B.; Gao, M.; Silver, D.; Cooper, S. P.; 1991; "Emergence of coherent patterns in vortex stretching during reconnection: A scattering paradigm"; *Phys. Rev. Lett* **18**, 2469-2472.

Zabusky, N. J.; Townsend, W.; 1997; "Naruto: Past and present"; *Phys. Fluids* **9**, pS8.

Zabusky, N. J.; Zeng, S.-M.; 1998; "Shock cavity implosion morphologies and vortical projectile generation in axisymmetric shock-spherical F/S bubble interactions"; *J. Fluid Mech.* **362**, 327-346.

Figures from

Scientific Computing Visualization - a new venue in the arts

Norman J. Zabusky

Figure 1: Naruto old and new. At left is Hiroshige's print and at right upper is an aerial photograph by M. Nanjo which appeared in [Zabusky et al., 1997]. At right lower is a zoom on a vortex whirlpool, an excerpt from the film by Townsend and Zabusky, shown at the APS/DFD Gallery of Fluid Motion in 1996.

Figure 2: Vorticity at four times produced when shock of Mach = 2 strikes a 60 degree inclined planar interface between air and freon (density ratio of 3.0). Note, the unstable roll-up of the blue/negative vorticity on the interface and merger at later times of the localized vortex structures. Also note the initially red-yellow/positive near-sheet of vorticity-emanating from the triple point in the heavier medium. Figure by R. Samtaney and published with his permission.

Figure 3: Space-time diagram of the y-integrated vorticity following a shock of Mach number = 1.2 striking an inclined interface. See [Hawley et al. 1989]. *Left*: slow/fast or heavy-to-light interaction and density ratio of 0.14. *Right*: fast/slow or light-to-heavy interaction and density ratio of 3.0.

Figure 4: A space-time diagram of the solution of the Korteweg-deVries equation with a cosinusoidal initial condition that produces nine solitons. These are manifest as the red-orange to blue streaks from the region of the x-axis associated with the maximum of the cosine function.

Figure 5: Vorticity isosurfaces showing roll up of a weaker vortex tube (initially vertical) by a stronger (initially horizontal and orthogonally-offset). Above, are shown four slices of the three dimensional volume (as one moves from left to right) through the region where the tubes are almost antiparallel.

Figure 6: *Upper left*: Evolution toward collapse of a single-filament, initially a Lissajous-elliptic vortex ring. Time proceeds from bottom to top. The ring is colored by $\omega \cdot \nabla \mathbf{u} \cdot \omega$; *Right*: 64^3 domain from a 512^3 data set showing an antiparallel configuration fitted in places with ellipsoids. The lines trace the vortex field from the center slice through the ellipsoid (five from each, colored red in one direction and blue in the other.) *Lower left*: Vorticity isosurfaces of tracked objects following shock passage of an inclined interface with a sinusoidal perturbation in the direction orthogonal to the incline.

Figure 7: Vorticity at two times from an axisymmetric simulation following and $M = 10$ shock passing over a heavy spherical bubble of density ratio ten. Only half the field is shown with blue positive and red negative. In the top image we see the complex near-axis supersonic ring and in the lower frame we see the complex

Disassembly of the structure near the interface above. This figure was provided by R. Samtaney from simulations with his EFM code and is published with his permission. See [Ray et al., 1999].

Arkady B. Tsinober

Double-Diffusive Christmas Tree

Figure 1; *Various stages of formation of the Christmas tree flow pattern:* a) -12s, *with a close up of the cap, b)* - 185s, *c)* - 1020s, and d) - 2110s. *Shearing motions, vortices and salt fingers:* e) - 2805s, f) - 2950s, *g)* - salt fingers in the second layer.

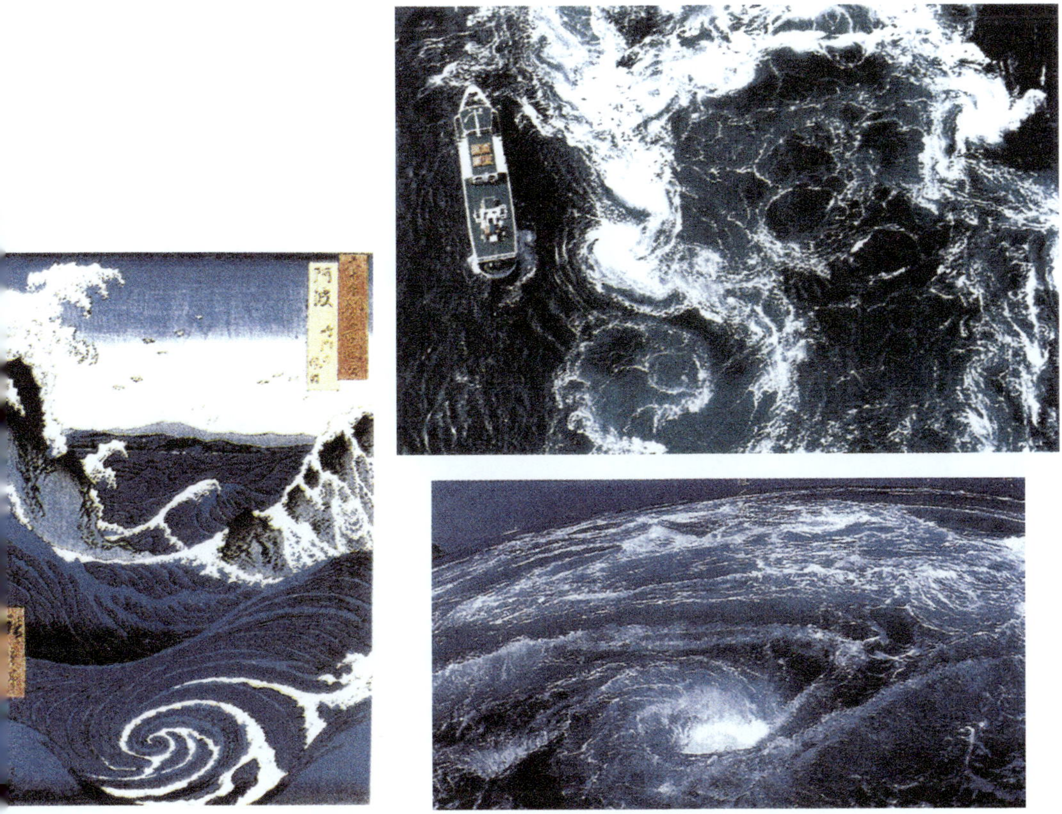

Figure 1

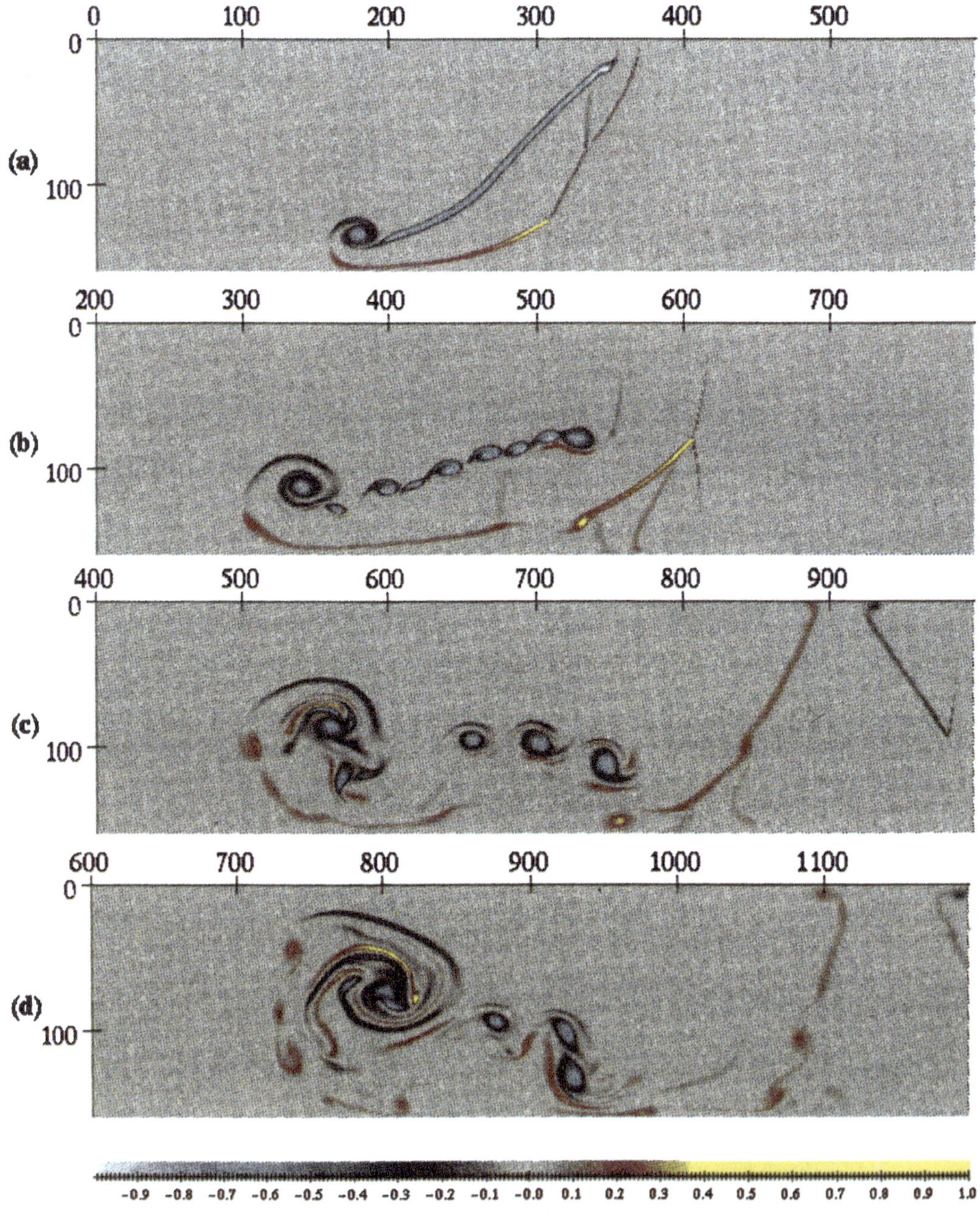

0
100
200
300
400
500
(a)
0
100
200
300
400
500
600
700
(b)
0
100
400
500
600
700
800
900
(c)
0
100
600
700
800
900
1000
1100
(d)
0
100
-0.9
-0.8
-0.7
-0.6
-0.5
-0.4
-0.3
-0.2
-0.1
-0.0
0.1
0.2
0.3
0.4
0.5
0.6
0.7
0.8
0.9
1.0

Figure 2

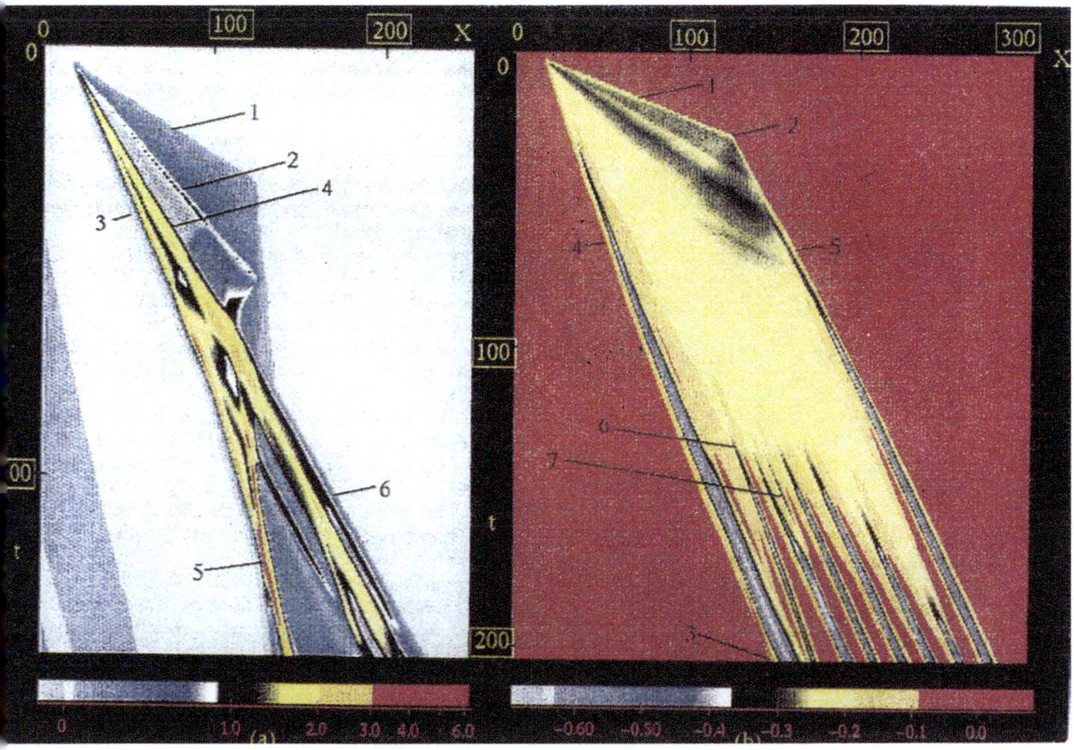

0
100
200
X
0
100
200
300
X
t
t
1
2
3
4
5
6
7
(a)
(b)
0
1.0
2.0
3.0
4.0
6.0
-0.60
-0.50
-0.4
-0.3
-0.2
-0.1
0.0

Figure 3

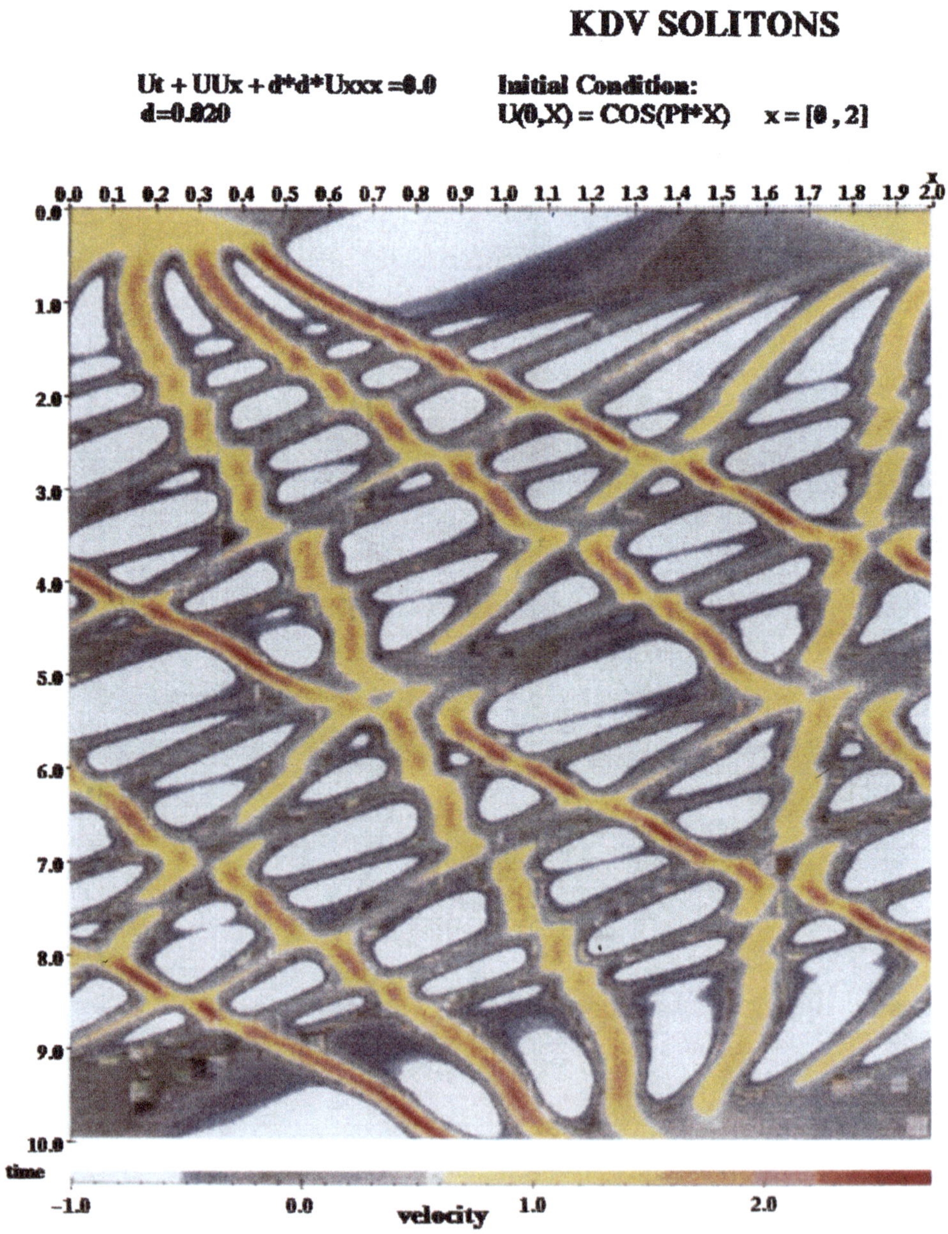
KDV SOLITONS
Ut + UUx + d*d*Uxxx =0.0
d=0.020
Initial Condition:
U(0,X) = COS(PI*X) x = [0 , 2]
0.0 0.1 0.2 0.3 0.4 0.5 0.6 0.7 0.8 0.9 1.0 1.1 1.2 1.3 1.4 1.5 1.6 1.7 1.8 1.9 2.0
x
0.0 1.0 2.0 3.0 4.0 5.0 6.0 7.0 8.0 9.0 10.0
time
-1.0 0.0 1.0 2.0
velocity

Figure 4

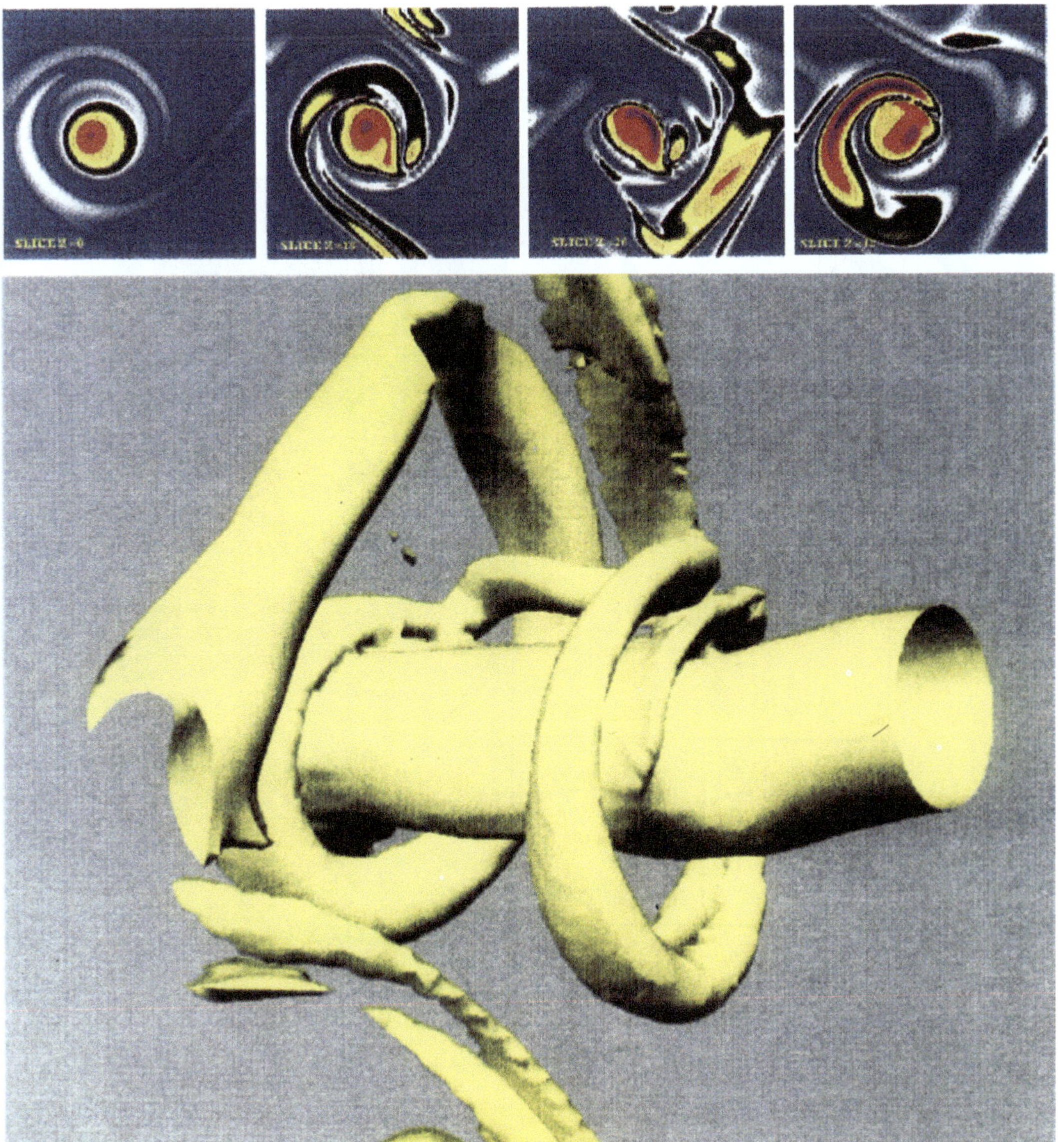

Figure 5

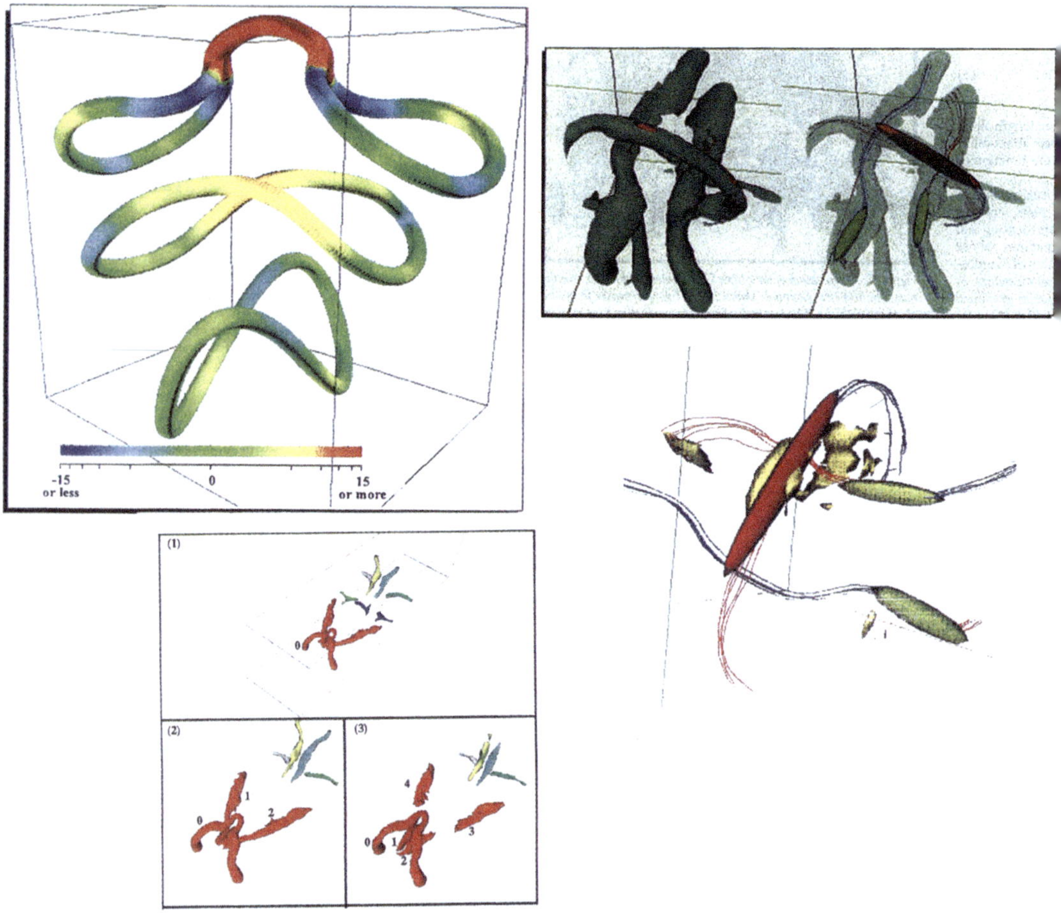

Figure 6

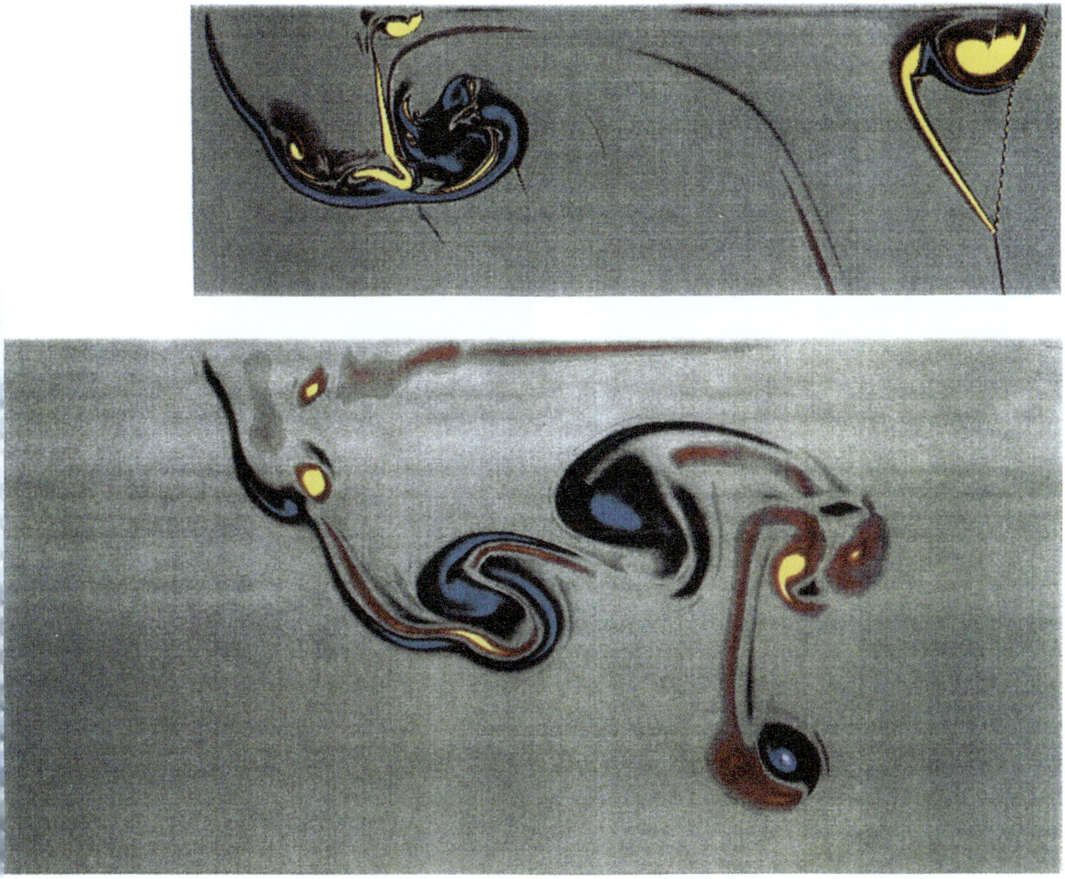

Figure 7

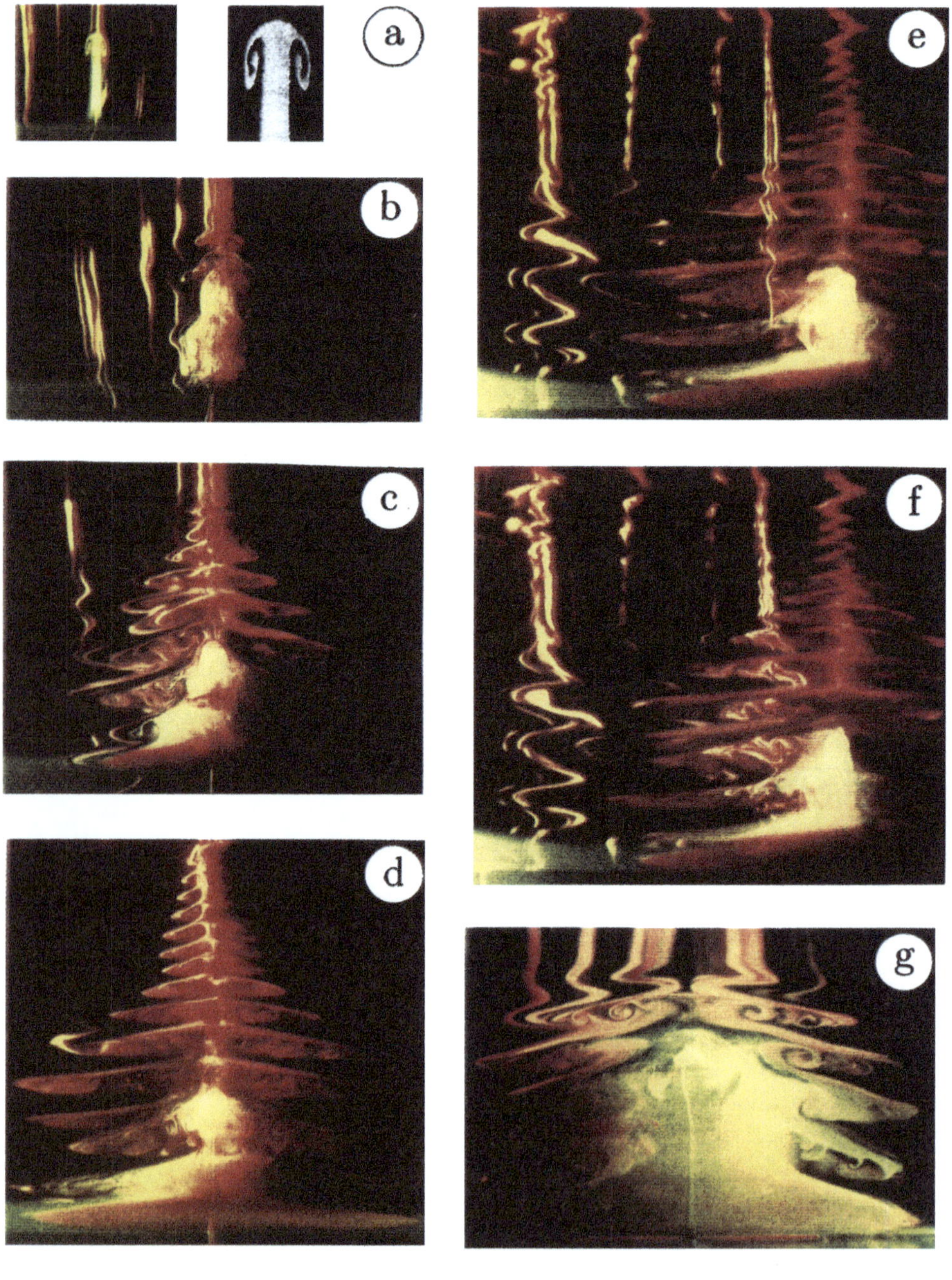
a
b
c
d
e
f
g

Vortex flows and the perception of movement in still images

L. Schouveiler[1], T. Leweke, M. Lansac[2] & M. Provansal
[1] Institut de Recherche sur les Phénomènes Hors Equilibre
Faculté de Saint Jerôme, Service 252
F-13397 Marseille Cedex 20, France
Lionel.Schouveiler@lrc.univ-mrs.fr
[2] Plastic Artist - L'Atelier de la Goutte d'Eau
Le Vallon du Frère, F-13114 Puyloubier, France

Abstract: Still images extracted from flow visualisations can sometimes convey a strong impression of movement. This feature is particularly evident for spiral dye patterns created by vortex flows and for periodic patterns. The confrontation of such scientific visualisations with artwork shows how artists can take advantage of this visual property for artistic purposes.
Keywords: perception of movement, spiral patterns, repetitive patterns.

1. INTRODUCTION

In this contribution, we discuss a visual property that can be found in scientific flow visualisations: the communication of motion in still images. This topic is related to a practical aspect of scientific work, namely the fact that, despite the increasing number of multimedia journals (on CD-ROM or via the Internet), the vast majority of scientific results are so far still published on paper. Since this "still" medium does not allow for animated images, the standard way to visually represent the motion of a fluid flow is to show a series of visualisations taken at successive instants in time. However, there exist certain flow structures for which a single visualisation image is enough to convey the information on the associated fluid motion. To illustrate this feature, we present flow visualisations extracted from diverse recent studies.

For the same reason of using a "frozen" medium, this notion of perception of movement can also be of concern to the plastic artist. In this context, the scientific images we present and their visual properties are discussed in parallel with artwork created by one of the authors (M. Lansac). These artworks show how expressing motion using visual flow properties can also be used for artistic purposes, and the confrontation reveals that the boundary between art and science can sometimes be very blurred. The artistic images presented in this paper are entitled "Turbulent landscape", they are part

A. Gyr et al. (eds.), Science and Art Symposium 2000, 21-28.

of the series of works: "Ecopoiesis 99, Water: adventurous coherences" and are introduced by the artist as follows:

"What touches me the most in the world is the perception of the movement engendering living forms.

Its constructive energy shows itself in its fluidity.

From this movement, always present, immense or tiny, forms and patterns are born, from the haziest to the most distinct, from the most secret to the most legible.

Fluids have properties, laws. Their dynamics seem to be represented by water on our planet.

Is water, carrier of the particles of life, the cradle of the world?

Cradle, place of dissolution and rest, and sculptor of life by its movement along its own river-bed."

The artworks were created using a technique inspired by the method of making so-called "marbled paper". A flat recipient is filled with a support fluid, namely rain water, whose surface is sprinkled with a few drops of colourings such as black ink. The artist then creates certain types of motion in the fluid using for example a thin needle. The unique dye pattern is captured, generally after all motion has died out, on a sheet of paper placed on the surface of the water. The scientific flow visualisations have been carried out in water and, excepted for Figure 3, they are obtained by injection of fluorescent dyes for which the Schmidt number is, in water, of the order of 2000. High values of the Schmidt number are also characteristic of the inks used on water by the artist, in such a way that the resulting patterns are more influenced by the artist's hand, and less so by the dynamic of the fluid. Moreover, they allow to freeze the pattern a long time after all motion in the support fluid has died out.

We present, respectively in section 2 and 3, two types of flows for which the perception of the movement is "self contained" in the resulting dye patterns, namely spiral and periodic patterns, both being due to vortical structures. For such flows, we illustrate also counter-examples where this perception of motion is misleading.

2. SPIRAL PATTERNS

2.1 Spirals with motion

Probably the most striking examples of patterns giving an impression of movement are spirals. This feature is clearly illustrated by the images of Figure 1. Figure 1 (left) shows a dye visualisation of an isolated vortex in a plane perpendicular to the vortex axis. The visual impression of rotation is very strong, and the general structure of the fluid flow is immediately apprehended. This is caused by the spiral structure of the dye pattern, which is an intrinsic feature of vortex visualisations, and which can be seen in a variety of more or less familiar situations involving rotation of a fluid: water swirls, tornadoes, galaxies, tea leaves on the bottom of a cup, etc. The flow field of a real (concentrated) vortex is characterised by differential rotation (except close to the vortex centre), i.e. the angular velocity of the fluid varies with the distance from the axis. This differential rotation and the associated strain deform any initially non-axisymmetric dye distribution into a spiral pattern (see e.g. [Flohr & Vassilicos, 1997]. The flow in Figure 1 results from the merging of two initially separate, parallel, co-rotating vortices dyed with different colours. These vortices were generated in water using two sharp-edged flat plates moved impulsively from rest [Meunier & Leweke, 1999]. It is interesting to note that the associated streamline pattern, which is often used to describe fluid motion in other situations with non-rotating flows, does not have such a strong effect on the observer here. For vortices, this pattern consists of closed concentric circles. Although the dye lines of the two images in Figure 1 are almost circular, the fact that they are not closed significantly enhances the impression of rotation in this image.

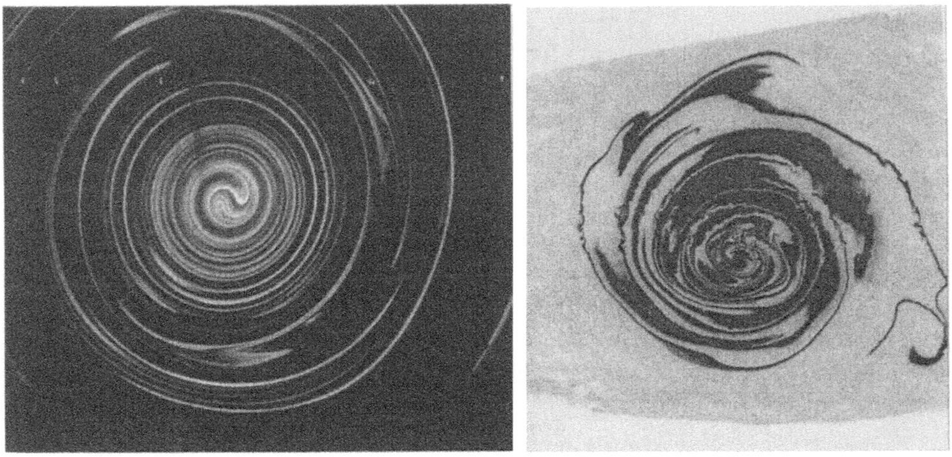

***Figure 1**; Spiral patterns giving a strong impression of rotating motion.*
Left: Visualisation of the flow resulting from the merging of two co-rotating vortices,
Right: Artwork, black ink on rice paper.

The expression of motion using vortices and spirals can also be recovered in Figure 2. In fact, among the different elements visible in this figure, the spirals are clearly the ones that most express movement. Figure 2 (left) shows an experimental dye visualisation, presented here as a negative image. The initial flow consists of two counter-rotating vortices extending from left to right and travelling downwards. The image visualises the plane between the two vortices, which initially contains no dye, after the development of a three-dimensional elliptic instability of the vortex pair [Leweke & Williamson, 1998]. This results in the formation of perpendicular secondary vortices visible here. The comparison and confrontation of the visualisation of this totally unforced flow with the creative artwork in Figure 2 (right) is quite remarkable. They show how the idea of expressing motion using vortices and spirals can also be used for artistic purposes.

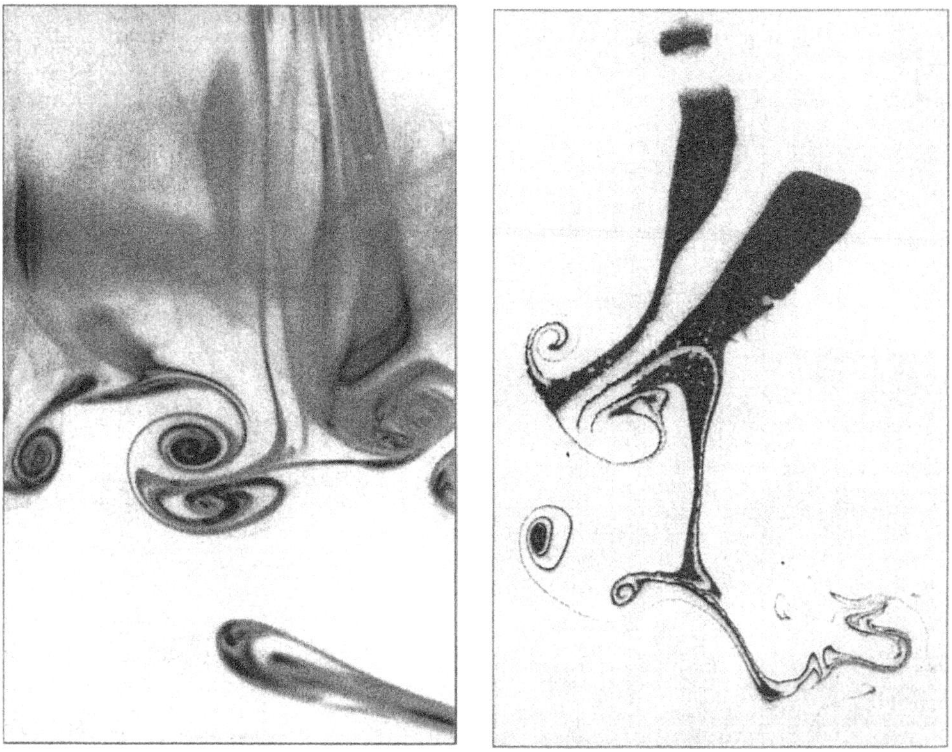

Figure 2*; Left: Visualisation of the late stages of a counter-rotating vortex pair undergoing elliptic instability, Right: Artwork, black ink on rice paper.*

2.2 Spirals without motion

The spiral patterns can, of course, sometimes be misleading. This can be illustrated with a more complex vortical flow, a visualisation of which is shown in Figure 3. The flow is confined between a stationary and a rotating disk, both of radius R, and separated by a gap of width h. Visualisation is achieved by adding anisotropic particles (small mineral mica platelets coated with tin oxide and titanium dioxide) to the water. In this way, the alternate dark and bright bands, in Figure 3, visualise vortices [Savas, 1985] whose curved axis lie in the plane of the image. These vortical structures result from instabilities of the boundary layer of the stationary disk [Schouveiler et al., 1999]. A large variety of vortex patterns in the form of spiral and circular rolls can be observed, sometimes simultaneously, in these flows depending on the parameters of the flow (Reynolds number Re, aspect ration h/R). Whereas the spirals again give a distinct impression of rotating motion, in the same direction (clockwise) as the moving disk, the circles do this to a much lesser extent. In addition, the latter certainly do not convey an impression of radial motion in this still image.

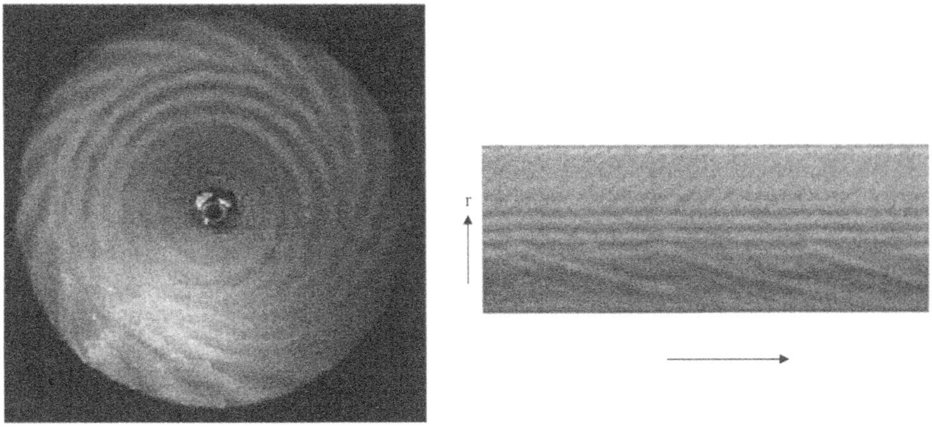

***Figure 3**; Left: Roll-like pattern due to instabilities of the flow between a stationary and a rotating disk (Re=17200, h/R=0.1143), Right: Space-time plot (for Re=46800, h/R=0.050).*

However, experiments have shown not only that the circular vortices always move radially inwards, but also that travelling as well as stationary spiral patterns can be observed. The construction of spatio-temporal plots, as presented in Figure 3 (right), allows to exhibit, and to analyse, the correct dynamics of such roll-like structures [Schouveiler et al., 1998]. These plots are obtained by extracting a line of pixels, taken on a radius, from successive digitised video images. The space-time diagram of the Figure 3 reveals a flow regime with a spiral and a circular roll-like pattern, travelling respectively outwards and inwards, which coexist with stationary spiral vortices whose the trace appear as bands parallel to the time axis.

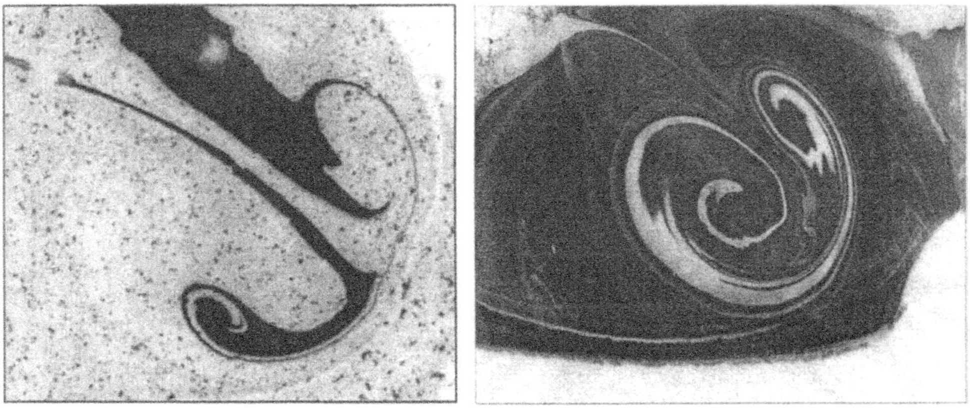

***Figure 4**; Paradoxical impression of movement. Artworks, black ink pattern on rice paper captured on water in motion (left), on still water (right).*

The perception of movement can also appear paradoxical in artistic productions as illustrated in Figure 4. In fact, although these two images seem very similar and convey an impression of movement due to the presence of spiral patterns, as discussed in section 2.1, it is noteworthy that the left one, contrary to the right one, has been captured with the fluid still in motion.

3. REPETITIVE PATTERNS

Repetitive, or periodic, structures are other examples of patterns in which the impression of movement is self-contained. Such a structure is obtained in fluid mechanics by the dye visualisation, as presented in Figure 5 (top), of the well known Kármán vortex street behind a circular cylinder placed in a uniform flow perpendicular to its axis. Another illustration of a periodic pattern is given in the lower image in this figure. It is extracted from series of experiments on the coupled wakes of two spheres, of same diameter *d*, whose centres are aligned perpendicular to a uniform flow and separated by a distance *h* [Leweke et al., 1999a]. The system is then controlled by two parameters: the Reynolds number *Re* and the non-dimensional spacing *h/d*. Whereas for large values of the spacing, we recover two separate sphere wakes characterised by an one-sided periodic shedding of vortex loops [Leweke et al., 1999b] above a critical *Re*, when *h/d* is less than 1.3 we observe the formation of a large scale wake with alternate shedding of the vortex loops which seems very similar to the Kármán vortex street.

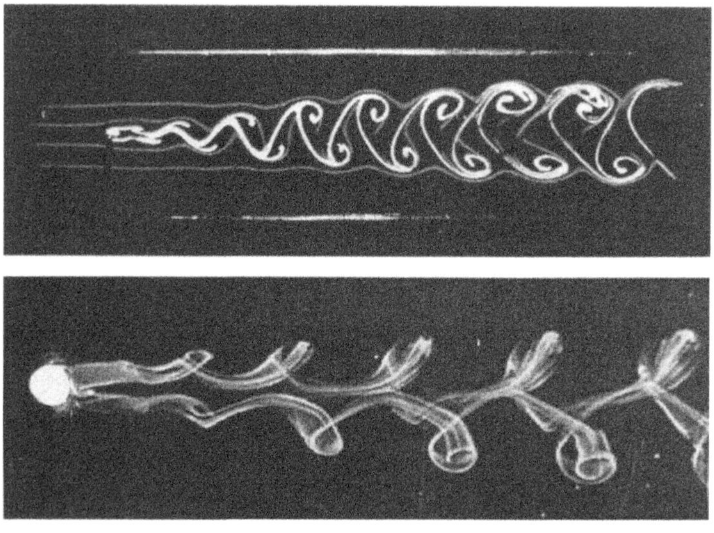

Figure 5; *Repetitive patterns. The von Kármán vortex street: Re=64 (top) and coupled wakes of two spheres: h/d=1.0, Re=360 (bottom).*

In these images, the perception of motion is due to the repetition and spreading of the same pattern, it is enhanced, in the upper one, by the visualisation of the outer streamlines, according to a process greatly used in comic strips.

The wake visualisations (Figure 5) seem to indicate that the vortices created in the near wake persist up to distances of several hundred cylinder diameters downstream. However, it is only the spiral pattern of the smoke lines that persists and which gives this impression. The actual vortices have diffused away well before they reach 100 diameters downstream [Cimbala et al., 1988]. The two scientific examples mentioned here, together with the artistic images of Figure 4, relate to a subtle and sometimes much debated issue in fluid mechanics: the correct interpretation of flow visualisations.

The perception of movement through repetition also appears clearly in the artistic illustration in Figure 6.

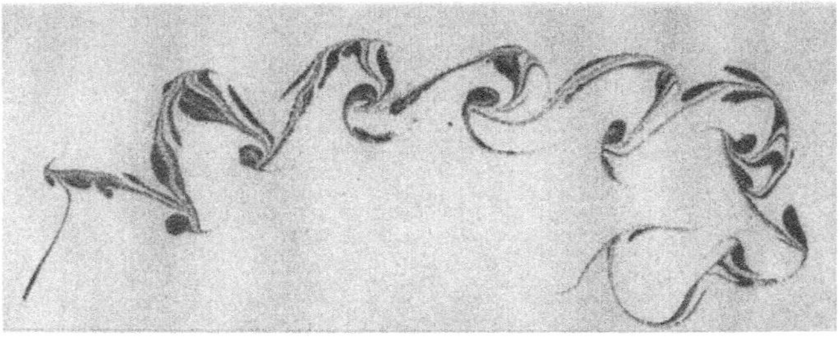

Figure 6*; Repetitive patterns in artworks: Black ink on water-colour paper.*

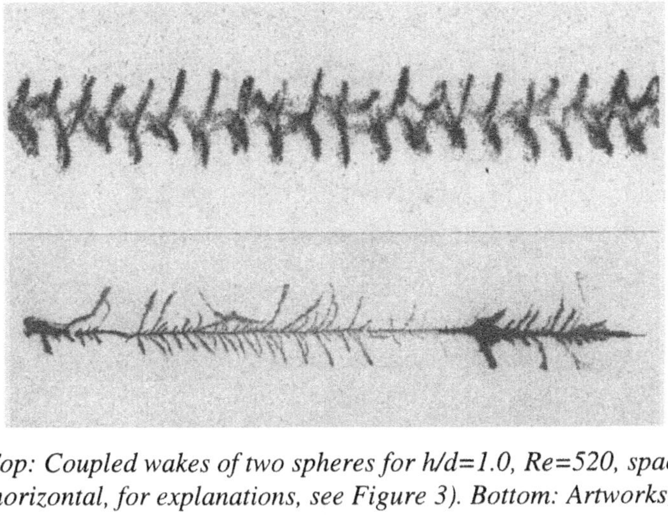

Figure 7*; Top: Coupled wakes of two spheres for h/d=1.0, Re=520, space-time plot (time axis is horizontal, for explanations, see Figure 3). Bottom: Artworks, black ink on water-colour paper.*

In contrast, the two images in Figure 7 do not give the impression of motion, in spite of the presence of a repetitive pattern. In particular, the lower artistic image exhibits plant-like structures, as found for example in ferns, which intuitively are not associated with self-motion.

4. CONCLUSION

In summary, we have tried to illustrate how the spiral and repetitive patterns typically generated by vortical flows create the strong impression (which is sometimes an illusion) of motion, even in a single still image, and that this feature can be used very effectively in artistic expressions.

On the artistic side, certain types of artwork contain ingredients similar to those found in visualisations of vortex flows, and which are deliberately used to express motion. Sometimes, actual fluid vortices can even be used in the artistic process as illustrated here.

REFERENCES

Cimbala, J.M.; Nagib, H. M.;Roshko, A.; 1988; "Large structure in the far wake of two-dimensional bluff bodies"; *J. Fluid Mech.* **190**, 265.

Flohr, P.; Vassilicos, J. C.; 1997; "Accelerated scalar dissipation in a vortex"; *J. Fluid Mech.* 348, 295.

Leweke, T.; Williamson, C. H. K.; 1998; "Cooperative elliptic instability of a vortex pair"; *J. Fluid Mech.* **360**, 85.

Leweke, T.; Ormieres, D.; Provansal, M.; Schouveiler, L.; 1999 a; "Visualisations des vortex déversés dans le sillage d'une ou deux sphères"; In *Visualisation et Traitement d'Images en Mécanique des Fluides (Hérbrard, P. ed.)* 103, ENSICA, Toulouse.

Leweke, T.; Provansal, M.; Ormieres, D.; Lebescond, R.; 1999 b; "Vortex dynamics in the wake of a sphere" *Phys. Fluids.* **10**, Gallery of Fluid Motion, S12.

Meunier, P.; Leweke, T.;1999; "Etude expérimentale de la fusion de tourbillons corotatifs"; *Visualisation et Traitement d'Images en Mécanique des Fluides (Hérbrard, P. ed.)* 183, ENSICA, Toulouse.

Savas, O; 1985; "On the flow visualization using reflective flakes"; *J. Fluid Mech.* **152**, 235.

Schouveiler, L.; Le Gal, P. Chauve, M. P.; 1998; "Stability of a travelling roll system in a rotating disk flow", *Phys. Fluids.*, **10**, 2695.

Schouveiler, L.; Le Gal, P. Chauve, M. P.; 1999; "Spiral and circular waves in the flow between a rotating and a stationary disk"; *Exp. Fluids* **26**, 179.

Pairing and Acoustics of Azimuthally Perturbed Vortex Rings

C. C. K. Tang[1*], R. C. K. Leung[2] and N. W. M. Ko[1+]

[1]*Department of Mechanical Engineering, The University of Hong Kong, Pokfulam Road Hong Kong, People's Republic of China.*

[*]*clief.tang@graduate.hku.hk*

[+]*nwmko@hkucc.hku.hk*

[2]*Department of Mechanical Engineering, The Hong Kong Polytechnic University, HungHom, Kowloon, Hong Kong, People's Republic of China.*

mmrckl@polyu.edu.hk

Abstract: The study of the interaction of the perturbed thin vortex rings with phase differences of $\Delta\gamma = 0$ and π for the wavenumber $m = 4$ are investigated numerically. The flow is assumed inviscid and incompressible. The model of the thin vortex rings is based on the vortex blob method, which gives the global dynamics of the vortices. By using acoustic analogy, the far field sound pressure of the evolution of the vortex system is found. Results show that, the phase difference of the two perturbed vortices causes different development of the vortex patterns of the vortex rings, resulting in instability and higher the sound pressures of the vortex system.
Keywords: vortex ring, sound and vortex method.

1. INTRODUCTION

The study of vortex flow has been given much attention in the present decade. In order to understand the importance of the coherent structures and their effect on the global statistical properties of the flow, it is desirable to analyse their flow dynamics in basic flows containing small number of interacting vortices (Leweke & Williamsin 1998). For low Mach number flow in unbounded fluid the source of aerodynamic sound is related to the flow vorticity which distribution is assumed to be confined in a finite

A. Gyr et al. (eds.), Science and Art Symposium 2000, 29-34.

region (Powell 1964 and Kambe & Minota 1981). It is commonly known that in a slip through or leapfrogging process the leading vortex ring expands and decelerates, while the trailing one contracts and accelerates. The trailing vortex ring catches up and slips through the leading ring. After the slip through, they exchange their role and repeat the slip through process.

Leung & Ko (1997) and Ko et al. (1999), based on the vortex blob method (Knio & Ghoniem 1990) and acoustic analogy (Möhring 1978), studied numerically the interactions of two coaxial vortex rings of thin cores with and without azimuthal instability of wavenumbers $0 \leq m \leq 11$, where m is the number of waves along the vortex. The far field sound generated by these interacting vortex rings of weak axial and radial sinusoidal perturbations was also studied. As the studies of Leung & Ko (1997) and Ko et al. (1999) concerned only the vortex rings with no phase difference between them, thus, the aim of this paper is to understand the interactions of the perturbed vortex rings at phase differences between the vortex rings and their far field sound generation.

2. MATHEMATICAL FORMULATION AND NUMERICAL METHOD

The flow is assumed inviscid and incompressible. The model is to investigate the effect of weak three-dimensional perturbation on the slip through of two coaxial vortex rings. The thin vortex ring is modelled by using a slender vortex filament in unbounded space, ignoring the role of core structure dynamics during the vortex evolution. It only gives the global dynamics of the vortex (Knio & Ghoniem 1990). The three-dimensional sinusoidal perturbation of each thin vortex ring is taken to be

$$\mathbf{Y} = (R + \rho_r e^{im(\theta+\gamma)})\mathbf{e}_r + (z + \rho_z e^{im(\theta+\gamma)})\mathbf{e}_z, \tag{1}$$

where $\boldsymbol{e}_r$ and $\boldsymbol{e}_z$ are the unit vectors in the radial and streamwise directions, θ the azimuthal angle, γ the phase of the perturbation wave, R the radius of an unperturbed ring, z the initial streamwise distance from a reference frame and m the integer wavenumber of complete wave along the filament (Figure 1). The respective radial and streamwise perturbation amplitudes, ρ_r and ρ_z, are assumed very small in compared with the ring radius, i.e. $|\rho_r|$, $|\rho_z| << R$.

Based on the three dimensional vortex blob method (Knio & Ghoniem 1990 and Winckelmans & Leonard 1993), the discretized Biot-Savart Law is

$$\mathbf{u}_w(\mathbf{y},t) = -\frac{1}{4\pi}\sum_\alpha \Gamma_\alpha \frac{(\mathbf{y}-\mathbf{y}_\alpha)\times\delta\mathbf{s}_\alpha}{|\mathbf{y}-\mathbf{y}_\alpha|^3}\chi\left(\frac{|\mathbf{y}-\mathbf{y}_\alpha|}{\delta}\right), \tag{2}$$

where

$$\chi(r) = 4\pi \int_0^r \zeta(\eta)\eta^2 d\eta = \frac{r^3(r^2+5/2)}{(r^2+1)^{5/2}}. \tag{3}$$

The discretized Möhring's sound pressure fluctuation formula (Möhring 1978) is

$$p'(\mathbf{x}) = \frac{\rho_o}{12\pi c_o^2 |\mathbf{x}|^3}\frac{\partial^3}{\partial t^3}\left\{\sum_\alpha \Gamma_\alpha (\mathbf{x}\cdot\mathbf{y}_\alpha)\mathbf{x}\cdot(\mathbf{y}_\alpha \times \delta\mathbf{s}_\alpha) - \frac{1}{3}\delta^2 C\mathbf{x}\cdot\boldsymbol{\Omega}\times\mathbf{x}\right\}, \tag{4}$$

where $C = 4\pi \int_0^\infty \xi(r) r^4 dr = 3/2$ and $\Omega = \sum_\alpha \Gamma_\alpha \delta s_\alpha$, which is zero in unbounded inviscid flow. The details of the numerical techniques have been presented by Leung & Ko (1997).

The two identical perturbed vortex rings have the following properties: the circulation $\Gamma = 1.0$, the core to unperturbed ring radius $\sigma/R = 0.2$, the in phase perturbation $\rho_r/R = \rho_z/R = 0.02$ (Figure 1). The suffixes L and T denote, respectively, the initially leading and trailing vortex rings. The wavenumber, m, for the present investigation is 4, the initial separation is $\Delta\lambda = 1R$ and the angle between the vortex ring $\Delta\gamma = 0$ and π. The far field pressure is evaluated at a radial distance of 120 R_T. The normalized length $x' = x/R_T$, time $t' = \Gamma_T t/R_T^2$, pressure $p' = p/(\Gamma_T/R_T)^2$ for unity density and circulation $\Gamma' = \Gamma/\Gamma_T$. The allowable error of the total vorticity and the linear vorticity impulse of potential vortex motions are kept within two percent.

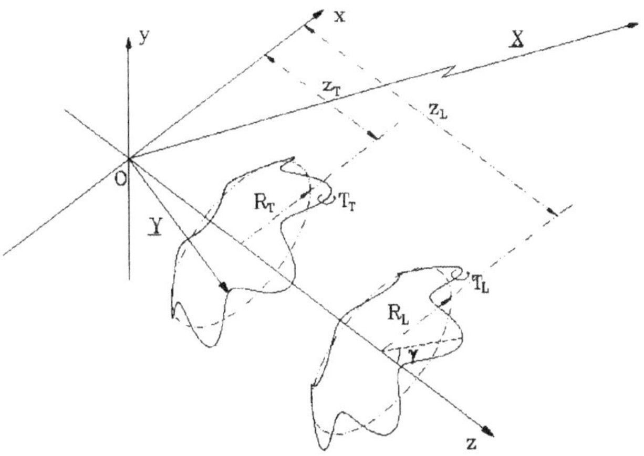

***Figure 1**; Schematic diagram of two perturbed vortex rings.*

3. RESULTS AND DISCUSSION

Within the allowable error, the computation was terminated at the non-dimensional time $t' \leq 30$ and 22 for $\Delta\gamma = 0$ and π respectively. Both cases have three slip throughs at $t' \approx$ 3.7, 11.2 and 18.7 (Figure 2). For the first and second slip through (Figures 2a, 2b 2e and 2f), the vortex rings maintain their basic waveforms. However, during the third slip through at $\Delta\gamma = 0$ (Figure 2c), the initially leading vortex ring doubles its wavenumbers, while the initially trailing vortex ring has more than eight waves. For $\Delta\gamma = \pi$ (Figure 2g), the vortex pattern of initially leading vortex ring maintains its basic waveforms, while the initially trailing vortex ring starts to roll-up and forms a complicated shape (Figure 2h). Moreover, there is a slight change of the azimuthal angle of both vortices. It, thus, shows that the phase difference affects the subsequent shapes of the vortex rings.

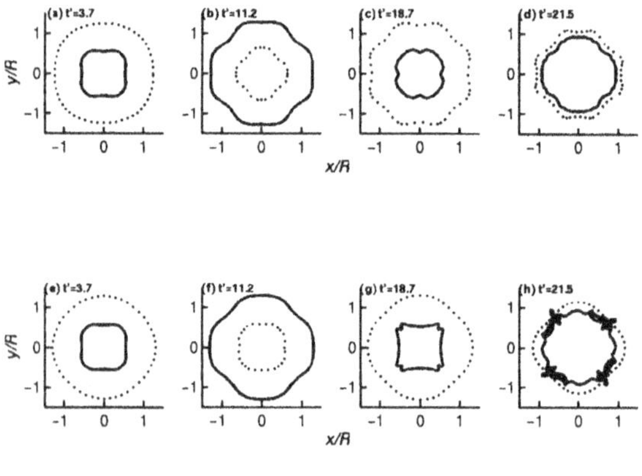

Figure 2*; Interaction of two vortex rings at m = 4. ———, Initially trailing vortex ring; - - -, initially leading vortex ring. (a-d), $\Delta\gamma = 0$; (e-h), $\Delta\gamma = \pi$.*

Leung & Ko (1997) and Ko et al. (1999) show the vortex system becomes unstable when the normalized streamwise peak to peak amplitude exceeds 5, beyond which exponential growth occurs further downstream. The presented normalized streamwise peak to peak amplitudes for $\Delta\gamma = 0$ and π are shown in Figure 3. For $\Delta\gamma = \pi$, only the initially trailing vortex develops into unstable mode, at $t' > 19$, as its vortex pattern changes significantly (Figure 2h). For $\Delta\gamma = 0$, both vortices are in stable mode even higher wavenumber waves are formed. It shows that the phase difference causes different behaviour during their interaction.

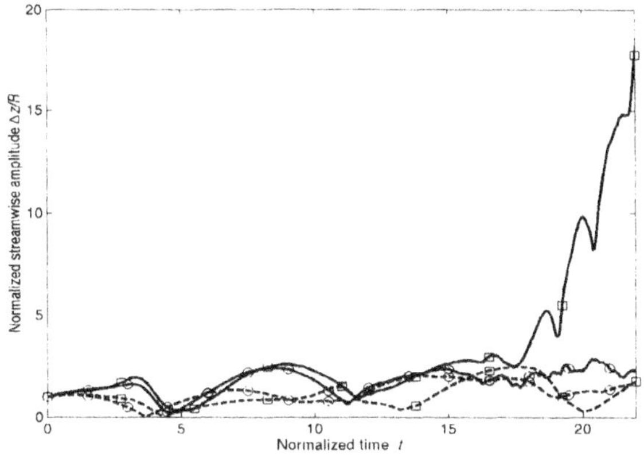

***Figure 3**; The normalized streamwise peak to peak amplitudes at m = 4.* ——— *,*
Initially trailing vortex ring; - - - *, initially leading vortex ring.*
• *,* $\Delta\gamma = 0$*;* • *,* $\Delta\gamma = \pi$*.*

Figure 4 shows the far field sound pressure fluctuations. For $\Delta\gamma = 0$, there are three peaks associated with the three slip through processes. The magnitudes of the first two slip throughs are the same of, $p' \approx 0.03$. The third slip through has a higher value of $p' \approx 0.035$. For $\Delta\gamma = \pi$, the pressures of the first two slip throughs are the same as those of $\Delta\gamma = 0$ (Figure 4), while the third slip through has the highest. Moreover, during the third slip through, high pressure fluctuations occur, revealing the vortex instability effect after the third slip through. It also indicates that unstable vortex system generates higher sound pressure.

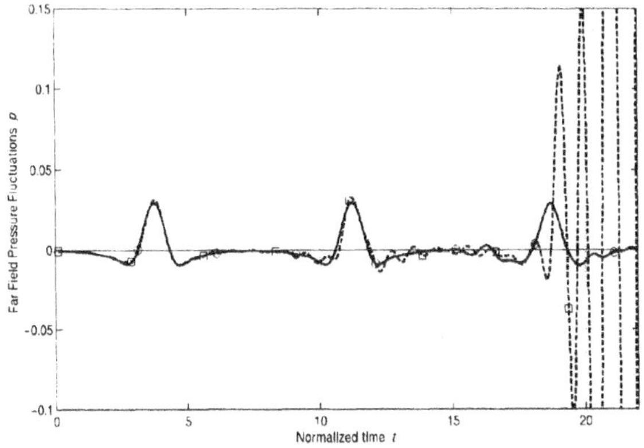

***Figure 4**; Sound pressure fluctuations on the major axis at m = 4.* ——— *,*
Initially trailing vortex ring; - - -*, initially leading vortex ring.*
• *,* $\Delta\gamma = 0$*;* • *,* $\Delta\gamma = \pi$*.*

4. CONCLUSIONS

A numerical study of the interactions of two perturbed vortex rings of $m = 4$ with phase differences of $\Delta\gamma = 0$ and π have been investigated. Results show that during the first and second slip through, the sound pressure fluctuations of the two cases are the same. Due to the change of the vortex patterns in the final development of the vortex rings, higher sound pressure is generated. The phase difference is, thus, one of the factors that affect the instability and the sound generation.

5. ACKNOWLEDGEMENT

This work was supported by a grant from the Research Grants Council, Hong Kong.

REFERENCES

Kambe, T.; Minota, T.; 1981; "Sound radiation from vortex systems"; *J. Sound & Vibration*, **74**, 61-72.

Knio, O. M.; Ghoniem, A. F.; 1990; " Numerical study of a three-dimensional vortex methods"; *J. Comp. Phys.*, **86**, 75-106.

Ko, N. W. M.; Leung, R. C. K.; Tang, C. C. K.; 1999; "The interaction of perturbed vortex rings and its sound generation Part II"; *J. Sound & Vibration*, to be published.

Leung, R. C. K.; Ko, N. W. M.; 1997; "The interaction of perturbed vortex rings and its sound generation"; *J. Sound & Vibration*, 202, 1-27.

Leweke, T.; Williamsin, C.H.K.; 1998; "Cooperative elliptic instability of a vortex pair"; *J. Fluid Mech.* **360**,85-119.

Möhring, W.; 1978; On vortex sound at low Mach number"; *J. Fluid Mech.* **85**, 685-691

Powell, A.; 1964; "The theory of vortex sound"; *The J. Acoust. Soc. America*, **36**, 177-195.

Wickelmans, G.S.; Leonard, A; 1993; "Contributions to vortical particle methods of the computation of three-dimensional incompressible unsteady flows"; *J. Comp. Phys:*, **109**, 247-273.

Air entrainment by a wavy plate entering a pool of stagnant liquid

René Delfos & Jennifer Kockx
J.M. Burgers Centre, Delft University of Technology
Lab. for Aero- and Hydrodynamics
Rotterdamseweg 145, 2628 AL Delft, The Netherlands.
r.delfos@wbmt.tudelft.nl

Abstract: We suggest a mechanism for the entrainment of air by a wavy plate or liquid film entering a pool of stagnant liquid. We have tested our hypothesis in a scaled experiment. With this set-up we show the detailed motion of the pool surface, clearly revealing the entrainment mechanism. We found that the onset of entrainment coincides reasonably well with the proposed model.
Keywords: entrainment, slug flow, liquid film, wave

1. INTRODUCTION

Entrainment of gas into liquid at the interface between the two plays an important role in many technical and environmental processes. In nature entrainment is a phenomenon, that for instance can be seen at spilling wave crests on the sea, or at 'white water' below waterfalls. Atmospheric air is torn below the surface as a result of the dynamic process of water plunging into water.

In technical applications, there are numerous examples of the occurrence of entrainment. There, such flows are of importance because they highly enhance gas transfer. For instance, in a chemical process where oxygen is required often air is brought into the liquid inside the reactor vessel by jetting some of the liquid onto the surface, from which by entrainment the air bubbles are brought in.

A second example is a vertical gas-liquid pipe flow. When the gas loading is high, small gas bubbles tend to coalesce to form large slug bubbles (called 'Taylor bubbles'). Usually, the liquid film flowing down around the slug bubble drags small bubbles with it while impinging into the wake below the slug bubble. The rate of entrainment has a large influence on the gas distribution, hence on the flow pattern. In experiments on a single slug bubble, we observed entrainment to occur only when the liquid film flowing along the slug bubble becomes turbulent, which is characterised by undulations on the film surface (Delfos 1996). Also, we have observed that very large wave lengths do not provoke entrainment. A visual study of the film entering the wake is rather difficult, because of the very fast process. Presently, we try to find out whether there is a minimum film velocity for entrainment to occur, and, if so, what it depends on.

A. Gyr et al. (eds.), Science and Art Symposium 2000, 35-40.

2. MODEL

When the surface of the film entering the pool is *flat*, viscous forces may tear gas into the liquid if the film velocity is high enough. For a low-viscous fluid like water, very high velocities (typically tens of m/s) are needed to cause such viscous entrainment. We consider the entrance of a *wavy* film surface. For such a surface, entrainment is known to occur already at a much lower film velocity. In that case the entrainment rate is about equal to the volume enclosed in between the crests of the disturbances present on the film surface, as was shown for a circular turbulent jet by McKeogh and Ervine (1981). For a wavy film surface we found a reasonable agreement as well (Kockx 1999).

This does not explain yet *why* entrainment occurs. If the film had a flat surface, it would smoothly join with the pool surface. Things change when there are (sinusoidal) disturbances present on the entering film surface. It is often suggested that the wave *crests* impinging upon the pool surface cause the pool surface to be splashed away and therefore leave a gap in the pool surface. We argue however, that it is the retreating wave crest, thus the approaching wave *trough* that is responsible for the gap.

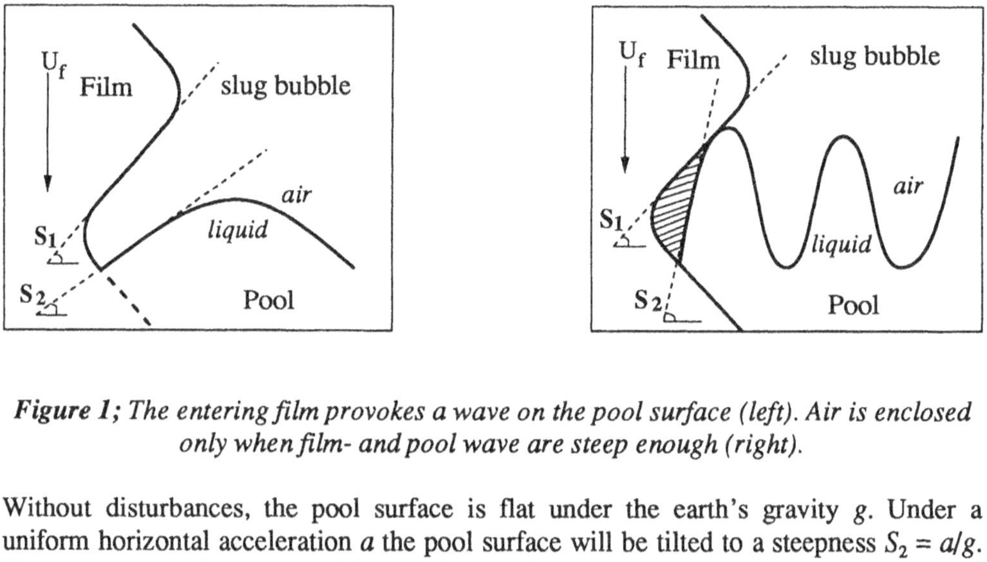

Figure 1; *The entering film provokes a wave on the pool surface (left). Air is enclosed only when film- and pool wave are steep enough (right).*

Without disturbances, the pool surface is flat under the earth's gravity g. Under a uniform horizontal acceleration a the pool surface will be tilted to a steepness $S_2 = a/g$. If we start to accelerate one of the side boundaries, a complex wave is sent out, built up of waves with different wave lengths λ_W and velocities U_W bound by the dispersion relation for deep water waves, $U_W^2 = g\lambda_W/2\pi$.

Now take a film falling with velocity U_F into the pool. We assume that sinusoidal waves with wave length λ and amplitude h_W are present on the film. The waves move with the film. Relative to the pool surface, the horizontal position x_E of the entrance point oscillates following the film surface, $x_E = h_W \sin(2\pi U_F\, t/\lambda)$. Then the transverse acceleration at the pool surface is given as $a_F = -(2\pi U_F/\lambda)^2\, x_E$. Even for small U_F this parameter gets large when λ is small.

We hypothesise that entrainment occurs when the pool surface is not capable of responding to these oscillations. In other words, an entering wave provokes a disturbance on the pool surface. Its steepness S_2 near the impinging point depends on gravity g and the (horizontal) acceleration a of the wave. When the provoked steepness S_2 is higher than the inverse steepness S_1 of the incoming wave, the trough being formed will close off to form a bubble. This is sketched in figure 1. We consider two limits: Firstly, that the pool surface is in equilibrium with the acceleration imposed by the entering film, and secondly, that on the pool surface only waves are formed with the same frequency as that of the incoming waves. Quantitatively, we then get a lower and an upper limit:

$$1 < \frac{U_{ON}^2 \cdot h_W^2}{g \cdot (\lambda/2\pi)^3} = N_{ENT} < \frac{3}{2}\pi$$

Onset of entrainment based on this number N_{ENT} is discussed by Prosperetti (see Zhu *et al.* 1999, Ohl *et al.* 1999). They studied an axi-symmetric vertical jet with a wave on it by means of both simulations and experiments. They suggested that entrainment is caused by the wave *crest* impinging onto the water surface, instead of the wave *troughs* sucking in the water as we suggest.

3. EXPERIMENT

To investigate the occurrence of either mechanism experimentally, we constructed an experiment with a much larger length scale than that in a turbulent film flow.

A scheme of the experiment is shown in figure 2. The receiving pool consists of a glass water tank of 0.4*0.25*0.2 m^3 (l*w*h). The wavy water film is mimicked by a non-deformable corrugated plastic plate. It is driven with a constant velocity through the water surface by a piston.

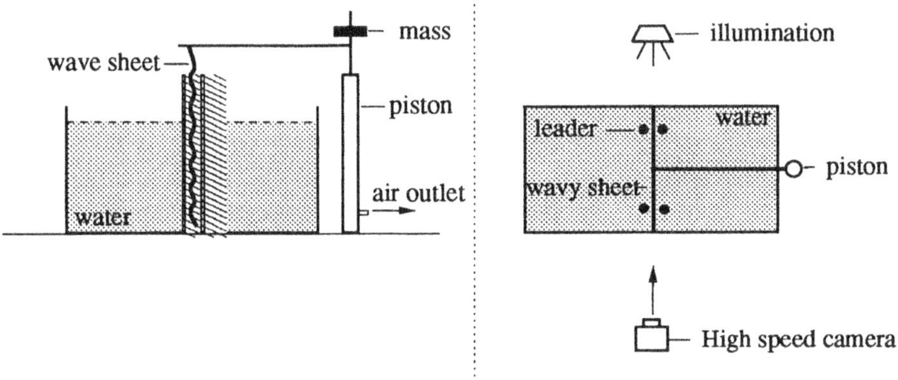

Figure 2; *Front (left) and top (right) view on the experimental set-up.*

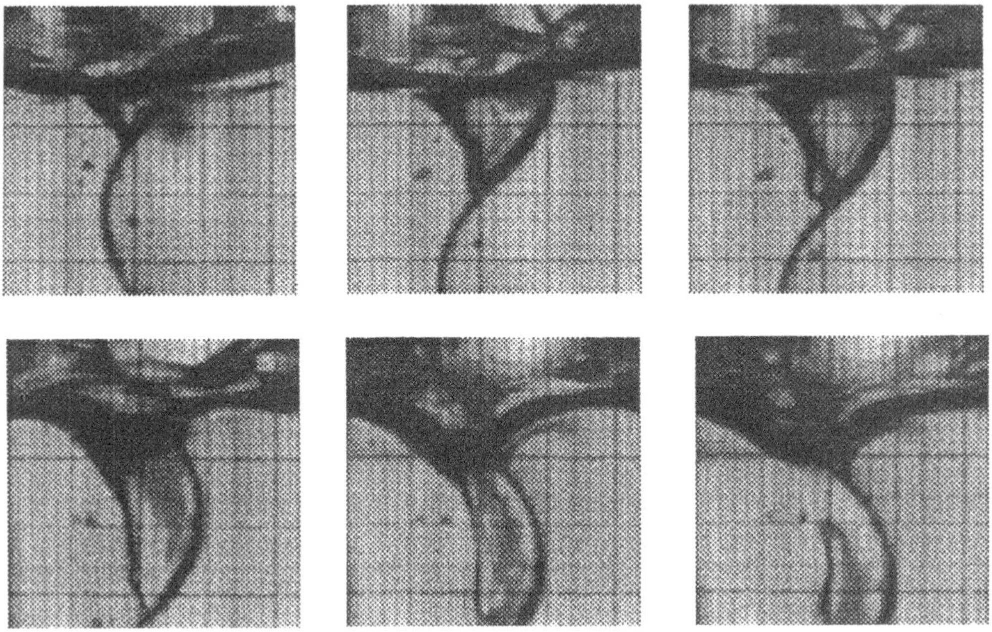

Figure 3; *Motion of the pool surface in case of the '**non-wetting**' surface, at $t' = tU/\lambda$ = 1.90; 2.00; 2.15; 2.23; 2.34; 2.38 (from left to right).*

The two largest deviations from the wavy plate with respect to a 'real' liquid film are that the liquid is deformable, and that there is three-phase contact line on the plate. The first effect can not be overcome but is thought to be small because of the high momentum of the liquid film; the latter is investigated by using two surfaces with different contact angles.

The variables in the experiment were the following:
- The fall speed of the wavy plate U (by different loads on the piston): (0..1.5 m/s)
- The wave height h_w (by different wavy plates): (2.2 & 3.9 mm)
- The contact angle at the plate (by different coatings on the plate): (~90° & ~0°)

The wave length λ was (nearly) the same for the two wavy plates, some 33 mm. Even at this 'large' length scale, the typical time scale of a wave passing by is a few tens of milliseconds only. Therefore, after preliminary experiments with a normal video-camera, we made high-speed recordings using a 750 frames/s CCD-camera storing 256*256 pixel^2 images directly in digital format. For this the whole experiment was placed in the "Kramers Laboratory for Applied Physics" of Delft University, to whom we are greatly indebted. The region near the entering plate was diffusely lit from the back with a 500 Watt halogen lamp. For the short exposure times (typically 0.25 ms) an LC-shutter was used with an infrared light blocking filter.

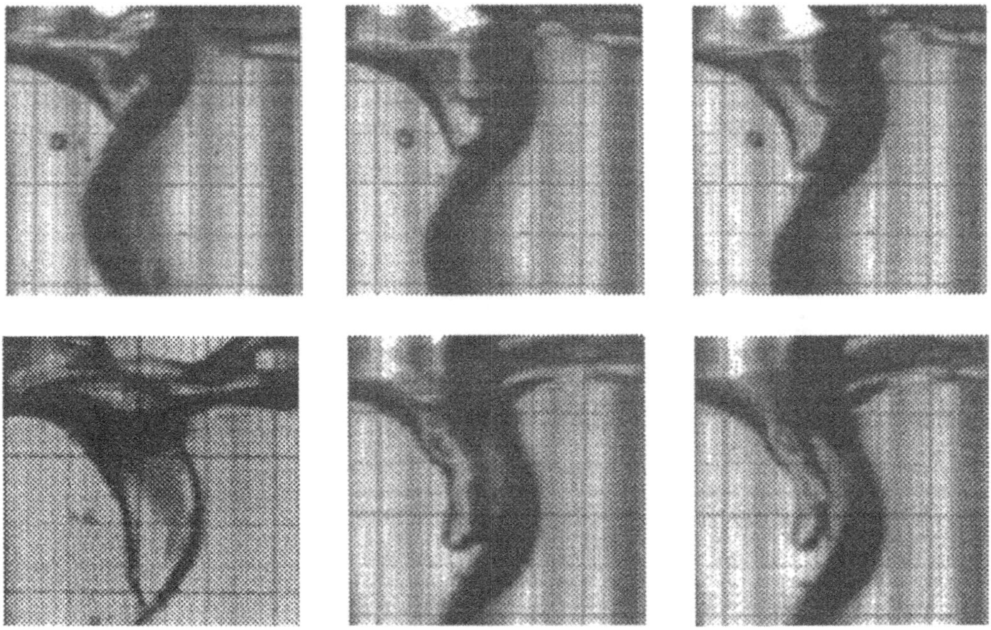

Figure 4; *Motion of the pool surface in case of the '**wetting**' surface, at $t' = tU/\lambda = 2.00; 2.15; 2.21; 2.42; 2.48; 2.52$ (from left to right).*

4. RESULTS

The images reveal the shape of the responding pool surface in detail, and make the process clearly visible. In figure 3 the 'non-wetting' is shown with the 'high' wavy plate at a velocity just beyond onset of entrainment ($U = 0.5$ m/s). We see that the contact line moves down, nearly stuck on the retreating side of the wave. The cavity gets deeper while the wave moves further down. When the next wave crest touches the water surface, the cavity gets separated from the air above, and a bubble is formed. In figure 4 the same conditions are used for the 'wetting' surface. Here there is no sticky contact line; instead a thin layer is easily scooped op on the entering wavy plate. However, the trough is finally closed off very similar to that in figure 3. Thus we see that the details of the contact line depend much on the wetting property of the surface, but that the entrainment process is not much changed by it. We conclude that it is the low pressure during the retreating wave on the film that causes the air to be sucked below the average water surface, and that this air is enclosed by the next wave on the sheet coming in.

At lower velocities, also a trough is formed, but it is not steep enough to get enclosed by the next wave. We also observed wave crests plunging into the water that produced a splashing motion, but without entraining air into the water.

We measured the minimum velocity U_{ON} needed to form air pockets under the different experimental conditions; the results are shown in table I. The wetting surface seems to mimic somewhat better an entering liquid film than the non-wetting surface, where the contact line seems to help dragging down the air. The differences between the two with respect to U_{ON} are small, however. The values determined for N_{ENT} are somewhere in the middle of the two boundaries as predicted by the model, which encourages our hypothesis on the mechanism we proposed. We are yet working on a more refined experiment and improvement of the model by a more complete simulation.

h_w (mm)	λ (mm)	Wetting	U_{ON} (m/s)	N_{ENT} (-)
3.9	32.0	yes	0.50	2.9
3.9	32.0	no	0.40	1.9
2.2	33.7	yes	0.80	2.0
2.2	33.7	no	0.75	1.8

Table I; *velocities measured for onset of entrainment and the corresponding* N_{ENT}.

5. CONCLUSION

In the experiments presented, we have scaled the process of air entrainment by a periodically disturbed surface. The length and time scales make high speed filming to a quantitative visualisation technique to study details of the air entrainment process. We have measured the minimum velocity for entrainment to occur, and have shown that it corresponds reasonably well with the proposed model value. We conclude that the entrainment is caused by the air being sucked into the water layer because of the transverse acceleration of the incoming film flow being too large for the interface to respond to the oscillations. This results in the formation of a wave on the receiving surface that is steeper than the inverse steepness of the wave on the incoming film.

REFERENCES

Delfos (1996) "Experiments on air entrainment from a stationary slug bubble in a vertical tube". Ph.D.-thesis Delft University of Technology; ISBN 90-370-0149-1

McKeogh, E.J. and Ervine, D.A. (1981) "Air entrainment rate and diffusion pattern of plunging liquid jets" *Chemical Engineering Science* **36**, 1161-72

Kockx, J.P. (1999) "Experiments on the gas exchange between a Taylor bubble and its liquid slug in a vertical tube". Ph.D.-thesis Delft University of Technology; ISBN 90-370-0173-4

Ohl, C.D., Oğuz, H.N. and Prosperetti, A. (1999) On the mechanism of air entrainment by a disturbed liquid jet, submitted to *Physics of Fluids*

Zhu, Y.G., Oğuz, H.N. and Prosperetti, A. (1999) Air entrainment by impinging liquid jets with surface disturbances, accepted by *J. Fluid Mech.*

Numerical simulation and manipulation of the vortex dynamics in a confined flow

Iraj Mortazavi
Mathématiques Appliquées de Bordeaux UMR 5466 CNRS, Université Bordeaux 1, 351 Cours de La Libération, 33405 Talence Cedex, France, mortaz@math.u-bordeaux.fr

Abstract : The dynamics of vorticity in a bluff-body geometry for an incompressible, unsteady, plane, high Reynolds number flow is studied using a vortex method. The numerical scheme corresponds to a hybrid Random Vortex (RVM)-Finite Element (FEM) method. Various techniques of instantaneous visualization are applied to characterize such flow's behaviour. Dynamic visualization techniques that are focused on the mechanisms of formation and transport of eddy structures represent a powerful tool for the analysis of such complex flows. The effect of the different inlet velocity ratios is studied to assess the evolution and interaction of vortical structures downstream of the bluff-body's mixing zone.
Keywords : Vortex simulation, vorticity dynamics, flow manipulation, visualization.

1. INTRODUCTION

Vortex methods are applied to simulate 2D incompressible steady and more particularly unsteady fluid flow in a wide range of Reynolds number regimes. The first successful presentation of vortex methods, containing a non-singular discretization of the Navier-Stokes equations has been proposed by Chorin [1] and named the Random Vortex Method. Later, Chorin modified his method to account for wall boundary conditions, and named it the Vortex Sheet Method [2]. During the last decade, different classes of vortex methods have been proposed to solve for the Navier-Stokes equations with a random or deterministic diffusion operator.

Vortex methods deal with non-primitive variables and their main well known advantage is related to the optimized computational efficiency arising from the fact that vortex elements are accounted for only in regions on non-zero vorticity, and partially from the fact that, dealing directly with vorticity, we have immediate physical insights on the behaviour in vortical sub-domains such as those where separated flows structures occur. Hence the essential feature of this class of simulations methods is related to its unsteady nature and its suitability to vortex dynamics studies.

In the present work the RVM is used to describe the strong dynamics of the flow due to the interaction between an inner jet and an external flow separated by two symmetrical flat plates. The flow is unsteady, two-dimensional (plane) and

A. Gyr et al. (eds.), Science and Art Symposium 2000, 41-50.

incompressible. The study is oriented toward the characterization of the formation and transport of eddy structures in the flow domain. The interaction and mixing of the two entry flows downstream of the separator imposes an interesting and complicated dynamics to the flow behavior. Because of the symmetry, only half of the whole geometry is considered here (figure 1). It will be observed that, in spite of the random nature of turbulence, the creation of vortical structures downstream of the obstacle, their mutual interactions just before the separation from the obstacle and their transport by the main flow follows a quasi-coherent logic which can be analized by instantaneous observations of the fluid patterns.

The main objective of this work concerns the effect of the entry velocity ratio on the vortex formation and transport. For this study, three velocity ratios are compared : $V_j/V_e = 1.$, $V_j/V_e = 0.5$ and $V_j/V_e = 2.0$, where V_e and V_j correspond respectively to the external flow velocity and the jet velocity. Numerical results show that events are dominated by the flow having the larger velocity and consequently larger absolute vorticity. Numerical flow visualization techniques are applied to complete this study.

2. NUMERICAL METHOD

The Random Vortex Method is a numerical technique to solve for the incompressible, 2D, unsteady Navier-Stokes equations converted to a rotational non-primitive formulation. The equations to be solved are the Poisson Equation

$$\Delta\psi = -\omega \quad (1)$$

and the Vorticity Transport Equation (VTE)

$$\partial\omega/\partial t + \vec{u}.\nabla\omega = (1/Re)\Delta\omega \quad (2)$$

associated with their boundary conditions. Where ψ is the stream function, ω the no-zero component of the vorticity, $\vec{u}$ the velocity vector and Re the Reynolds number. To discretize the vorticity, the computational domain is divided into two regions : the "interior" region and the numerical boundary region. In the interior region the vorticity is discretized using vortex "blobs" [1], whereas in the boundary domain vortex "sheets" are used [2]. The VTE is solved by a two fractional step method : the first step corresponding to the convection and the second one to the diffusion.

The mathematical scheme for the advection and diffusion calculations in the boundary and interior domains are the same (the viscous splitting method). For the convection step vortex sheets and vortex blobs are advected by their own local velocities.

For the diffusion step the Random Walk algorithm [1,3,5] is used. This is based on the rule that a large collection of particles undergoing Brownian motion provide a good approximation of the heat equation.

Finally, the total transport of vortex elements is obtained adding the two fractional movements

$$\vec{x}_j^{k+1} = \vec{x}_j^{k} + L(\vec{u}.\Delta t) + \vec{\eta}_j \quad \text{with} \quad \vec{\eta} = (\eta_x, \eta_y) \qquad (3)$$

Where, L is the convective tranport operator (in this work, the Heun's second order method and Euler's first order technique are used for the time integration of vortex particles, respectively, in the interior and boundary regions).

The precision of the computational procedure is influenced by : the time step (Δt), the intensity of the elements (their strength or discretisation in circulation terms, Γ_m), the spatial discretization values of sheets or blobs (h or δ), and the mesh size. The accuracy and numerical convergence of the scheme for complex flows, as a function of the above mentioned parameters, has been already studied by Ghoniem & Gagnon [3], Sethian & Ghoniem [4] and Mortazavi et al [5]. Numerical parameter values implemented in this work are : Δt=0.05, Γ=0.0033 and h=0.05. These parameters fulfill all requierments for numerical convergence of computational results.

4. ANALYSIS & VISUALIZATION OF NUMERICAL RESULTS

Quantitative and qualitative flow visualizations are applied to identify various fluid structures and characterize their spatial dependencies. Once defined, the spatial relationship between these structures can be compared by visualizing them. This visualization is also helpful to study their generation, evolution and mutual interactions. Furthermore, the statistical significance and relevance of each class of structures, their spatio-temporal relationship and dynamical behavior can be studied using this class of visualization methods. This flexibility is accompanied by a huge amount of information on the structural features of the flow behavior and so it permits us to select the statistics convenient to the description of the reproducible sequences of events. The accrued interest for the case of flows with large Reynolds number, where vortical structures have an important role in the turbulence production and in the momentum transfer, should be noted [6]. In this work, the generation of two vortex layers of opposite signs that merge in two sides of the plate, just after the separation zone, causes instabilities and deforms the eddy structures in the mixing zone [7]. On the one hand, since Re is large, the vortical zones are very precisely defined and concentrated in sub-vortical regions. The analysis of complicated fluid mechanisms, e.g. "backs (velocity discontinuities)", needs a powerful numerical visualization tool. Another difficulty that must be overcome is the study of the background regions of very low vorticity. The weak vortical intensity of these regions, can make their influence disappear from the computational field. This elimination can be the origin of a false interpretation of the vortical behavior. Vortex methods are powerful tools for such critical studies.

4.1. FLOW DYNAMICS WITH EQUAL INLET VELOCITY VALUES

The evolution with time of ω, the vorticity, is an important source of information to study the vorticity dynamics. Figure 2 correspond the granular vorticity ω. Two important series of events detected in the vortex behavior concern essentially the formation and transport of eddies. This kind of visualization shows the quasi-periodic creation of two important vortical structures : a clockwise rotating eddy in the external edge of the bluff-body (that moves toward the axis, with negative circulation values) and a positive counterclockwise eddy created by the inner jet flow which has an outward motion. The jet flow which has a momentum equal to : ρV_j^2, provides a positive circulation in the flow : $\delta\Gamma_{inj} = V_j^2.\Delta t/2$.

The formation phase - The mutual interaction of the jet and the external flow is the principal origin for the formation of vortical structures. It is easy to see that the upper vortical structure dominates generally the motion of the lower one. This aspect limits the growth of the vortex which seats underneath, which finally is separated from the normal wall by a vertical upside motion. Furthermore, when these structures come together to create an instantaneous region with twice the length of the average recirculation zone, a breakdown with the creation of a vortex pair is observed. These events are described as follows : the convergent motion of these two eddies creates for a short time a composite vortical structure and initiates progressively the vortex shedding. Later, a separation of these eddies from each other occurs, approximately at the same time that the bigger external vortex separates from the bluff-body wall. The vortex pairs created are immediately injected in the flow. Few steps later, the rest of the inner structure separates from the wall, to be replaced by two new vortical structures which form in front of the bluff-body (the separated lower vortex is moving in fluid following two general behaviors : either it is advancing along the axis toward the exit boundaries, or it is propagating in the domain closer to the upper wall). This procedure represents a quasi-periodic sub-cycle which occurs about every 200 steps, in the statistically steady regime. Nevertheless, in a high Reynolds number flow this periodicity is never fixed with an identical repetition of fluid motion (e.g. sometimes, during the creation of two vortex pairs, several very small structures with less important circulations are formed and dispersed along the axis of symmetry) ; however, it constitutes a capability of prediction of global fluid macro-structures behaviour (e.g. vortical structures motion and interactions) during the time.

The transport phase - Once imparted in the flow, vortex pair structures have various trajectories (figure 2). Either they are attracted toward the upper wall and correspond to the so-called mechanism of dipole ejection, with the negative vortex that is advancing in the vicinity of the wall downstream of the positive vortex, or they are accelerated attaining upstream eddies and creating more complex structures that contain several vortices mixing and spinning around each other. Also, other behaviors can be detected : some particles of the negative structure spin around the positive vortex and the remaining of that structure accelerates toward the exit boundary across the upper wall. It is apparent that, globally, negative structures move faster than positive vortices. As shown in the figure 2, some eddy agglomerations are formed from smaller particle patches which advance in time translating and spinning around each other. However, some isolated patches with individual behavior are noticeable in different parts of the flow. These isolated agglomerations, that were previously detached from the large eddies, correspond to low-vorticity regions. Later, they can reattach again to other large scale structures or continue their individual movement.

4.2. EFFECT OF THE VELOCITY RATIO VARIATION

This comparison is done using two different entry velocity ratios corresponding to $V_j/V_e = 0.5$ and $V_j/V_e = 2.0$ (V_e is the external entry velocity and V_j represents the jet entry velocity). For the second velocity ratio, respecting the Courant conditions [6], the following parameters are used (Δt=0.025, Γ=0.0033 and h=0.05). For larger values of the external flow we observed complex interactions of the large eddies essentially dominated by the main stream. Nevertheless, the evolution of these complex vortex interactions is predictable by two principal series of quasi-cyclic events in the flow map. For the opposite case study (larger jet velocities) we observed a sensible decrease in the vorticity formation that leads directly to a decrease of the interactions between vortical structures of internal and external flow.

first case : $V_j/V_e = 0.5$ - Figure 3 shows the evolution of vortical structures. One can easily observe that the inner jet is completely blocked by the external flow which stops its motion along the axis. Therefore, the jet flow that is the origin of the creation of the positive vorticity, either goes up on the vertical face of the bluff-body toward its upper edge, or is dispersed downstream of the obstacle intensifying the shedding procedure of negative structures. This positive eddy that is extended along the wake of the obstacle is also broken in two parts by the external flow. The shedding of large negative structures is then followed by the separation of an ensemble of positive vortices (broken in two parts). These two large eddy structures which show a reciprocal progressive creeping motion are convected together in the flow.

These shedding events are repeated about every 200 steps in a quasi-cyclic manner. At times a kind of artificial blockage is created by a part of an eddy structures, just downstream of the formation region, which waits for the joining of some other detached eddies from the formation region. Larger eddies due to this process are then convected in the flow. An important portion of the large vortical structures are formed from small positive vortex patches which often orbit around the large negative structures until they exit the domain.

second case : $V_j/V_e = 2.0$ - In this case we observed the permanent difficulty of eddies for merging and mixing. This is directly related to the weaker energy level of the structures.

Figure 4 confirms this observation. The jet flow maintains his individuality across the axis of symmetry ; and the influence of the negative structures (that have an injection rate of circulation four times smaller) on the jet flow is weak. However, the accumulation of the eddies issued from the upper flow, just on the top side of the jet flow, permits the formation of a large negative structure and its interaction with neighboring positive vortices entrains a mutual simultaneous shedding of positive and negative structures. The *'roll up'* behavior of the negative vortex around the positive eddy and its development by the last one can be considered as a particular case of the *'eddy suction'* phenomenon. Moving toward the exit boundary of the domain, the two structures remain always separated and grow gradually. The negative eddy that continues to grow blocks the channel flow by reaching the top of it, and the positive structure is diluted slightly. Negative vortices are not sufficiently energetic to break down positive structures. Therefore, these structures continue systematically their motion along the axis of symmetry. A portion of the particles issued from the jet flow rise along the bluff-body's vertical wall, but this phenomenon does not generate a direct shedding circumstance. Essentially, positive and negative structures which advance and develop progressively in the channel, remain almost parallel together. They do not have enough dynamics to mix and merge and therefore, no dipole creation is noticed in the flow. The higher velocity of the jet flow and its larger momentum seems to be at the origin of this behavior.

5. CONCLUSION and PERSPECTIVES

In the present work, we used the Random Vortex Method to study the flow downstream a bluff-body geometry. The effect of the inlet velocities ratio on the physics of the flow and the vortex dynamics was analyzed. The variable of interest , ω, is directly connected to the vortical structures ; hence, the computational task is adapted and refined in regions of high vorticity, where vortex shear events, the physics of vorticity generation, and the dynamics of evolution are quite well depicted by this vortex method.

Visualization techniques are applied to achieve a better understanding and generalization of instantaneous flow events. They allow us to classify the various steps of the fluid flow. The post processing of the results occurs during a large time sequence compared to the evolution scale of the biggest eddies.

First, we chose a velocity ratio between two streams equal to one in order to increase the topological interest of the phenomenon. Two vorticity layers on the two edges of the separation feed positive and negative vortical structures, that grow, interact, and then merge inside the channel as their length scale increases. The dynamics is governed initially by the shedding of large vortex eddies from the inner and outer sides of the bluff-body. Mixing between the two streams is enhanced by the merging of these eddies downstream the bluff-body and the formation of composite structures. The entrainment mechanism is the merging of the two eddies. The evolution of these complex vortex interactions is predictable by two principal series of quasi-cyclic events in the flow map. The formation and shedding procedure of eddy structures occur approximately on a cycle of dimensionless period equal to about 10. These events lead to the creation of vortex pairs which evolve and advance in the flow. Then, the influence of differential inlet velocity on the evolution and interaction of vortical structures for two opposite cases is studied. In both cases, the flow mechanisms are essentially dominated by the flow having the largest inlet velocity value.

ACKNOWLEDGEMENTS

Most of the calculations and image processing were performed at the "Institut du Developpement et des Ressources en Informatique Scientifique (IDRIS)" of "CNRS-Orsay, France". The author thanks Sandrine Layrisse for her help to prepare this paper.

REFERENCES

1- Chorin, A. J., 1973, "Numerical Study of Slightly Viscous Flow" *Journal of Fluid Mechanics*, pp. 273-289.

2- Chorin, A. J., 1987, "Vortex Sheet Approximation of Boundary Layers, " *Journal of Comput. Physics*, Vol. 27, pp. 428-442.

3- Ghoniem, A. F., And Gagnon, Y., 1987, "Vortex Simulation Of Laminar Recirculating Flows, " *Journal of Comput. Physics*, Vol. 68, pp. 346-377.

4- Sethian, J. A., And Ghoniem, A. F., 1988, "Validation Study of Vortex Methods, " *Journal of Comput. Physics*, Vol. 74, pp. 283-317.

5- Mortazavi, I., Micheau, P., & Giovannini, A., "Numerical Convergence Of The Random Vortex Method For Complex Flows," *European Series in Applied and Industrial Mathematics*, Vol. 1, (1996), pp. 571-589.

6- Robinson, S. K., Kline, S. J., And Spalart, P. R., 1988, "Quasi-Coherent Structures In The Turbulent Boundary Layer : Part II. Verification And New Information From A Numerically Simulated Flat-Plate Layer," *Zoran P. Zaric' Memorial International Seminar On Near-Wall Turbulence*, Dubrovnik, pp. 218-242.

7- Giovannini, A., Mortazavi, I. & Tinel, Y., "Numerical flow Visualization in High Reynolds Numbers Using Vortex Method Computational Results," *ASME Summer Meeting, FED 218* (1995), pp. 37-43.

FIGURES

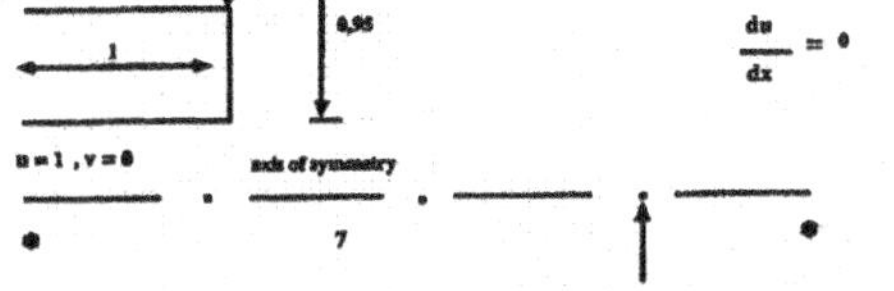

Figure **1**: Geometry of the computational domain with inflow and outflow boundary conditions.

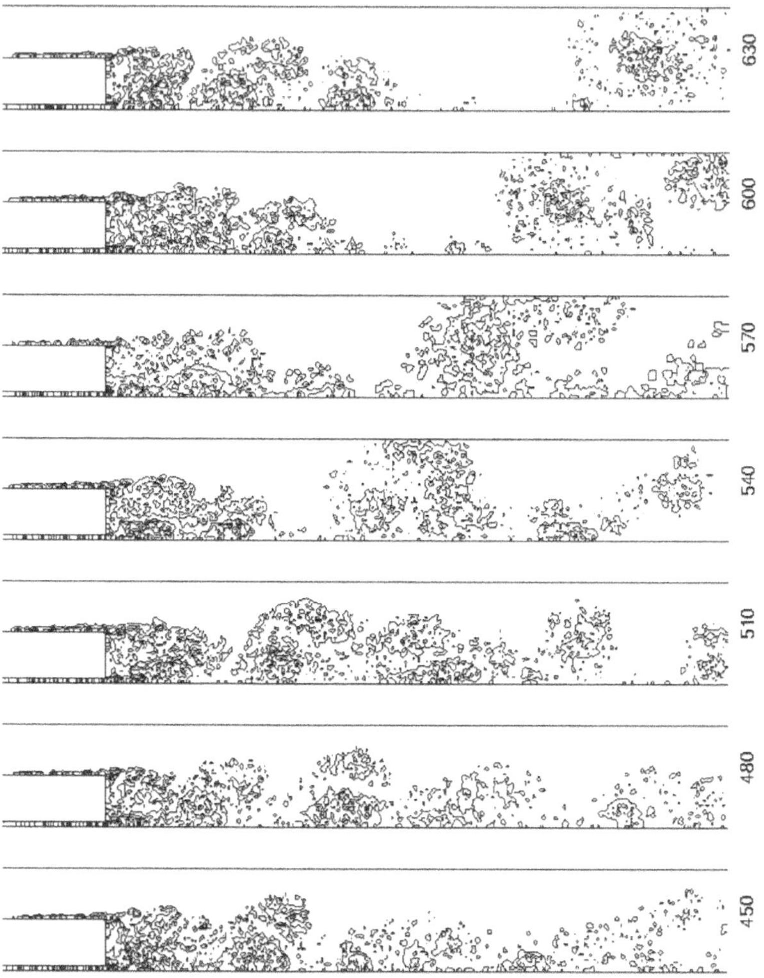

Figure **2**: Instantaneous vortex particle evolution with granular representation (Γ=0.0033, Δt=0.05 and h=0.05, Re=7000 ; equal inlet velocity ratio).

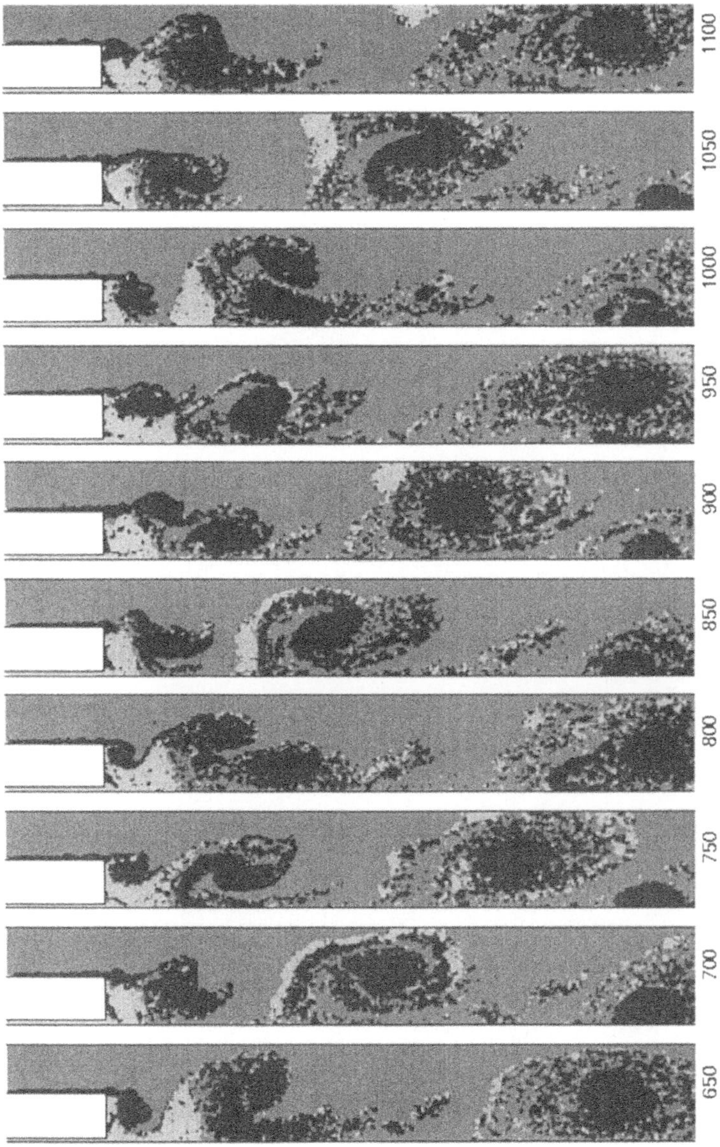

Figure **3**: Instantaneous vortex particle evolution ($V_j/V_e = 0.5$, Γ=0.0033, Δt=0.025 and h=0.05, Re=7000).

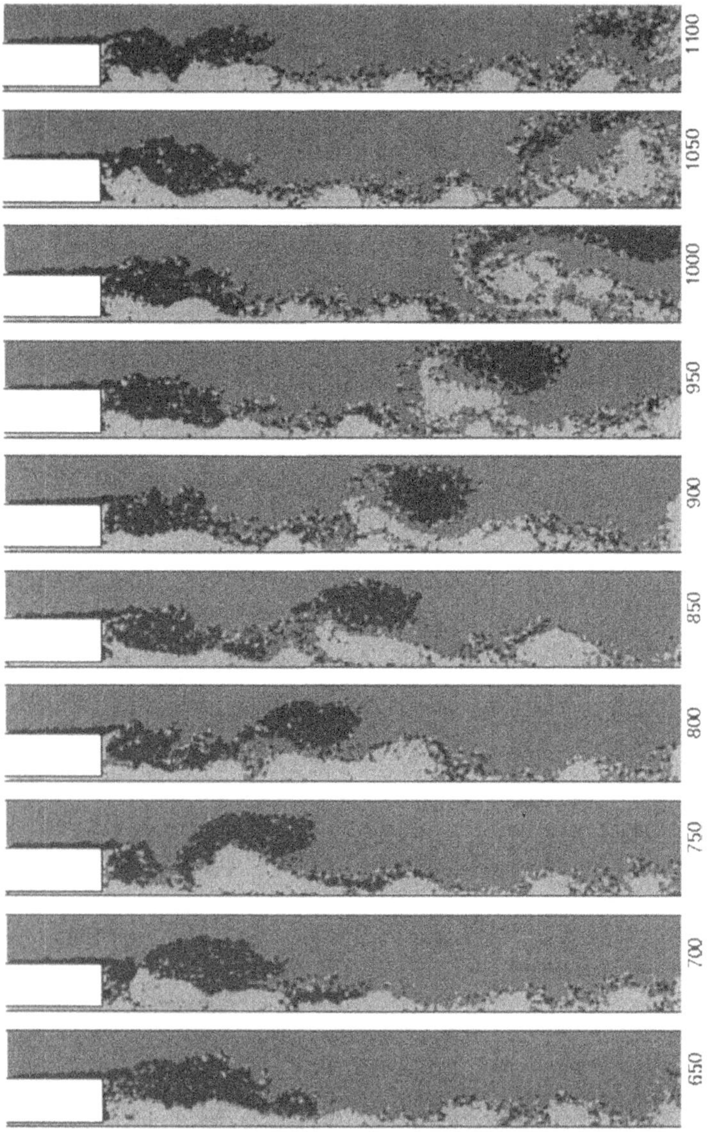

Figure **4**: Instantaneous vortex particle evolution ($V_j/V_e = 2.0$, Γ=0.0033, Δt=0.025 and h=0.05, Re=7000).

Double-Diffusive Christmas Tree

Arkady Tsinober
Department of Fluid Mechanics, Faculty of Engineering
Tel Aviv University, Tel Aviv 69978, ISRAEL
tsinober@eng.tau.ac.il

Abstract: Double- and multi-diffusive phenomena occur in fluids containing more than one component contributing to their density (e.g. salt and heat), and exhibit unusually beautiful patterns. Such a pattern in the form of a Chrismas tree is observed when a fluid with stable salinity gradient is heated at a single point. The pattern consists of a primary rising hot plume capped by a vortex dome, a falling downwards annular plume, and a series of layers around and above. A characteristic feature of the layers is the existence of systematic shearing motions and vortices driven by the horizontal shear in the layers, the upward buoyancy flux and the horizontal density gradient. This type of flow can be regarded as a basic individual convective element, and to some extent it plays the same role in the double-diffusive convection problems as the plume in one-component convection problems.
Keywords: Double-diffusion, instability, layers, shearing motions, vortices.

1. INTRODUCTORY NOTES

1.1 Historical overview

The subject of double-diffusive phenomena - surprisingly for a branch of classic science - has its beginning only at 1956 when Stommel et al. published a paper entitled 'An oceanographical curiosity: the perpetual salt fountain'. The real start, however, has to be attributed to Stern who published in 1960 the first paper explaining in physical and mathematical terms the phenomenon of double-diffusive convection as an instability resulting from the coupling between gravity and diffusion, when the fluid contains at least two components with different diffusivities and having comparable contribution to fluid density. Since then the subject aroused great interest amongst oceanographers, but soon it was realized that double-diffusive phenomena have far broader implications and are important in astrophysics (e.g., helium-hydrogen systems), geophysics (e.g., magma chambers and Earth core), chemistry (ternary systems, polymers), material science and metallurgy (solidification and crystallization), iceberg melting, solar ponds, liquid gas storage and transportation, sewage and other disposals, thermosyphons and some other possible applications. For a review and references see [Huppert and Turner, 1981], [Stern 1985], and [Turner, 1985].

A. Gyr et al. (eds.), Science and Art Symposium 2000, 51-57.

1.2 What is double diffusive convection - basic mechanisms.

Double (multi)-diffusive convection occurs when the fluid contains two (or more) components making opposite contributions to its density and the diffusivity of the components is *different*, e.g. heat and salt(s) in water. In such a case the gravitational potentional energy of the system is not minimal.[1] Strong convective motions can arise even when the net density decreases upwards, i.e. the density distribution is statically stable. The flow patterns depend on the relation between the concentration gradients of the components contributing to the fluid density and on the relation between their diffusivities. There are two basic configurations of statically stable stratification when the components make opposing contributions to the fluid density. The first one is when hot salty water overlies cold fresh water. This is the so called salt finger regime analogous to the ordinary Rayleigh Bernard instability. The second configuration is when cold fresh water overlies hot salty water. This is the so called diffusive regime, corresponding to the oscillating instability mode. In both cases the motions evolving from an initial disturbance are driven by the potential energy in the top-heavy component (salt in the salt fingers regime and heat in the diffusive regime). This energy is released by the differential diffusion of heat and salt. In both cases a parcel displaced vertically from its equilibrium will loose (gain) its heat much faster than salt due to much smaller diffusivity of salt. Under certain conditions this leads to instability with a direct mode in the salt finger regime and with an oscillatory mode in the diffusive regime.

Double diffusive instabilities occur also in many other configurations, e.g. in case of a homogeneous fluid with horizontal gradients of temperature and salinity, but no horizontal density gradient.

There exist multidiffusive systems having more than three agencies such as with several components contributing to the fluid density (two or more) in the presence of rotation and/or magnetic field. Such systems are relevant to astrophysics, material science and in chemical and biological contexts.

The double-diffusive instability of a fluid with initially smooth distribution of temperature and salinity leads at high Rayleigh numbers to formation of a series of layers separated by diffusive and/or finger interfaces. The process is such that the convection in the layers (driven by transports through the interfaces) keeps the interfaces sharp and the density differences across the interfaces tend to increase in time. Therefore, once formed, a system of layers is a long lived pattern and its evolution is of considerable interest in many contexts. This property seems to be universal for double-diffusive systems and is exhibited in a great variety of situations

Systems of layers form via several routes all of which exhibit a numer of most beautiful features. One of such routes is described below. We describe here mostly the observational evidence directly related to the aesthetic aspects. More technical details and references can be found in [Tsinober et. al., 1983].

[1] It is assumed that the cross-diffusion effects – Soret and Dufour effects – are negligible.

2. THE EXPERIMENT

2.1 The system

The system under consideration consists of a point source of buoyancy via supplying constant heat flux at the bottom of a stably stratified fluid with a constant salinity gradient. The diffusivity of salt ($NaCl$) in water is 80 times smaller than that of the injected buoyancy (heat). This system develops in time into a complicated system consisting of a rising plume (mostly laminar), a falling annular plume (mostly turbulent) around the rising one and, at later stages, of a system of double-diffusive layers which are partially turbulent and partially laminar. The physical quantities which completely specify this system are the buoyancy flux, the initial density gradient, the kinematic viscosity, and the thermal and salinity diffusivities. These physical quantities can be reduced to three dimensionless parameters: the effective Rayleigh number, the Prandtl number and the diffusivity ratio, i.e. the ratio of salt to heat diffusivities. In the experiments described below, the Prandtl number (≈ 7) and the diffusivity ratio ($\approx 1.25 \cdot 10^{-2}$) remain fixed and thus the dependence on the Rayleigh number was investigated with the emphasis on the variety of patterns arising as a result of double-difusive instabilites and their subsequent development.

2.2 Apparatus and techniques

Constant-salinity gradients were produced by the "double-bucket" method, filling the 40 X 40 X 30 cm high aquarium. These were heated from below by a point-source heat injection. This was achieved by passing a 5k Hz electric current from a small electrode (1 mm diameter and 1 mm high). The maximum heat flux was limited by bubble formation on the electrode.

Shadowgraph and dye techniques were used either simultaneously or separately. The light source for the shadowgraph was a 15 mW He-Ne laser. For the dye visualization, a 3 W argon laser beam in the form of a 0.5 mm thick vertical sheet illuminated a section containing the symmetry axis. Since fluorescent dyes were used, only the dye in this section was visible. The salt solution below the electrode was tagged with Fluorescein dye (yellow-green) while vertical lines in the illuminated plane were tagged orange-red by dropping particles of Rhodamine B in the solution. Also were used ciné film and small particles dyed by special fluoresecent dye. The latter was yellow and was excited by the blue light of the argon laser. Use of a yellow filter enabled to obtain flow details with very small particles ($3\mu m$ in diameter).

It was found that some important details of the flow pattern were lost in the shadowgraph pictures, since the image in this case is an integrated one along the path of the light and since the fluid density in the upper region of the layers is almost homogeneous. Therefore, most of results to be presented will be based on the dye visualization.

3. OBSERVATIONS

3.1 The initial stage

Immediately after the heating is turned on, a vortex cap forms at the front of the starting plume as it rises (figure 1a). Owing to the excess of momentum the plume overshoots the level of zero buoyancy. It forms a dome on top as it reaches the maximum height, falls downwards in an annular region forming a double plume structure, and then spreads out horizontally at its own density level. Depending on the Rayleigh number the outer falling annular plume can be laminar or turbulent. The salty water raised from the heat-source level (green liquid) spreads out in one layer slightly above its original density level. Further on, layers form successively above this one, since it is warmer (and more saline) than the fluid above. The layers also develope horizontally, and systematic shearing motions set up in the layers. An important detail of the layer development is also the formation of vortices (rolls) near the first plume top and their slightly sloping advance into the ambient fluid.

After the initial stage, a mixing-layer-type instability develops in the shear region between the annular plume and the ambient fluid (see figure 1*b* and figure 4 in [Tsinober et al., 1983]). Vortices form as a result of this instability. These vortices distort the outer boundary of the annular falling plume. The horizontal density gradient between the plume and the ambient fluid contributes to the vorticity and enhances the shear-generated vortices. The outer boundary of the plume becomes strongly convoluted and breaks down, and the annular plume becomes turbulent. We note, however, that the rising circular plume remains laminar during the whole observation time (about 3 h).

3.2 Double- difusive layers

As the turbulent annular plume develops, layers of three types form in a Christmas tree flow pattern (figure 1*a*-*d*). The layers of the first type (type I) form on top of the basic plume, since the warm top of the basic plume acts as heat source for the fluid situated above the plume. In such away a hierarchy of plumes forms successively on top of the basic one, which later develops into a system of layers. An example of this hierarchy is seen figure 1*d*, where at least eleven successive secondary plumes can be seen. The number of layers in this hierarchy (at a particular time) depends upon the Rayleigh number in a surprisingly non-monotonic manner. A maximum of this number was observed at Rayleigh number around $2 \cdot 10^4$.

The layers of the second type (type II) are formed around the upper part (roughly one-third) of the basic plume. These layers are driven by the horizontal temperature difference between the warm basic plume and the cold ambient fluid. Therefore, these layers are similar to those developing in the side-wall heating experiments. For references see [Huppert and Turner, 1981], [Tsinober et al., 1983] and [Turner, 1985]. In most cases, one layer of this type was formed.

Finally, the layers of the the third type form around the lower part of the basic plume. These layers are driven by the same mechanism as the layers of type II. However, they contain green (warmer and more saline) liquid, originating from the heat-source level, and in this respect are similar to the double-diffusive intrusions studied by Turner in 1978 and to the layers developing at fronts described by Ruddick & Turner in 1979. In contrast with the number of the type-I layers the number of layers of types II and III together was independent of the Rayleigh number and equal to three, and at larger times to four.

An example of these three types of layers is shown in figure 5 in [Tsinober et al., 1983]. They are seen also in figure 1*e-g*. Note the salt fingers in the right part of the second layer of figure 1*g* (also figure 5*b* in [Tsinober et al., 1983]).

It should be noted that the layers of the type I formed successively, whereas the layers of types II and III formed simultaneously.

3.3 The structure of the layers

The layers of all three types consist of a mixed upper region (orange for types I and II and green for type III), and of a non-mixed dark lower region. The dark lower region consists of the ambient fluid, which moves slightly upwards towards the warm plume where, due to heating from the basic plume, it becomes lighter and flows out into the interior as the mixed upper region of the layer. Thus systematic shearing motions are set up in the layers, which is the motion characteristic of layers formed by horizontal (lateral) temperature and salinty gradients.

The shearing flow extends beyond the mixed upper regions of the layers into the ambient fluid, which is seen from the deformation of the dye streaks shown in figure 1*e-f*, see also figure 2. This is generally attributed to the upstream influence produced by internal waves with small vertical wavelength.

In addition to the radial motions, systematic azimuthal shearing motions are observed in the layers. The sense of the motions in the mixed upper part of the layer is opposite to the one in the lower dark part of the layer. In addition, the sense of these motions at a fixed level is not the same for different polar angles, which indicates the existence of horizontal cells. No azimuthal motions are observed beyond the mixed upper regions of the layers (and corresponding dark regions) in the ambient fluid, where radial motions still persist. We have not been able to develop an appropriate visualization technique for these motions, which were observed simply by dropping a dye particle and following the development of their traces. An example and interpretation of this kind of observation is shown in figure 5*b* in [Tsinober et al., 1983].

Another important feature of the structure of both type-I and type II layers is that the upper region of the layers consists of a system of vortices or rolls, which are seen in figures 1*e-g* and figure 2. The time-lapse motion-pictures observation shows clearly that the vortices in the layers of type II are created near the plume when the dark lower part of the layer approaches the plume, and then are advected away from it. The vortices in the layers of type I are the caps of the secondary plumes separated from it and advected into the ambient fluid. The tilting of the type-I and II layers is due to double-diffusive convection driven by the buoyancy flux across the diffusive interfaces, separating the mixed part of one layer and the non-mixed part of the next layer above it. The creation of the vortices is due to heating of the approaching ambient fluid in the lower nonmixed part of the layers and to strong horizontal velocity gradients near the plume core. Both causes contribute to the vorticity in the same sense. Once created, these vortices do not become passive after leaving the plume. They remain active since they are driven by the three following factors. The first one is the shear across the interface between the mixed upper part of the layer, and the lower unmixed part of the layer moving toward the plume. The second process is the upward buoyancy flux due to the negative temperature gradient produced by the larger horizontal heat flux in the lower layer. Finally, the third factor is the horizontal density gradient. All the three factors contribute to the vorticity in the vortices in the same sense.

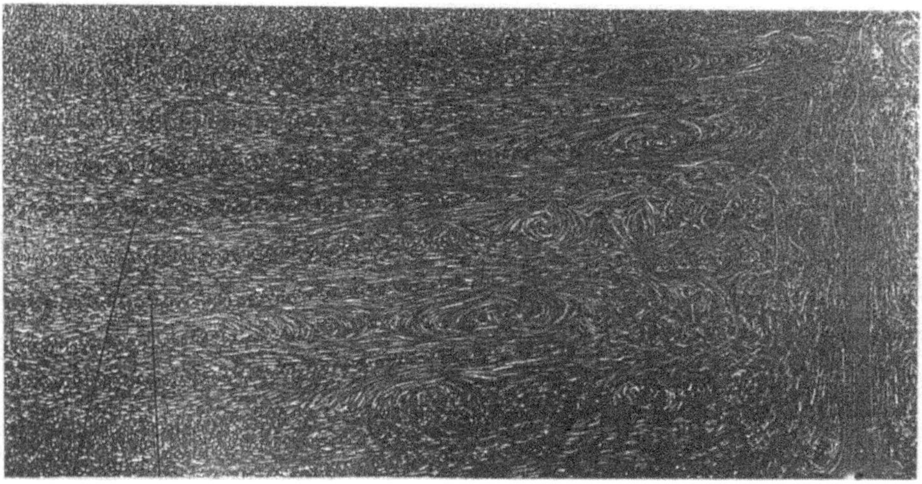

Figure 2; *Structure of the flow in double-diffusive layers as visualized by 3 μm particles dyed by a special fluorescent dye (see the text). The solid lines indicate the upper and the lower interfaces of a layer, and the tip of the heating electrode appears as a black dot in the lower right corner of the picture. The height of the figure corresponds to 19*cm *in the experiment.*

The layers of the type III (the most lower ones) consist mostly of the fluid originating from the heat source level. Therefore, they contain an excess of salt and temperature, so that a diffusive interface forms above and salt fingers below these layers. An example of such (sheared) salt fingers is seen in the left second layer from below in figure 1*g*.

4. CONCLUDING REMARKS

Visual observations of the development of an *initially* simple system consisting of a salinity gradient heated from below reveal quite a *complicated* and *beatuful* subsequent pattern in the form of a Christmas-tree. It consists of a rising primary plume capped by vortex dome, a hierarchy of secondary plumes on top of the primary one, a falling downwards annular plume and a system of layers.

The layers consist of a mixed upper region moving away from the plume and non-mixed lower region of ambient fluid moving towards the plume and forming in this manner a systematic radial shearing motion. The upper mixed region of the layers contains vortices driven by the horizontal shear in the layer, the upward buoyancy flux and the horizontal density gradient. In addition the lower layers (of type III) contained salt fingers.

In addition to the radial systematic azimuthal shearing motions were observed in the layers. The patterns of these latter were quite intricate and really impressive. From above the lower layers look as multi-petal flowers, while the upper ones have much finer structure. Visualization of their beauty is not a trivial matter, but definitely deserves to be realized. Among the difficulties in producing a beautiful visualization form above is that the layers are not horizontal – they are tilted, another one is that the dye is introduced from above: its remnants are not that beautiful.

REFERENCES

Huppert, H.E.; Turner, J.S.; 1981; "Double diffusive convection"; J. Fluid Mech. 106, 299-329.

Stern, M.E.; 1975; "Ocean circulation physics, Chapter 11; Academic Press.

Turner, S.J.; 1985; Multi-component convection"; Ann. Rev. Fluid Mech. 17, 11-44.

Tsinober, A.B.; Yahalom, Y.; Shlien, D.J.; 1983; J. Fluid Mech. 135, 199-217.

POWER OF BEAUTY: STREAMWISE VORTICES

N. Yurchenko

Institute of Hydromechanics NASU, 8/4 Zheliabov St., Kiev, Ukraine
Phone: +38 044 441-74-52, Fax: +38 044 446-4229, E-mail: yur@rdfm.freenet.kiev.ua

Abstract: Streamwise vortices are studied as an intriguing feature intrinsic to the vast variety of fluid motion types. The studies are focused on mechanisms of the vortical structure evolution both under natural and controlled conditions of its formation. Special attention is paid to practical applications of the fundamental knowledge and advantages of this structure for flow control. Experimental and numerical results reveal peculiarities of the vortex scale transformation and possibilities to favorably modify the vortical structure, in particular, for heat transfer enhancement.
Keywords: streamwise vortices, motion scales, flow control.

1. INTRODUCTION

Many faces and names of fluid motion of a similar type (streaky structure, longitudinal/Taylor/Goertler vortices, coherent vortices, etc.) display its attractiveness for observers and researchers. Elegance of elongated shapes connected with some kinds of motion even in a "frozen", naturally visualized, form can be seen in Figure 1 though being of a speculative nature if to refer to longitudinal vortices. Streamwise vortical structure can also be visualized and observed naturally due to sand, vapor, dust, and fish-scale or in reflected works of art. It is an important structural element typical for various flows, such as

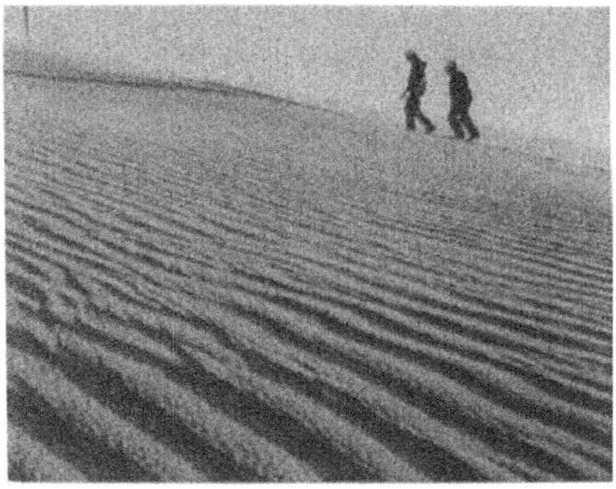

***Figure 1**; "Streaky" structure.*

- laminar-turbulent boundary layers;
- flows over heated surfaces;
- turbulent spot environment in a form of a "fringe" from streamwise vortices;
- sublayers of turbulent boundary layers;
- rotating flows, e.g. those in a gap between concentric cylinders (Taylor vortices);
- secondary flows in curved channels and river-beds;
- MHD flows with imposed transverse magnetic fields;
- Langmuir circulation in ocean...

A. Gyr et al. (eds.), Science and Art Symposium 2000, 59-64.

A universal form of a fluid motion supposes similar mechanisms of its development as well as similar approaches to its control. In this connection, the research problem was formulated as follows.

Analysis of emergence and development of streamwise vortices generated in boundary layers affected by centrifugal (body) forces, i.e. in flows where they are a naturally dominating vortical structure: space-scale selection process of evolving vortices generated with a given scale; vortex scale transformation downstream and across boundary layer with an emphasis to possible applications of the obtained fundamental results.

2. INVESTIGATION APPROACH. FACILITIES.

It is known that evolution created a streamwise skin structure of fast marine animals, the more orderly, the faster a creature is. This finding stimulated a number of research works dealt with its fluid dynamical interpretation, modeling and applications (in particular, for drag reduction).

As long as a hint to advantages of streamwisely organized motion was given by marine animals, their skin properties were exhaustively studied and served as a prototype for engineering solutions, e.g. in a form of riblets. However the present strategy of research and approaches to the boundary-layer control is based on available streamwise vortices with certain parameters not depending on a way of their formation. Therefore the main issue of the work and, consequently, its resolution consists in generation of a streamwise vortical system using whatsoever means and further analysis of this system behavior. Figure 1 shows how vortices were practically generated, as well as a general scheme of an experimental/numerical approach to the problem solution.

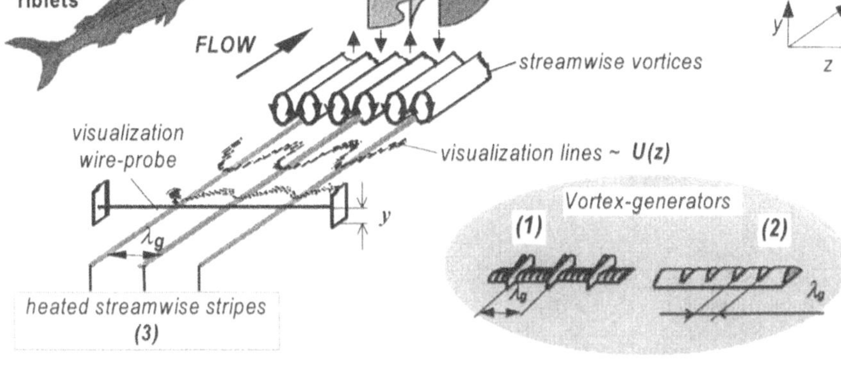

***Figure 2**; Sketch of experiment formulation and arrangements.*

Another important issue of the work refers to modification of a streamwise vortical structure to control flows where it is their inherent feature, a dominating vortical element under certain conditions. Here, these are boundary-layer flows under body forces, such as centrifugal forces and buoyancy. Combined experimental and numerical investigations of boundary layers over concave surfaces [Yurchenko et al., 1999] were carried out in a frame of the Goertler stability theory [Saric, 1994]. To show practical benefit of streamwise vortices for heat transfer enhancement, the experimental part of the work was extended to studies of a turbulent boundary layer over a heated flat plate [Yurchenko, 1998] with transversely mounted vortex-generator arrays of (2) and (3) types as shown in Figure 2.

Most experiments were carried out in a low-turbulent water channel with an open test section of 10 x 25 x 300 cm at the length-based Reynolds numbers, Re=(0.6-1.0)$\cdot 10^5$. The studied boundary layer developed on its bottom, 25 x 300 cm, which could be flat or contain a concave section (with a curvature radii, R = 1.0, 4.0 and 12 m). Streamwise vortices were generated using mechanical vortex-generators (1) and regularly spaced along z longitudinal flush-mounted heated stripes (3). Scales, λ_g, of generated vortices covered a range from neutral to most amplified vortices according to the Goertler diagram. A typical wavy shape of visualized U (z) velocity profiles gave clear evidence of available streamwise vortices in a boundary layer and enabled to analyze a flow response to any external excitation and, subsequently, to the choice of control parameters. The matched numerical approach to studies of generated streamwise vortices was based on their thermal excitation that was formulated as a constant boundary condition in a form of a periodic temperature gradient $\Delta T(z)$. A code developed for the direct numerical simulation of a laminar-turbulent transition in compressible subsonic boundary layers was applied here with the inclusion of body force terms into the Navier-Srokes equations.

The second part of the experiments was carried out in a turbulent boundary layer under buoyancy, i.e. over a uniformly heated horizontal test plate in a wind tunnel. It aimed for comparative studies of heat transfer enhancement downstream standard two-dimensional transverse ribs and specially constructed three-dimensional ones (see Figure 2, (2). Regularly (λ_g) spaced grooves along them provided conditions for a z-periodic flow separation and downstream reattachment and, accordingly, stimulated the development of a regular vortical structure in the reattached flow.

Local surface temperature was measured using a set of thermocouples flush-mounted over a surface to register an effect of generated streamwise vortices on heat transfer. Wave-like distribution of Stanton number along z (Figure 3) similar to U(z) shown in Figure 2 displays the embedded streamwise vortical structure.

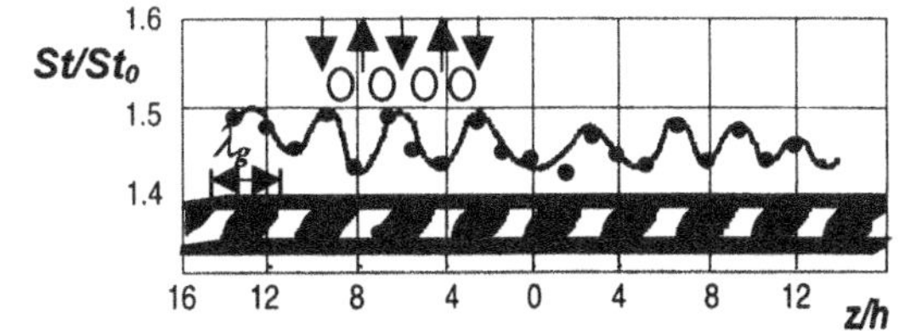

Figure 3; *Heat transfer distribution downstream of a transverse 3D rib:* $h=5mm$, $\lambda_g=12mm$, $x_0 = 0.7m$, $Re=1.2\cdot 10^6$.

3. RESULTS AND DISCUSSION

To get an insight into vortex dynamics under conditions of streamwise vortical structure generation, first of all, natural evolution of the vortices from their emergence to breakdown in a process of a laminar-turbulent transition was considered. Good agreement was found between experimental and numerical results that are resumed in a form of a sequence of flow field patterns taken during the evolution process (Figure 4, case 0) [Yurchenko et. al., 1999]. It was found to be typical that in the very beginning of the vortical system formation, the overall vortex size and the vertical position of the vortex core remain constant. This stage is followed by rapid growth of the vertical scale of the vortices and, consequently, by the development of strong shear layers separating adjacent counter-rotating vortices. Finally it results in the breakdown of this initial vortical system and formation of a new one with different space scales.

Forced development of streamwise vortices was analyzed using experimental and numerical results overlapped for two considered scales, λ_g, of generated vortices. They are shown in Figure 4 as Cases 1 and 2. Case 1 describes excitation of a purely second mode with a non-dimensional vortex scale $\Lambda=84$ linearly amplified according to the Goertler stability diagram; Case 2 corresponds to generation of a slightly irregular vortical structure that means additional stimulation of the most amplified first mode, $\Lambda=236$. Practically, Case 2 was realized due to one of the heated stripes insignificantly shifted along z. Both experimental and numerical visualization demonstrated fast growth and breakdown of the most amplified vortices in spite of small amplitudes of initially introduced disturbances and in spite of the imposed constant boundary condition $\Delta T(z)$. Generation of vortices with scales close to neutral (having zero growth rate) and even with $\Lambda\approx84$ slowed down the process of the vortex structure development.

The vortical structure transformation was observed not only in a downstream but also in a normal to the surface direction. Generated vortices gradually change their shape and scale giving rise to a smaller scale vortex pair near a wall if their scales are different from a scale which can be supported by the flow under given conditions. It is well seen in the first patterns of Case 1, Figure 4. The same conclusions were made from simultaneous flow visualization in xz-plane at two different distances from a wall. Visualization-wire probes oriented along z emitted colored lines propagating downstream and taking a form of U(z) velocity profiles, as it is shown in Figure 1. The wavy pattern adjacent to the wall had almost twice shorter wavelength compared to one in the outer region of the boundary layer.

These fundamental results should be taken into account if applied to control laminar-turbulent transition and transport processes near a wall. In particular, they can help to determine parameters of vortices (correlated with basic flow parameters) with longer lifetime after they have been generated in a boundary layer.

The advantage of regular streamwise vortices compared to non-organized vortical motion is demonstrated in Figure 5 for heat transfer enhancement using the transverse 3D rib shown in Figures 1 and 3 versus a 2D rib typically applied for this purpose (both in laboratory experiments and as an element of heat exchangers).

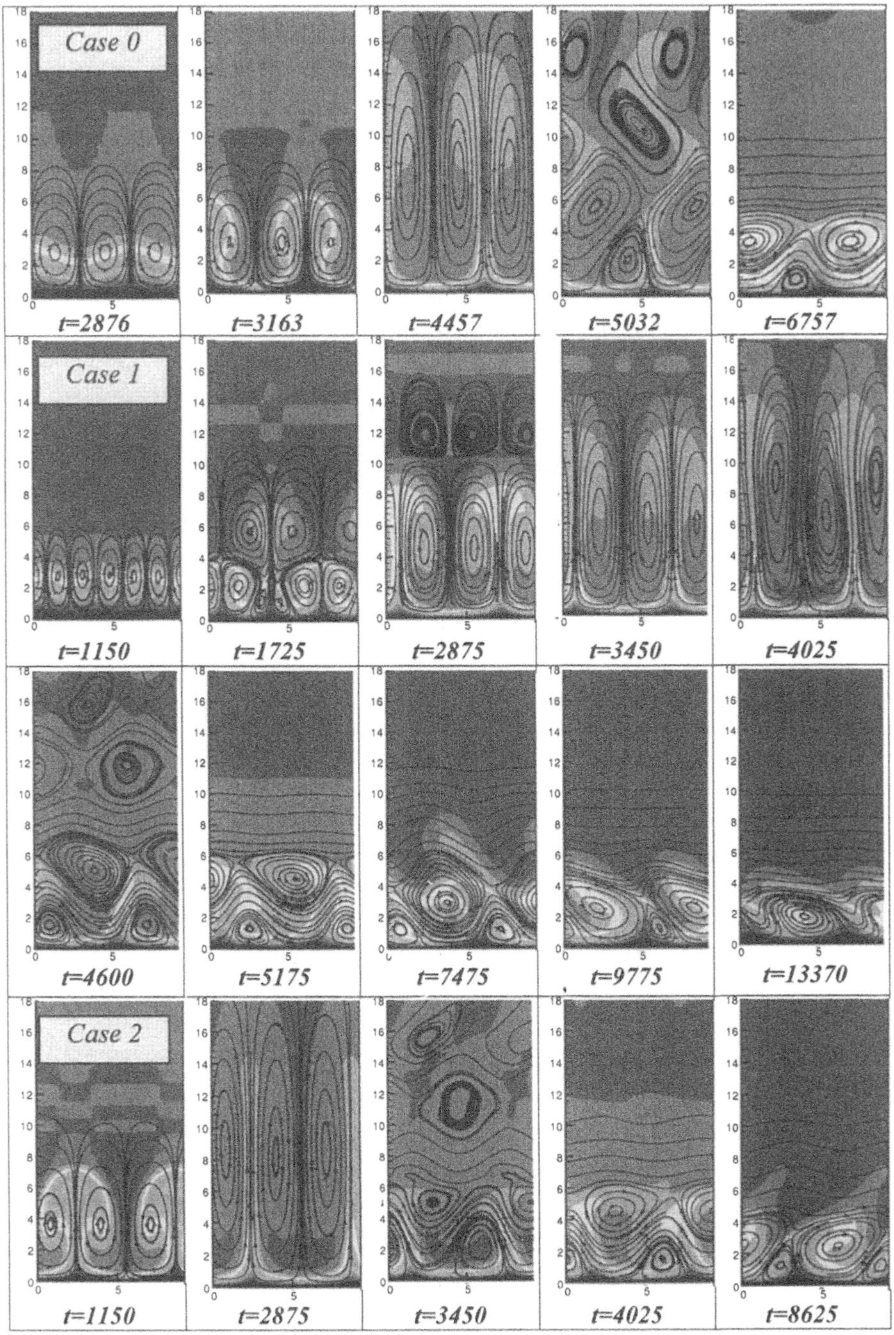

Figure 4; *Contour maps of iso-U-velocity traces in yz-plane.*

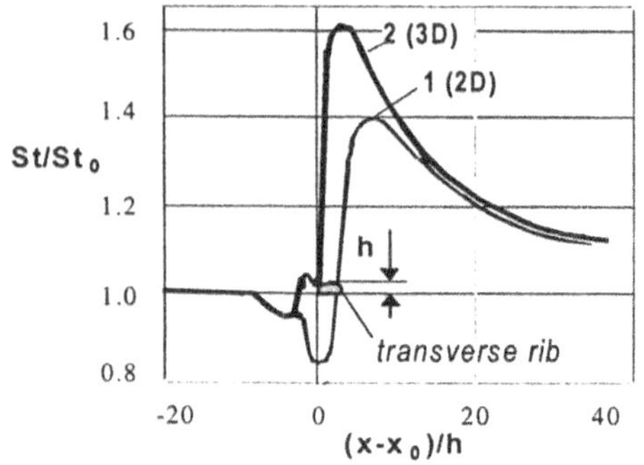

Figure 5; *Heat transfer enhancement around (1) a conventional transverse 2D rib and (2) a 3D rib generating streamwise vortices (experimental conditions are as in Figure 2).*

4. CONCLUSIONS

Alluring prevalence of streamwise vortices in various flows was considered to reveal its mechanism and efficient ways of flow control. A balanced complex of experimental and numerical studies of natural and forced vortical structure evolution showed its scale transformation downstream and across a boundary layer depending on correlation between parameters of generated vortices and the flow. This process can be controlled resulting e.g. in delay of the transition to turbulence or in heat transfer enhancement.

Rush hours...days, and months. Rush YEARS!
We are used to life in haste
as if punished by some nervous devils
insisting that we see no colours, feel no taste.

Dragged through life by deadlines, terms and duties
never stopping, never time to rest,
to look about and find the greatest beauty,
in simple things, to feel them blessed....

REFERENCES

Saric, W.S.; 1994; "Goertler vortices"; Ann. Rev. Fluid Mech. 26, 379.

Yurchenko, N.F.; 1998; "Optimization of heat transfer control based on a receptivity approach"; In: Proc. Turbulent Heat Transfer Conf. Manchester, UK, P72-P81.

Yurchenko, N.F.; Delfs, J.W.; 1999; "Optimal control of boundary layers under body forces"; In: Proc. IUTAM Symp. On Laminar Turbulent Transition, Sedona, USA, to be published.

ACKNOWLEDGEMENTS

This material is based on work supported by the European Office of Aerospace Research and Development, Air Force Office of Scientific Research, Air Force Research Laboratory under the contract F61775-99-WE075. The author acknowledges with thankfulness the encouragement of Dr. Charbel Raffoul during the whole course of the work implementation.

Interaction Between the "Two Cultures": Art & Science

Abraham Tamir

Department of Chemical Engineering, Ben-Gurion University of the Negev, Beer-Sheva, Israel. E-mail: atamir@bgumail.bgu.ac.il

Abstract: This paper presents an approach demonstrating the following interaction between art & science: art is used as a means to illustrate science, on the one hand, and science serves as an instrumentality in creating art, on the other. Its major objective is to educate people to see science in the art and art in the science.

A demonstration of the subject appears in *The museum of Art & Science* founded by the author in 1998 at Ben-Gurion University of the Negev in Israel (www.bgu.ac.il/museum) as well as in the book *The One Face of Art & Science* by the author containing about 500 pictures demonstrating the above interaction. This book may also serve as a useful text book for a new introductory course in schools or universities on the *Interactions Between Art &* Science as well as a source of ideas for establishing exhibitions or museums on *Art & Science* for the benefit of the society and the future generation.

Keywords: Art and Science

INTRODUCTION

Leonardo da Vinci, consummate genius: artist, sculptor, architect, urban planner, inventor, scientist, writer, musician - not exactly a slouch in creative or intellectual pursuits - was probably, the first to utilize the close interrelationship between art & science. His famous *Mona Lisa* painted around 1503-1506, the epitome of art, represents scientific principles imbedded in the picture through the use of perspective and optical illusion, that usually escape the casual observer.

It was Cheng-Dau Lee, Nobel laureate in physics, who expressed the following about the close interaction between art & science. He said: "Both, science and art are the presentation of the creature of the human being. They are not separated from each other. There is even a similarity between them as they help us observe nature. With the help of science we can find out routines of nature. On the other hand, by means of art we can describe the emotions of nature."

Since science and art and the interaction between the two are major conceptions in the presentation, elaboration of these is made in the following.

A. Gyr et al. (eds.), Science and Art Symposium 2000, 65-76.

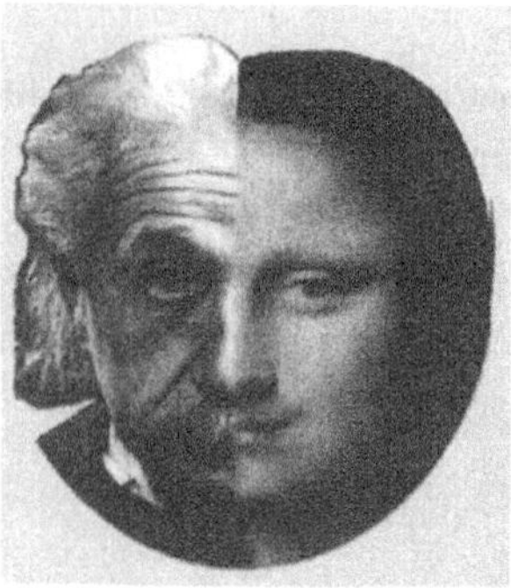

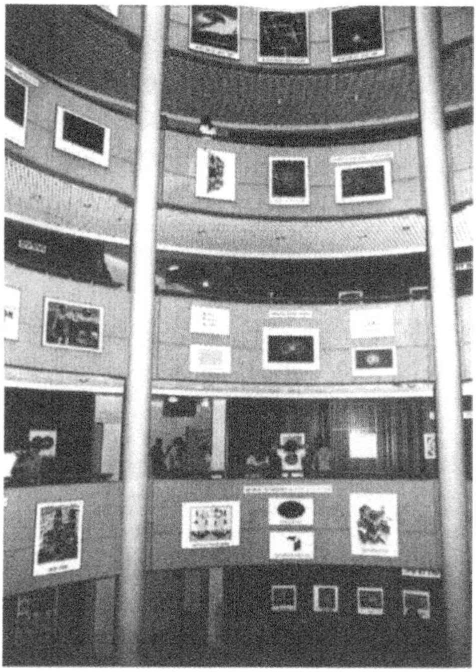

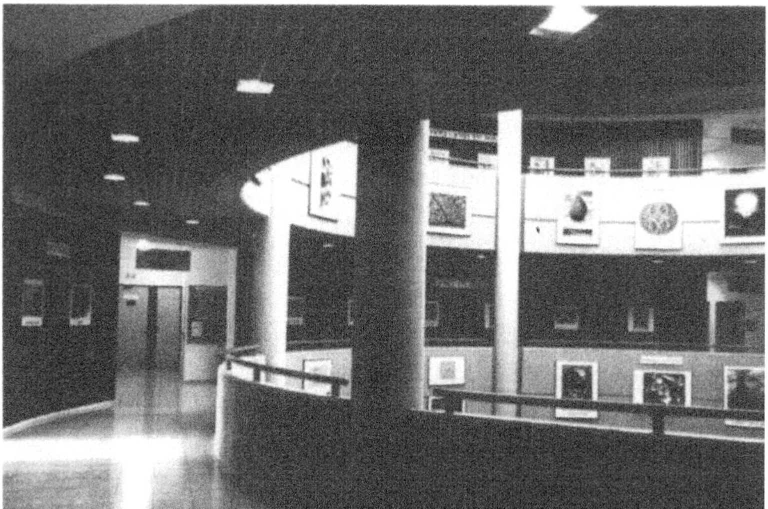

Horizontal view

Fig.1. The Museum of Art & Science at Ben-Gurion University of the Negev

WHAT IS SCIENCE?

Science to the layman is, usually, identified with something precise, such as mathematics or physics. Science is much easier to define in comparison to art because it is exact, it does not depend on one's emotions, and has a very limited, but essential scope to reveal the laws of nature which govern our life every second. Many definitions for "science", the Latin for "knowledge", are available; however, only a few are spelled out below; a more comprehensive list is given elsewhere [Tamir, 2000]. For example, the *Oxford Advanced Learner's Dictionary* and Even-Shoshan's *New Hebrew Dictionary* definitions are summarized as follows: "Science is a systematic research in a certain subject, such as in mathematics, physics, botany, and history, organized and based on facts, observations or experiments and summarized into laws, rules and axioms". Thus, "a scientist is a scholar occupied in research of a certain subject". Or, in an early edition of the *Encyclopedia Britanica* "science is certainty" while in the new edition "science is the knowledge of the world of nature". The latter is universal among mankind, and this is why it has exited since the dawn of human existence. Campbell's [Campbell, 1953] definition is very interesting. He says that nature means, practically, the part of the world which man regards as external to him. Thus, "Science is that branch of pure learning which is concerned with the properties of the external world."

Other interesting questions are: What is science for, and how is science acting? There are two aspects of science. First, practical science, which is a body of useful and practical knowledge and methods of obtaining it. In its second aspect, pure science is a branch of pure learning which aims at intellectual satisfaction. Science is not an individual experience. It is a shared knowledge based on a common understanding of some aspects of the physical or social world. Science studies laws of nature and we seek to discover laws in order to make nature intelligible to us.

WHAT IS ART?

"Art" like "love" is a word that every one knows what it means, most of us will tell you that it is necessary, that it is also good for the soul and that they enjoy it, as well as that it is an activity which manifests beauty. On the other hand, people will have difficulties in defining art.

The volume of definitions of art is enormous where only a few are spelled out. Picasso expressed once the following idea: "We all know that art is not truth. Art is a lie which teaches us how to comprehend the truth, or at least that truth we human beings are capable of comprehending" while Duchamp claims that: "A product becomes an art product if it enters a museum." Tolstoy brings a long list of definitions of different aestheticians based on the concept of "beauty", for example: according to Batteux: "Art consists in imitating the beauty of nature, and its aim is pleasure" while Vernon expressed that: "Art is the manifestation of emotions, conveyed externally by a combination of lines, forms, colors, or a sequence of gestures, sounds, or words subject to certain rhythms." If we discard the concepts of beauty, the definition according to Tolstoy reads: "Art is that human activity which is based on the fact that man, as he receives through hearing or sight the expressions of another man's feelings, is capable of experiencing the same feelings as the man who expresses them." Art, therefore, is a means of human communition, necessary for life and for the movement towards the good of the individual man and of mankind, uniting them in the same feeling.

My understanding of what is art has been crystallized along my scientific career as well as my interest in art, mainly painting. "Art is all what affects our senses and emotions resulting from different stimuli which penetrate into our brain creating there the"work of art" what we eventually see or hear. These stimuli may be optical, acoustic, touch, feel, smell and others." The means for creating art, for example, paintbrush, computer, knife, telecommunication technologies, hammer, internet, etc., are immaterial. An interesting question is what does art do to people? According to Dissanayake [Dissanayake, 1995], quote, "Art is said to be both pleasurable and advantageous because it is therapeutic: it integrates for us powerful contradictory and disturbing feelings; it allows to escape from tedium or permits temporary participation in a more desirable alternative world; it provides encouraging illusions; it promotes catharsis of disturbing emotions, and so forth."

INTERACTIONS BETWEEN ART & SCIENCE

It is my belief that investing in the future generation is the most promising way towards achieving a fruitful interaction between art & science. And indeed, efforts in this direction have been made in recent years. One attempt, realized in California, is described elsewhere [Kimberley, 1994]. The underlying philosophy behind this experiment was, quote: "Art & science are two different ways of understanding and knowing the world around us ... Through the connection of art and science, students increase their appreciation of the natural world and develop an openness to the wonder and joy of life."

Another attempt, made in Israel, was the foundation by the author of this paper, in 1998, the *Museum of Art & Science* at Ben-Gurion University of the Negev in Beer-Sheva . The philosophy of the museum, which aims at educating the future generation in science through art, and vice versa, is that art is used as a means to illustrate science, and science is instrumental in creating art. Such a combination between Art & Science in a museum is unique, making it the first of its kind. The museum is located in the central lobby of the four-story Kreitman-Zlotowski Building which reminds the famous Guggenheim Museum in New York. A vertical and horizontal sections of the museum are depicted in Fig.1, where there are totally five vertical sections. The building is usually crowded with students, faculty, school children, as well as the general public, who are, thus, continuously exposed to famous art paintings. Guided tours are also conducted. The museum displays holograms and approximately 200 works of famous artists to demonstrate the above duality between Art & Science.

DEMONSTRATING SCIENTIFIC SUBJECTS THROUGH PAINTINGS

Fig.2 demonstrates how art may be used as a means to illustrate science. In the book *The One Face of Art & Science*, approximately 500 pictures demonstrate *science via art* and *art via science* for the following sciences: Natural sciences, Life sciences, Technological sciences, Social sciences and Science fiction. In the following, a description is given of a scientific subject related to the corresponding picture. The titles appearing are my interpretation of the scientific concept demonstrated; the artist's name appears as well. It should be empasized that in the majority of the cases, artists did not even think about the scientific idea imbedded in their painting.

The first conception is *Mass*. This basic characteristic of a material body, which is the constant of proportionality in Newton's law $F \propto a$, is demonstrated fantastically in Fernando Botero's painting "The bath".

Black hole is a hypothetical region of space possesing a *gravitational field* so intense that no matter as well as radiation can escape from it. This is also the reason why its name is black hole. Such regions are believed to form as a result of the *gravitational collapse* of a star. There are many pictures demonstrating a black hole in which an impression of suction is transferred by the artist to the observer, as the one presented here.

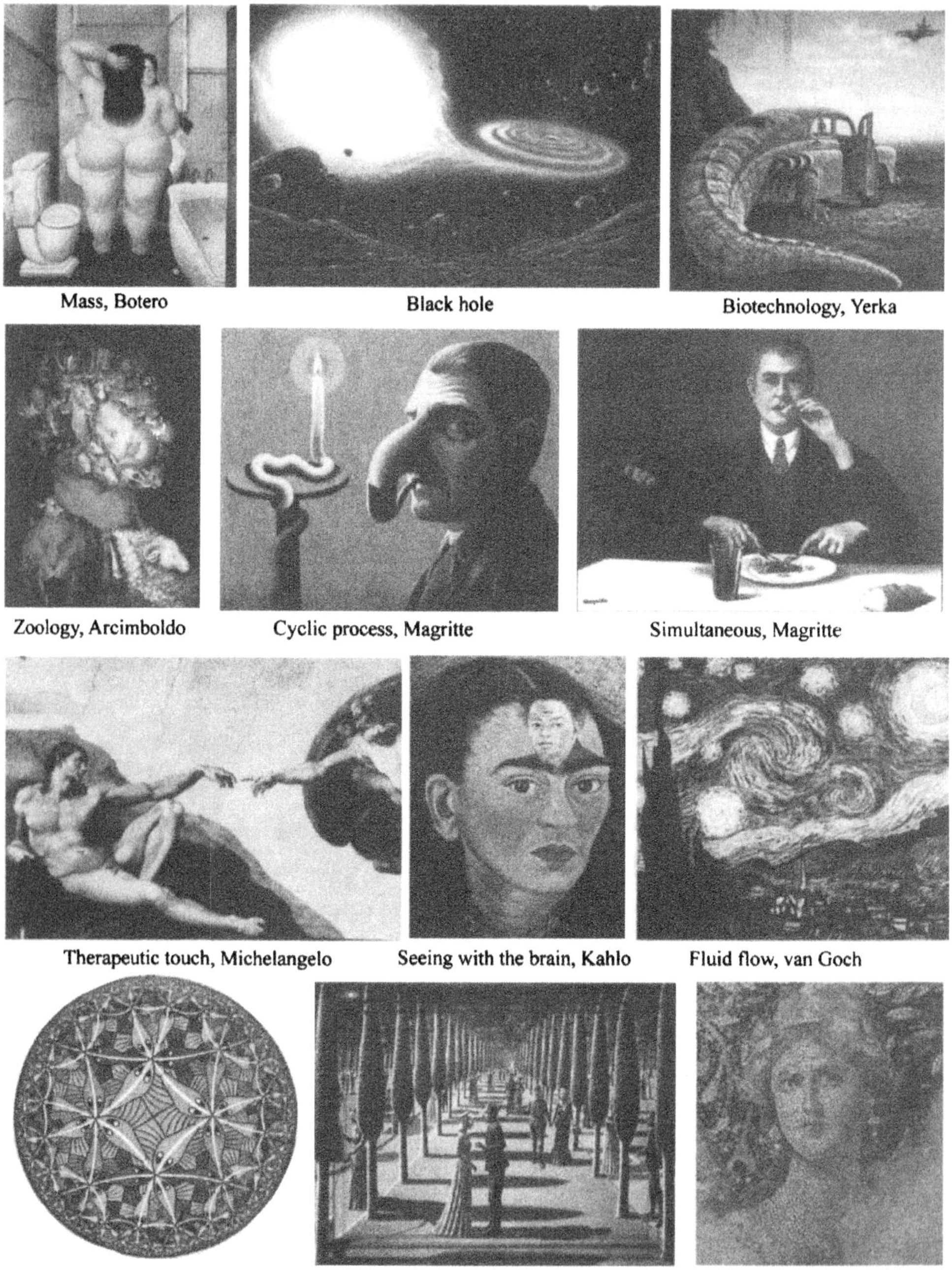

Mass, Botero | Black hole | Biotechnology, Yerka

Zoology, Arcimboldo | Cyclic process, Magritte | Simultaneous, Magritte

Therapeutic touch, Michelangelo | Seeing with the brain, Kahlo | Fluid flow, van Goch

Infinity, Escher | Perspective, Crook | Archeology

Fig.2. Science via Art

Biotechnology, incorporates biology and technological processes to produce such substances as cheese, antibiotics, beer, etc. The painting "Attack at Down" by Jacek Yerka's, a talented Polish artisrt, demonstrates this combination in a most convincing way.

Zoology, the scientific study of animals, is demonstrated by Guiseppe Arcimboldo in his ingenious painting "Eart" of 1570. As seen, many animals create the profile of the head such as: Indian gazzle, a fallow doe, a leopard, a dog, a deer, an ibex, a rhinoceros, a mule, a monkey, a bear, a lion, a horse, an elephant, a donkey, a mouse, a fox, a wolf, a cat as well as a lizard coming out of the open mouth.

Cyclic process is that in which a system is restored periodically to its original state. Such processes are intimately related to production of work from heat, the major job of thermodynamics, and specifically to the Carnot cycle with the highest possible efficiency. "The Philosopher's Lamp" by Magritte demonstrates such processes most impressively. By the way, almost every book that reports the death of Carnot, indicates that he died of Cholera. However, the truth is that he passed away in an asylum for rich people, probably of Gastroenteritis [Wisniak].

Simultaneous - different events occuring concurrently - is a conception widely encountered, such as simultanous equations, users, translation, or engineering. Magritte, in his fascinating painting "The magician", gives this conception a wonderful demonstration.

Alternative medicine - or more specifically - *the therapeutic touch* is demonstrated in *"The Creation of Adam"* by Michelangelo. Such a touch, which doesn't involve physical contact at all is based on the premise that when a person is ill, the body's energy forces become interrupted or unbalanced. Skilled practitioners believe that they can detect and influence these energy fluctuations by passing their hands above a person's body from head to toe to create an interaction between their own and patient's energy field. In the picture, Adam is seen with breath of life, already physically complete and alive; so he was already created by God. However, one may note that his face look, somehow, not in best shape - may be he is sick - and God, near him, is using *the therapeutic touch* to heal him.

Seeing with the brain, related to *psychology* which is the science that studies behavior of the soul, is demonstrated by the painting *Diego and I* of Frida Kahlo. She painted this self-portrait during the period when Diego Rivera, her husband, was having an affair with a film star, a relationship which provoked a public scandal. Wretched and weeping, she looks mournfully, seeing the betraying husband in her brain.

Fluid flow, a science concerned with the motion of fluids and the conditions affecting that motion, is demonstrated by van Goch in the most famous of all his paintings *Starry Night*. One sees vortices, interaction between streamlines formed in the flow over the stars as well as the boundary layer created above the mountains.

Infinity, a quantity greater than any assignable quantity, can never be experienced by humans. However, Escher, demonstrated infinity by means of regular division of a plane, or by reducing the size of some basic configuration. While looking at *Circle Limit III*, infinity is demonstrated by outward size reduction. The fishes keep diminishing, each time to half their preceding size, where the finite circle encloses an infinite number of fishes.

Perspective, representing a three-dimensional space on a flat plane, is the realization of how the brain operates and the way we perceive space and its content. Brunelleschi, in the 15th century, was the first to apply perspective in painting. The most important consequence of linear perspective is that if we look on an object of a constant width from different distances, the longer the distance the shorter will be the length of the object seen in the brain. The above concepts are demonstrated convincingly by P.J. Crook in his *The Tryst*.

And finally, *Archeology*, which is the study of the human past. The picture demonstrates a detail of a spectacular mosaic floor from the Roman period, 3rd century C.E., which is displayed in the Israel Museum.

SCIENCE SERVES AS INSTRUMENTALITY IN CREATING ART

Fig.3 demonstrates "the 180 degrees" of the previous chapter, i.e. art via science, where an extended presentation appears in the book *The One Face of Art & Science* [Tamir, 2000]

We begin with *Fractals* - patterns of chaos - extremely aesthetic patterns, generated by non-linear equations, that consist of an identical motif which repeats itself on an ever diminishing scale. It was Benoit Mandelbrot, the inventor of fractal geometry, who at the early 1980s used simple iterative equations, which produced spectacular patterns such as the one depicted in Fig.2 [Briggs, 1992]. Such patterns, and numerous others, may be considered as art while emphasizing that the means for

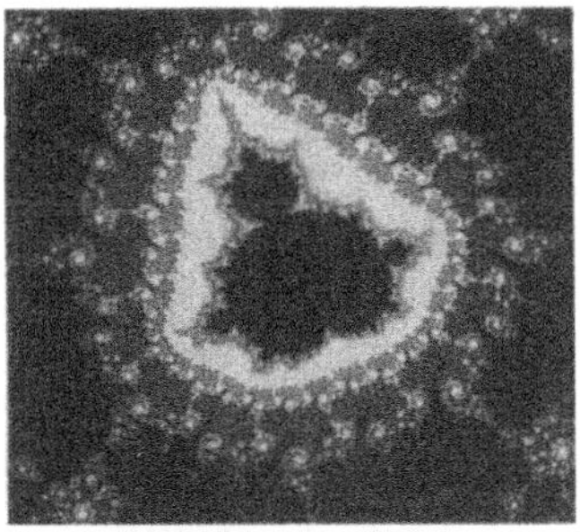

Fractals, Mandelbrot

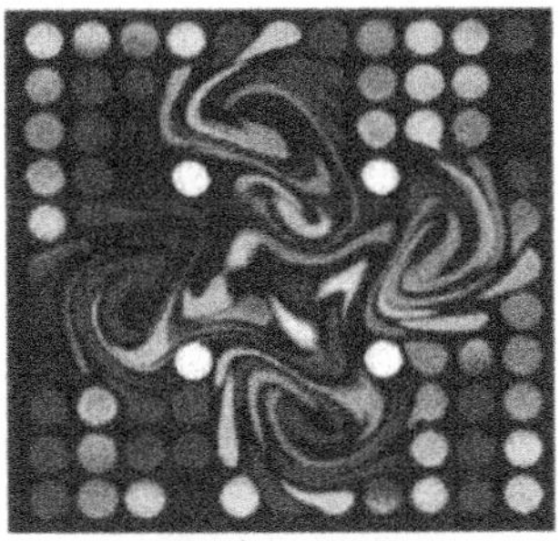

Rheo-Art, Takaki

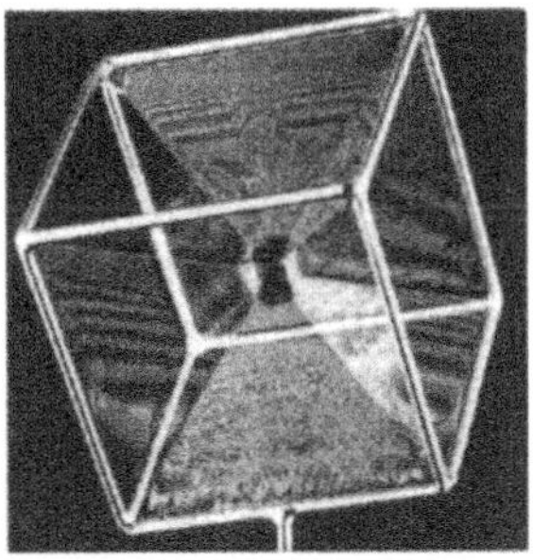

Surface tension, Deckart

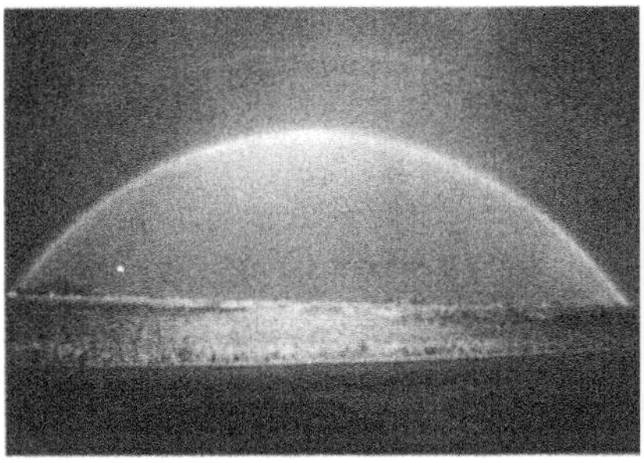

Rainbow

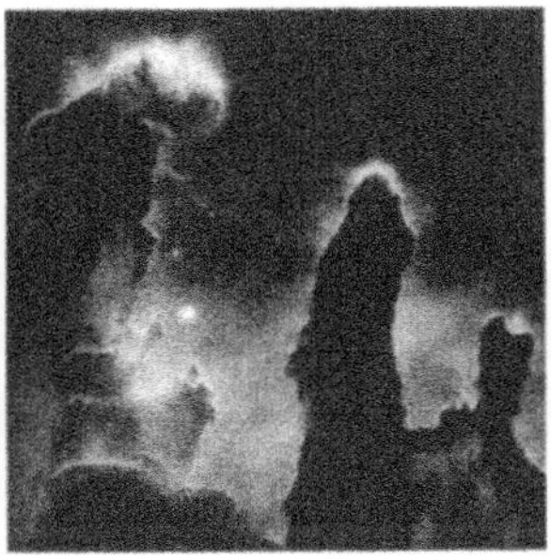

Pillars of creation

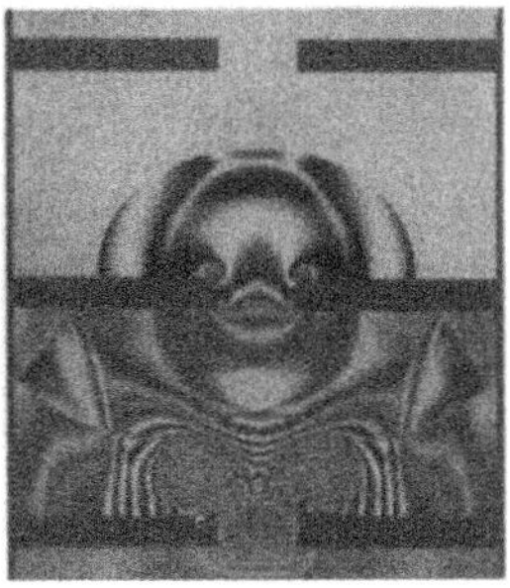

Shock wave

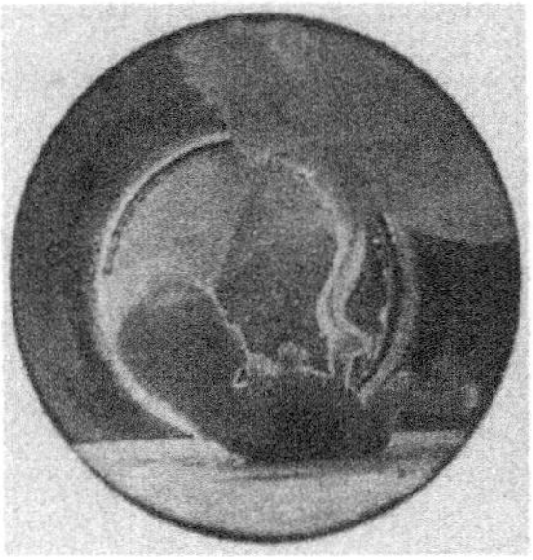

Fetus, Goro

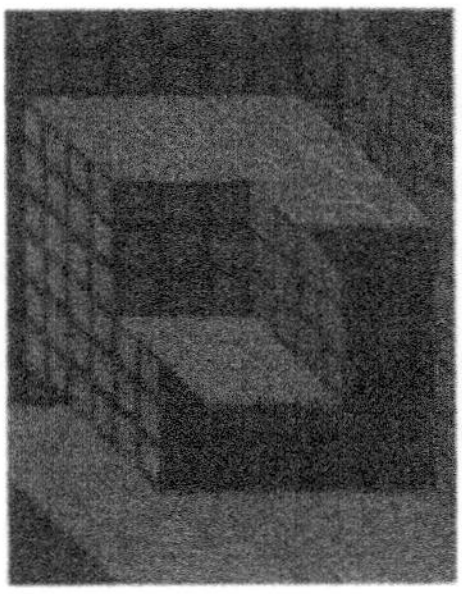

Virtual motion, Vasarely

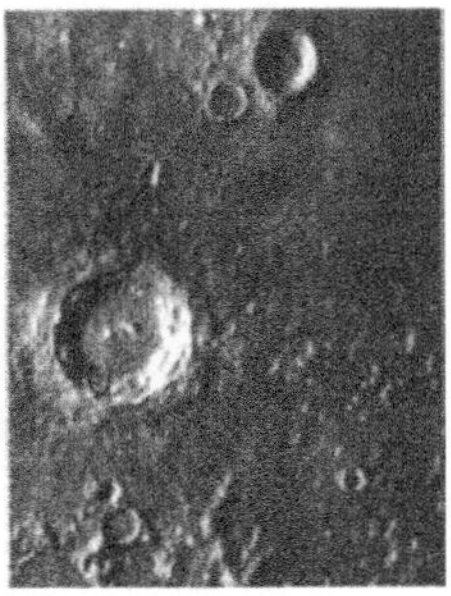

Concave-Convex

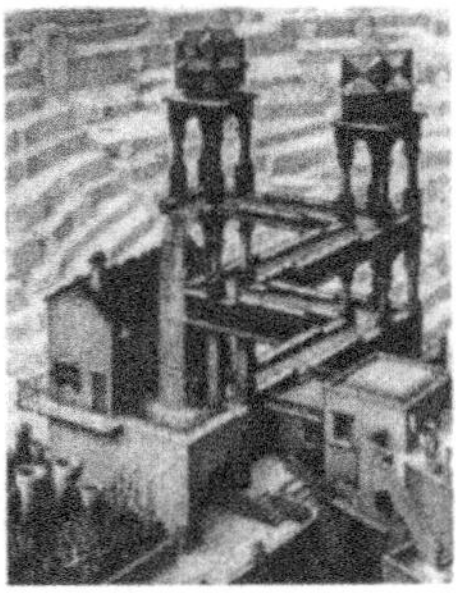

Waterfall, Escher

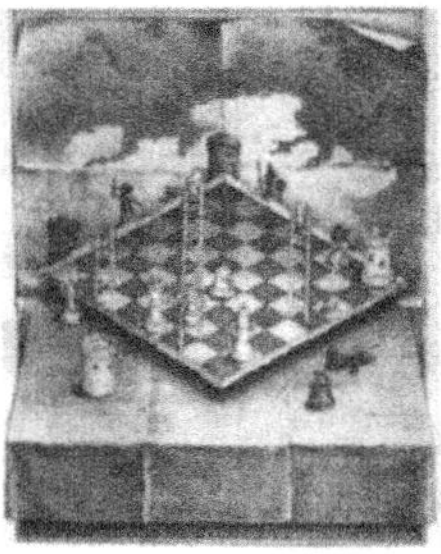

Chessboard, Del Prete

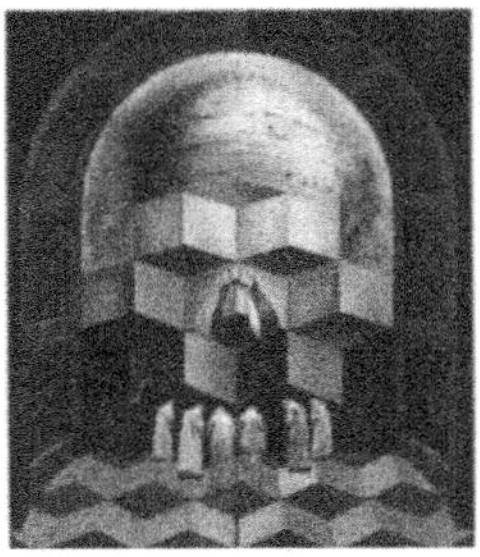

The skull of Zurbaran, Dali

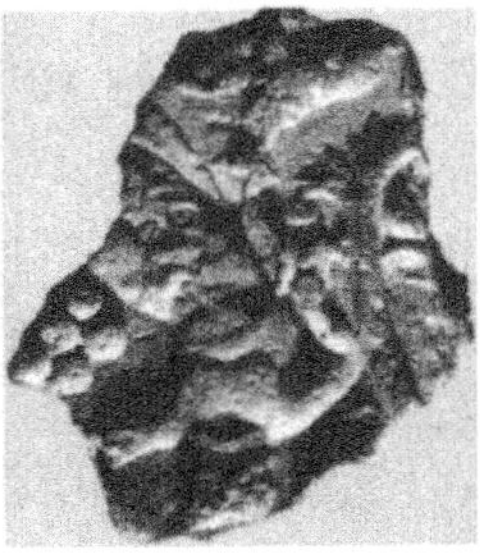

Front & profile

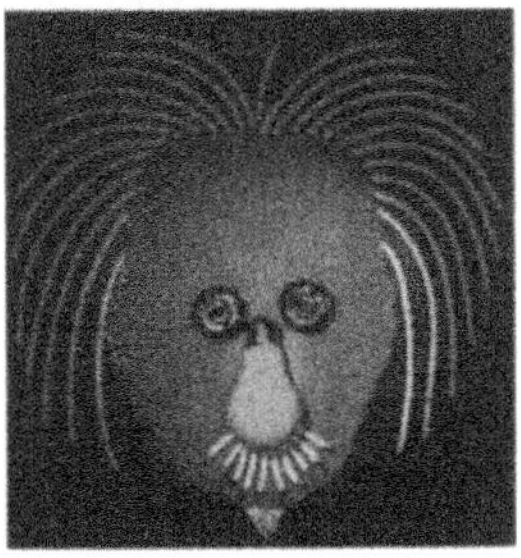

Einstein, Piven

Fig.3. Art via Science

creating art, for example, paintbrush, knife, telecommunication technologies, as well as computers are immaterial.

Rheo-Art, attained [Takaki, 1994] by applying laws of viscous fluid dynamics on an initial dye pattern put on a viscous fluid which is deformed by repeated processes of intermittent rotations of two cylinders inserted into the fluid. Consequently, the extremely nice pattern presented is created.

Surface tension is a basic scientific property of liquid surface and "the artist of nature". On a metal structure immersed in soap solution and taken out, thin films are formed due to surface tension forces. Their interaction with light, as captured on film by the German photographer Karl E.Deckart, creates, as seen in Fig.3, unpredictable aesthetic patterns.

Rainbow, a wonderful work of art, is related to weather sciences and light. It is a color effect produced by the internal reflection of sunlight in rain drops in the air.

Pillars of creation, related to space sciences, is the name given to Hubble's telescope incredible pictures of clouds of gas and dust where within these clouds, stars are forming.

Shock wave, related to fluid flow dynamics, is a narrow region of high pressure and temperature in which air flow changes from subsonic to supersonic. Consequenly, magnificent patterns may be generated, as the face-like one depicted here, created when air flows through a slit, and photographed by Prof. Heinz Reichenbach, Ernst Mach Inst., Freiburg, Germany.

Fetus is related to life sciences. Here, a miscarried fetus is kept alive in a glass and steel artificial womb, demonstrates [Goro, 1993] an unbelievable work of art of the initial stage of man's creation.

Virtual motion, related to the science of psychology, is the result of what the brain does with the infomation entering it through the eyes. Two situatios may be observed in the painting *Kotzka* by Vasarely, among the greatest of the Pop-Art . One can see a blue cuboid, on the one hand, and a "hole" in the big red cuboid of the blue cuboid-shape, on the other. It should be noted that it takes a few seconds to reveal the two situations; one has just to look with patience! Since, there are two possible situations - "two faces of same picture" - the picture seen in the brain will oscillate between these two. This is exactly what is meant by *virtual motion* which does not exist in reality, but, only, in the brain. What makes this picture an interesting work of art is that *added value* caused by the virtual motion, usually not observered at first glance.

Concave-convex. In the picture by Jean Dragesco, one sees the concave-shape Copernicus crater on the moon formed about 810 million years ago. However, if the picture is inverted, a convex shape is revealed. How this happens? The phenomenon is related to the cognition of vision by the brain and the effect of shadow-light on interpreting by the brain of the information it absorbs. If one does not know it a priori, what is the real situation on the moon, concave or convex? And finally, what makes this an interesting art is the above duality, concave-convex.

Waterfall by Escher, his most well-known and fascinating art work, is composed of impossible configurations due to forbidden connections in the construction elements. Psychological effects play here a major role, where at first glance the waterfall seems O.K., but afterwards, taking a few seconds, one observes that all water channels are at an identical level, thus, the waterfall is irrational. The clue here lies on three impossible triangles. Such a triangle was suggested, independently, by Oscar Reutesvard in 1934 and Roger Penrose in 1956, and, imbedded with genius by Escher in the painting.

Chessboard by Sandro Del-Prete, a famous Swiss artist, creates in the observer the so-called cognitive dissonance. This sensation is caused by seeing the chessboard, almost simultaneously, either from its top view and from its underneath. It is this sensation, not usually felt at first glance, which makes this painting so attractive and interesting.

The skull of Zurbaran by Dali belongs to the so-called *figure-ground reversibility* or *perceptual reversal* taking place in the brain. One can observe in the figure a gigantic skull, the figures - monks - forming the teeth, the arch forming the nasal cavity, the dark surfaces of the central row of cubes forming the eye sockets and so on. All these elements oscillate between each other. It is also posssible to read the top surface of some of the cubes as the bottoms of other - virtual motion - so that, instead of six cubes stacked in the form of a pyramid, the altar can also be read as seven cubes in a two-three-two configuration. All these effects provide an added value to the attractiveness of this picture in which many psychlogical effects play simultaneously a role.

Front & profile combination, shown at the bottom of Fig.3, serves creating in the brain a sensation of movement from the front to profile and vise versa. This happens due to the one pair of eyes which stand for the two eyes of the human face, but at the same time are a duplication of one profile eye. However, the fascinating thing about this picture is its inverted side, which, as a matter of fact, was the original side I found in some journal. What is seen there is a relief, about 1900 years old, of Dyonisos, God of wine, with a cluster of grapes to the side of him. So, what makes this picture a fascinating work of art are its "two faces", each showing a completely different picture. The practical conclusion is that one should get used to invert a picture, and many times, he might reveal a new "face", as the concave-convex effect mentioned before.

Einstein, by the Israeli artist Piven, demonstrates Gestalt doctrine saying that we see reality in "parts" and the "complete" perceived by the brain is greater than the sum of "parts". In other words, if we have in the brain the figure of Einstein, then if the brain receives only partial information, it will complete the picture.

REFERENCES

Botero F (1997) Cameo/Abrams, Harry N.Abrams, Inc., Publishers
Briggs J (1992) Fractals - the patterns of chaos. Touchstone
Campbell N (1953) What is science? Dover Publications, Inc., New York
Dissanayake E (1995) What is art for? University of Washington Press
Encyclopedia of Art (1971) Encyclopedia Britanica International Ltd., London
Even-Shoshan A (1975) The new dictionary. Kiryat Sepher Ltd., Jerusalem (in Hebrew)
Goro F (1993) On the nature of things. Aperature publishers
Kimberley T (1994) The art and science connection: hands-on activities for intermediate students. Addison-Wesley Publishing Company
Neret G (1998) Michelangelo. Benedikt Verlag GmbH Taschen, Koln
Oxford Advanced Learner's Dictionary (1989, 2nd ed.) Oxford University Press
Piven H (1999) Piven works 1990-1999. Am Oved Publishers, Tel Aviv, Israel
Takaki R (1994, vol. 9) Proposal of a new kind of art "Rheo-Art". Forma, 203-208
Tamir A (2000) The one face of art & science. In preparation
Tolstoy N (1995) What is art ? Penguin Books Ltd., London
Wisniak J, private communication
Yerka J and Ellison H (1994) Mind fields. Morpheus International

Vortex Packets and the Structure of Wall Turbulence
Extended Abstract

Ronald S. Adrian
Laboratory for turbulence and complex flow
Department of theoretical and applied mechanics
University of Illinois, Urbana, Illinois 61801 USA

In the classical view of statistically homogeneous turbulence large-scale eddies break down into smaller eddies which break down into still smaller eddies, and so on, thereby creating a cascade of energy from large scale to small scales. The cascade concept is incontestable for homogeneous flow, but spatially inhomogeneous turbulence it must be modified to account for the flow of energy between different regions in space. Turbulent flow created by fluid moving over a smooth wall is intrinsically inhomogeneous with small scales being favored close to the wall and large scales favored farther away from the wall. Moreover it is known that most of the energy production occurs in the near-wall region, so that energy must flow from smaller eddies to larger eddies. Recent work on the structure wall turbulence leads to a mechanistic picture of the eddies in this flow and the means by which the energy cascade process is modified. First it was shown that the long-suspected hairpin-shaped eddy is, indeed, the structural building block of wall-turbulence. Second, a mechanism was found by which small hairpins form close to the wall an assemble into much larger packets of hairpins. The hairpin packets are capable of assembling into still larger structures, creating energy flow from small to large. One formed, the large scales may undergo instability resulting in energy cascade from large to small. The simple scenario provided by the hairpin vortex packet paradigm conceptually explains many puzzling and heretofore unexplained aspect of wall turbulence, and it offers a substantially different view of fundamental characteristics.

A series of PIV measurements in planes parallel to a smooth wall in boundary layer wind tunnel was performed. The purpose is to see if the parallel plane data could be used to identify packets of vortices and to observe their spanwise growth, c.f. Fig 1 (plate). More refined methods of identifying the packets, including tilting the plain at the mean angle of packet growth, and extracting low momentum patterns were discussed.

A new direct numerical simulation code using pseudo-spectral methods were used to double the Reynolds number that was previously attained in simulations of hairpin packet growth (Re_τ=300 vs 150). It was found that slight asymmetry in the form of the starting hairpin evolves in complex, asymmetric multi-hairpin packet, Fig.2 (on plate), as opposed to the relatively well-ordered, symmetric packet that evolves from symmetric initial hair-pins. The complex packet looks much more like the complex fields that are seen in fully turbulent DNS results for channel flow, c.f. Fig 3 (ON plate).

A. Gyr et al. (eds.), Science and Art Symposium 2000, 77.

The Hidden Structure in a Round Jet

S. McIlwain, A. Pollard and S. Amalfi[1]
Department of Mechanical Engineering
Queen's University
Kingston, ON K7L 3N6
CANADA
pollard@me.queensu.ca

Abstract: A turbulent round jet at a moderate Reynolds number was calculated using large eddy simulation. Large scale structures in the flow were identified using vorticity and eigenvalue discriminant analysis techniques. The results show that the two techniques successfully identify essentially the same large scale structures in the flow.
Keywords: round jet, LES, vorticity, eigenvalue discriminant

1 MODEL DESCRIPTION

The near field of a turbulent round jet at a Reynolds number of 68,000 was calculated using Large Eddy Simulation (LES). In LES, the large scale structures in the flow field are resolved while the contributions of the smaller scales are included through a subgrid scale model. The following equations are used:

$$\frac{\partial u_j}{\partial x_j} = 0 \tag{1}$$

and

$$\frac{\partial u_i}{\partial t} + \frac{\partial u_i u_j}{\partial x_j} = -\frac{\partial p}{\partial x_i} + \frac{1}{Re_\tau}\frac{\partial}{\partial x_j}\left[(1+\nu_t)\left(\frac{\partial u_i}{\partial x_j} + \frac{\partial u_j}{\partial x_i}\right)\right] \tag{2}$$

where u_i are components of the velocity vector in the x_i direction corresponding to the x, y, and z directions, respectively; t is time; p is pressure; ν_t is the turbulence viscosity; and Re_τ is the turbulence Reynolds number. For the calculations reported here, the dynamic Smagorinsky subgrid scale model was used [Germano et al., 1991].

The calculations were performed on a grid consisting of $73 \times 56 \times 56$ points in the x, y and z directions, where x is the streamwise coordinate. The round jet inlet was approximated using 316 square control volumes. The computational domain size was $4.25D \times 3D \times 3D$ where D is the diameter of the round jet at the nozzle.

[1]DEA Student, ISTIL, Université Claude Bernard, Lyon, France

A. Gyr et al. (eds.), Science and Art Symposium 2000, 79-84.

Tests were performed to ensure that increasing the size of the computational domain from $3D$ to $4D$ in the y and z directions had little effect on the results. Time- and space-averaged velocity profiles agree well with the experimental data of Citriniti [Citriniti, 1996].

Since numerical calculations must be performed in a finite computational domain, particular attention must be directed towards the boundaries. The jet inlet was prescribed using a top-hat velocity profile upon which were superimposed temporal disturbances with a maximum value of 4% of the mean inlet velocity. The frequency of these disturbances was set to provide the greatest excitation of the turbulence structures. A convective boundary condition was used at the jet outlet. Zero normal velocity gradients were imposed at all other boundaries.

2 ANALYSIS TECHNIQUES USED

Large scale turbulence structures contain most of the turbulence energy of a flow field. It is important to be able to identify these structures as they control the mixing of the jet with the ambient fluid and therefore have a strong influence on combustion. Since the definition of large scale turbulence structures is a matter of contention, several methods of identification are in common use. Two are examined here.

2.1 Flow Vorticity

The flow vorticity, $\overline{\Omega}$, is a measurement of the rotation of a fluid element in the flow. It is defined as

$$\overline{\Omega} = \nabla \times \overline{V} \tag{3}$$

where $\overline{V}$ is the velocity vector at any point in the flow. The normalised vorticity can be obtained by taking the square root of the square of the components, $\Omega_n = \sqrt{\Omega_1^2 + \Omega_2^2 + \Omega_3^2}$. Areas of high vorticity are associated with large scale structures.

2.2 Eigenvalue Discriminant

The geometry of the streamline pattern close to any point in the flow can be classified by studying certain invariants of the velocity gradient tensor, $A_{ij} = \partial u_i / \partial x_j$, [Chong et al., 1998]. The characteristic equation of A_{ij} is

$$\lambda^3 + P\lambda^2 + Q\lambda + R = 0 \tag{4}$$

where λ represents the eigenvalues, and P, Q, and R are the tensor invarients,

$$P = -tr(\mathbf{A}) \tag{5}$$

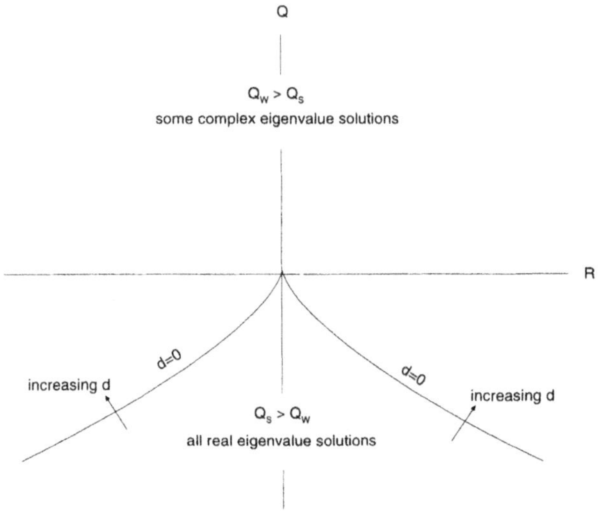

Figure 1: *$Q - R$ space for the velocity gradient characteristic equation.*

$$Q = \frac{1}{2}(P^2 - tr(\mathbf{A^2})) \tag{6}$$

and

$$R = -det(\mathbf{A}) \tag{7}$$

For incompressible flow, $P = 0$. The eigenvalue discriminant, d, which describes the surface that separates points with complex eigenvalues from those with all real eigenvalues is, [Chong et al., 1990]:

$$d = \frac{27}{4}R^2 + Q^3 = 0 \tag{8}$$

The tensor invarient Q can be used to relate d to the relative strain and rotational forces acting on a fluid element in the flow field. For incompressible flow, Q can be broken down into rotational, Q_W, and strain, Q_S, components:

$$Q = Q_W + Q_S = \frac{1}{2}(W_{ij}W_{ij} - S_{ij}S_{ij}) \tag{9}$$

where

$$W_{ij} = \frac{1}{2}\left(\frac{\partial u_i}{\partial x_j} - \frac{\partial u_j}{\partial x_i}\right) \tag{10}$$

and

$$S_{ij} = \frac{1}{2}\left(\frac{\partial u_i}{\partial x_j} + \frac{\partial u_j}{\partial x_i}\right) \tag{11}$$

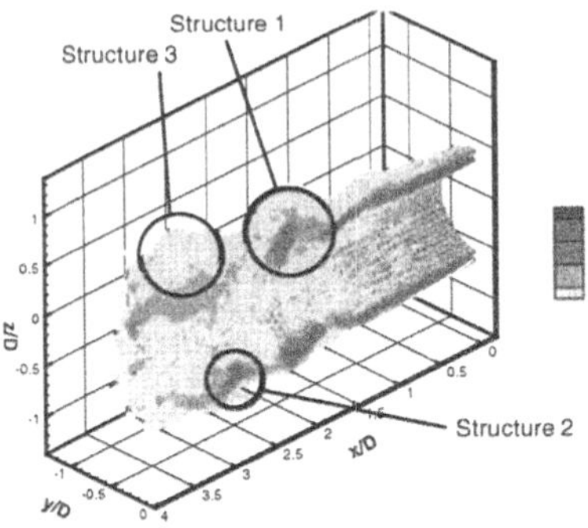

Figure 2: *Three dimensional plot of the normalised vorticity in a round jet, sliced along the $x-z$ plane at $y = 0$ (see Colour Plate).*

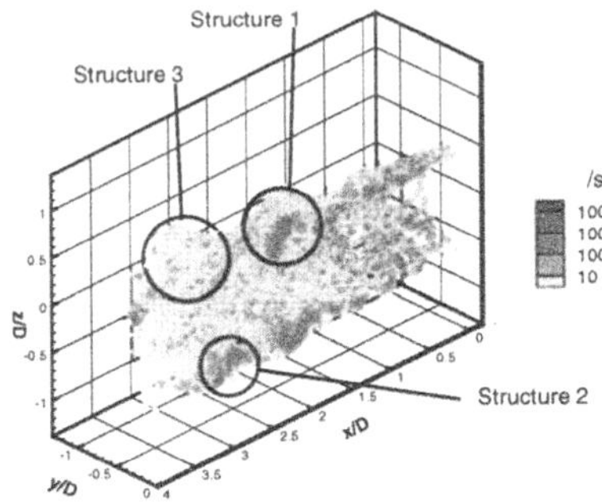

Figure 3: *Three dimensional plot of positive values of the eigenvalue discriminant in a round jet, sliced along the $x-z$ plane at $y = 0$ (see Colour Plate).*

Figure 1 is a plot of $Q - R$ space showing the $d = 0$ line. Regions of the flow where $Q_W > Q_S$ correspond to positive values of d; $d < 0$ where $Q_S < Q_W$. Higher positive values of d are associated with large scale structures as they correspond to areas of the flow where the rate of rotation of a fluid element greatly exceeds the strain rate.

3 RESULTS

The results presented here are for an instant in time at 24.6 large eddy turnover times (LETOTs) from the start of the calculations. The LETOT for a round jet is defined as the jet diameter divided by the mean jet inlet velocity. These figures are representative of data obtained throughout the simulation.

A three dimensional contour plot of the normalised vorticity, Ω_n, is given in Figure 2. The flow domain has been sliced in the $x - z$ plane at $y = 0$ so that only values for $y \leq 0$ are shown. This is to better illustrate the turbulence structures. The jet nozzle is at $x = 0$. A similar plot of the eigenvalue discriminant, d, is shown in Figure 3. Only positive values of the discriminant, *ie.* those areas corresponding to regions of high rotation, are shown. A two-dimensional slice through the flow field at $x/D = 3.0$ is provided for Ω_n and d in Figures 4 and 5, respectively. In Figure 5, both positive and negative values of the discriminant are given. In Figures 2-5, low values of Ω_n and $\mid d \mid$ are not shaded to reduce clutter. Thus an arbitrary low cut-off value has been introduced. While it is possible to use lower cutoff values, this does not highlight any additional flow structures in this case.

The three dimensional plots in Figures 2 and 3 clearly illustrate a vortex ring

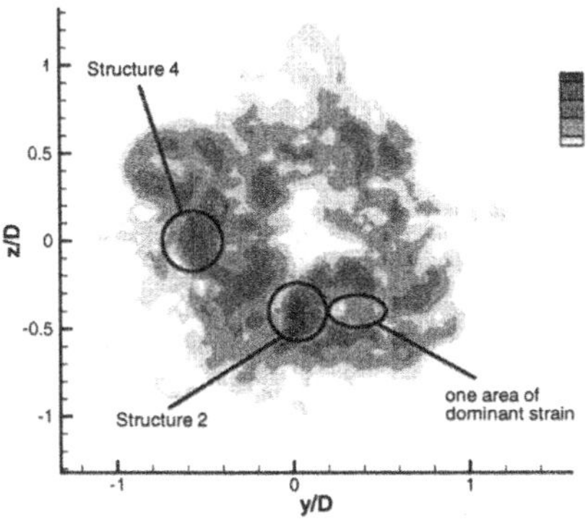

Figure 4: *Two dimensional plot of the normalised vorticity in a round jet, along the $x/D = 3$ plane (see Colour Plate).*

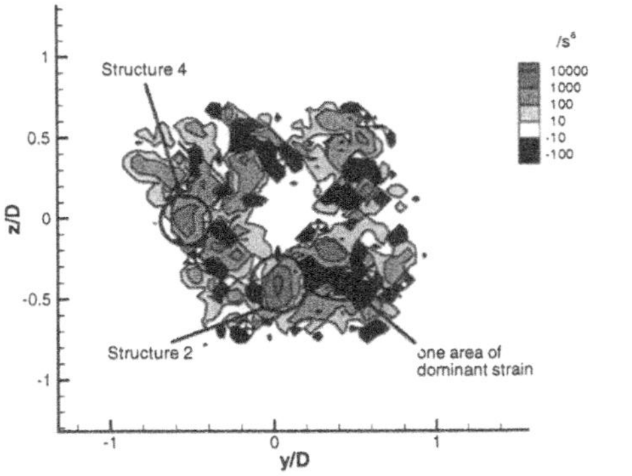

Figure 5: *Two dimensional plot of the eigenvalue discriminant in a round jet, along the $x/D = 3$ plane (see Colour Plate).*

surrounding the jet core at the nozzle and for the first 1.75 units downstream. The vortex ring is continuous in the Ω_n plot, but in the d plot it consists of isolated packets of regions with high rotation rates. Areas of high Ω_n or d are visible in the vortex ring along the slice through the $x - z$ plane. Farther downstream, the ring begins to break up as large scale structures appear next to the jet core. Areas of high Ω_n and d are visible along the slice through the $x - z$ plane in "Structure 1" and "Structure 2" at $x/D = 2.25$ and 3.0. The bulge in the contours at $x/D = 3.4$ also corresponds to a structure, "Structure 3", but the high values of Ω_n or d are not visible since the slice along the $x - z$ plane does not pass through this area. "Structure 2" is visible also in the two dimensional plots as indicated in Figures 4 and 5. A second structure, "Structure 4", which was hidden behind the contours in the three dimensional plots, is visible as well in the two dimensional plots. Areas of high strain rates ($d < 0$) are indicated in Figure 5, contained within the ring surrounding the jet core. These areas are indistinguishable from areas of low rotation in the vorticity plot, Figure 4. This is the reason why the three dimensional d plot in Figure 3 does not show a smooth ring around the jet core near the nozzle.

The results from the two analysis techniques correlate well. The same structures are identified in the same locations. The Ω_2 and Ω_3 components of vorticity contribute to most of the normalised vorticity, Ω_n, shown in Figures 2 and 4.

An arbitrary low cut-off level was required for the three dimensional plots to reduce clutter that would otherwise hide the turbulence structures. Although not required, the same cut-off levels were used in the two dimensional plots to provide consistency between the figures. This introduces a degree of arbitrariness into the analysis. The cut-off must be high enough so the resulting contours do not hide the important features of the turbulence, yet low enough not to remove those features altogether. The eigenvalue discriminant plots provide a definite advantage when attempting to choose an appropriate cut-off. Areas of high strain rates correspond to negative values of d, and the difference between low and high rates of rotation in the d plots is much greater than the difference in vorticity levels in the Ω_n plots. Although the two-dimensional plots do not require a cut-off level (the contours corresponding to lower levels of Ω_n or d will simply fill in the white space in Figures 4 and 5), it is difficult to accurately represent characteristics of a three dimensional flow field in a two dimensional plot.

4 CONCLUSIONS

The near field of a turbulent jet at moderate Reynolds number was calculated using LES. Large scale structures were identified around the jet core, starting 1.75 jet inlet diameters downstream from the nozzle, using vorticity and eigenvalue discriminant analysis techniques. Both techniques identified the same structures at essentially the same locations. An arbitrary low cut-off level had to be introduced when analysing and plotting the three dimensional results from either technique. This cut-off was easier to define when using the eigenvalue discriminant analysis technique.

ACKNOWLEDGEMENTS

This work was funded by the Natural Sciences and Engineering Council of Canada. The authors would like to acknowledge the support of the Rhone/Alps-Canada exchange.

REFERENCES

Chong, M.S.; Soria, J.; Perry, E.A.; Chacin, J.; Cantwell, B.J.; Nay, Y.; 1998, "Turbulence structures of wall-bounded shear flows found using DNS"; *Journal of Fluid Mechanics*, Vol. 357, pp. 225-247.

Chong, M.S.; Perry, A.E.; Cantwell, B.J.; 1990, "A general classification of three-dimensional flow fields"; *Physics of Fluids*, Vol. A2, pp. 765 - 777.

Citriniti, J.; 1996, *Experimental Investigation into the Dynamics of the Axisymmetric Mixing Layer Utilizing the Proper Othogonal Decomposition*, PhD. Thesis, State University of New York at Bufflalo, 161 pp.

Germano, M.; Piomelli, U.; Moin, P.; Cabot, W.H.; 1991, "A dynamic subgrid-scale eddy viscosity model"; *Physics of Fluids*, Vol. A3(7), pp. 1760-1765.

The interaction between longitudinal vortices in a channel flow visualised through seeding and hydrogen bubble techniques

Holger Kocherscheidt, Maren Schmidts and Venkatesa I. Vasanta Ram
Ruhr-Universität Bochum
Institut für Thermo- und Fluiddynamik
D–44780 Bochum, Germany
Kocherscheidt@lstm.ruhr-uni-bochum.de

Abstract: The present paper is a visualisation study of the patterns of a flow in a channel in which longitudinal vortices are generated by winglets. Subject of this paper is the presentation of vortex patterns visualised by the method of hydrogen bubble technique in comparison with a study in which the vortices were generated by the same geometric winglet configuration and the flow visualised by particle seeding. The hydrogen bubble technique shows certain states of spatial development of the vortices at the Reynolds numbers Re_H=320 and Re_H=370. The flow visualisation was recorded on photographs and video.
Keywords: flow visualisation, hydrogen bubble method, embedded vortices

1. INTRODUCTION

In many engineering applications there appears the phenomenon of shear flows with embedded longitudinal vortices. Some examples are : aircraft wings, diffusers and compact heat exchangers. In these and other applications the embedded longitudinal vortices are often generated by surface mounted winglets. The vortices created do not only exert a dominant influence on the overall character assumed by the flow but also determine the various quantities of engineering interests, like drag, power requirement and heat transfer. An understanding of the behaviour of these vortices in the flow field and their interaction with each other is therefore essential for a meaningful flow control. To this end flow visualisation is an indispensable tool. In this context the question of the most suited method of visualisation is also of great interest, hence this study.
In the preceding conference in this series, SCART 97 held 1997 in Berlin, [Lau et al., 1997] presented the visualisation of the interaction patterns of longitudinal vortices embedded in a channel flow. The visualisation of these patterns was realised by seeding the flow with Iriodin® particles. These are saucer shaped titanium dioxide coated particles (10-25 μm diameter, 3-4 μm thick) having nearly the same density as water. Due to their shape they align themselves in preferred directions with respect to the axes of the local strain rate tensor in the flow and produce, when illuminated from a suitable direction, a pattern of bright and dark patches, which renders the vortical structure in the

A. Gyr et al. (eds.), Science and Art Symposium 2000, 85-90.

flow visible. In the study of [Lau et al., 1997] patterns in a Reynolds number range of Re_H=300 up to Re_H=6700 were visualised, where, as in this study, the semi-channel height H (=5mm) has been chosen as the reference length for the Reynolds number.
In this study our interest lies in visualisation the same flow phenomenon by an alternative method, namely the hydrogen bubble technique and comparing the outcome with the earlier flow visualisation at the two Reynolds numbers Re_H=320 and Re_H=370.

2. THE EXPERIMENTAL FACILITY

Figure 1 shows a schematic diagram of the water channel facility specially built for the purpose of the recent flow visualisation studies. The facility is of the closed-circuit type. The channel walls are made out of plexiglass to enable visualisation of the flow. In its cross section the channel is rectangular with dimensions 10 mm x 200 mm. Denoting the space between the channel walls as 2 H (=10 mm), the salient dimensions of the test section (1) are 2 H x 40 H x 120 H. Upstream of the test section, the flow passage is a channel with the same cross section dimensions as the test section with a length of 800 mm (160 H). The contraction ratio ahead of this section is 6:1. The long uniform channel (2) upstream of the test section and its large aspect ratio of 1:20 assures the two-dimensionality of the oncoming flow in the channel section upstream of the winglets. Water is pumped up from the lower level against the steady head maintained in the reservoir at the top (3). The circulating water leaves the reservoir into the return piping (4), flowing over V-notches that are calibrated to measure the volume flow rate. The flow rate through the channel is regulated by a frequency divider for the electric motor, throttle valve (5) and a by-pass (6). The pump (7) of the system is able to maintain a steady flow rate through the channel in the range from 0.02 l/s up to 4 l/s, which corresponds to an average through flow velocity in the test section of 1 cm/s up to 2 m/s. Longitudinal vortices embedded in the channel flow are generated by winglets mounted on one of the channel walls. This wall will be called *rear* channel wall in the follow. Figure 2 shows the geometrical configuration, in which the winglets were arranged in both studies. The two winglets are thin (0.1 H) rectangular shaped sharp edged plates with dimensions 1 H x 4 H placed at an angle of 30° to the main flow direction.

3. THE METHOD FOR FLOW VISUALISATION

For the visualisation of the flow we use the hydrogen bubble method. In our facility this method is realised by inserting 9 very fine platinum wires (∅ 150 μm), acting as cathodes at different positions in the channel. The wires form three groups of three wires, each group being positioned perpendicular to the main flow direction at different distances to the winglets. The first group is located at a distance of 1 H upstream of the lower edge of the winglets, the second group is installed 1 H downstream of the top edge of the winglets and the third group is fixed in a distance of 15 H downstream of the

top edge of the winglets. Within a group, it is possible to distinguish the wires through their distance to the channel wall: they are located at 1/2 H, 1 H and 3/2 H from the front channel wall. The wires are consecutively numbered (figure 3). Each group of wires belongs to one out of three anodes consisting of small strips of high-grade steel, that are embedded into the rear channel wall near the corresponding group.
By using a transformer a voltage between 37,5 volts and 176 volts can be applied to any of the wires and the referring anode. Thus a regular layer of bubbles is generated at the wire (continuous run). In addition a frequency generator can be connected to the transformer in order to generate a pulsating voltage on the wire which then emits lines of bubbles (timelines) in a certain frequency rate (pulsating run).
The mode 'continuous run' gives a comprehensive picture of the general flow distribution showing the spatial formation of vortices very clearly.
The mode 'pulsating run' can be seen as a useful instrument to gain knowledge about the local and global velocity distribution by making full use of the timelines.
The flow visualised with this method has been recorded photographically and videographically. For the photographs a 400 ASA b/w film was taken and an exposure time of 1/60 seconds was chosen. For the video recordings we used a Sony Video H8 Handycam.

4. THE EXPERIMENT

Throughout the experiment, the flow was observed at two Reynolds numbers, Re_H=320 and Re_H=370. At this the state of the flow was analysed by generating hydrogen bubbles at each of the nine wires in both of the before-mentioned modes. Cause of the different positions of the cathodes, the used form of visualisation allows us to obtain a precise analysis of the spatial development of the embedded longitudinal vortices in the flow. So, e.g. it becomes obvious that at a Reynolds number of Re_H=320 the generated vortices show a smaller influence on the flow in direction of the channel depth as at the higher Reynolds number. Applying the continuous mode to wire 1, it becomes apparent the parts of the flow passing the wire are influenced by the vortices, which can be seen by a curvature of the layer of bubbles, but a rotation of the streaklines of the bubbles can not be observed. This curvature is formed like a furrow in the layer of bubbles that starts approximately at the centre of the winglets. In this furrow, the flow accelerates towards the rear channel wall (figure 4). This phenomenon is still visible in the section downstream of the winglets; it only shows very small wave-like disturbances along their inner edges (s. video). An increase of the Reynolds number to Re_H=370 leads to the same formation (figure 5). However, the disturbances along the edges of the furrow are more distinct both temporal and spatial. An even bigger difference in the spread into the depth of the channel can be made out by comparing the observations on wire 4 at the different Reynolds numbers. Applying the lower Reynolds number two furrows can be seen developing on the right and on the left of the centre of the channel. These furrows show a strong acceleration towards the rear channel wall (figure 8). Again, no rotations of streaklines can be detected. At the higher Reynolds number of Re_H=370 the parts of the flow passing the wire are dragged into the vortices right downstream of the corresponding wire (figure 9). Additional extremely distinct periodical structures develop, and it is conspicuous that this phenomenon can be observed both at the continuous run and at the pulsating run. These structures must be regarded as the rotation of bubbles around the vortex axis.

The vortex axis lies, at least in proximity to the winglets, in the central layer of the channel. Observations on wire 2 lead to that conclusion. At Reynolds numbers of both Re_H=320 and Re_H=370 projections of vortex-like patterns, which in both cases have their starting point at the lower edge of the winglets, appear. Obviously these patterns have relatively straight and little disturbed inner edges (figure 6) at the lower Reynolds number, whereas at Re_H=370 these edges are interspersed with spatially and temporally periodical disturbances (figure 7).

Tests with wire 5 and 6 each one shows a distinct internal shear layer between the outsides of the vortex-like pattern and the undisturbed flow in the outer section of the channel at Re_H=370. On the one hand this layer has a periodical wave form (figure 11), on the other hand it also produces a likewise periodical zigzag form (s. video). The structure of the vortex-like patterns itself is also highly interspersed with temporal and spatial disturbances. In contrast it is also possible to detect a small transitional area between the vortices and the outer section of the flow at the lower Reynolds number, but in this case the differences of the velocity between the flow at the outer areas and the vortices are not big enough for tearing asunder the layer of bubbles (figure 10). Observations at both Reynolds numbers in the continuous mode show that the spatial vortex structures appear up to a certain distance downstream the winglets. Observations of the flow by generating hydrogen bubbles at the wires 7, 8 and 9 reveal that at a distance of more than 15 H downstream of the upper edge of the winglets structures like the aforementioned do not occur anymore. At such distance only the conspicuous distribution of velocities hints to the disturbance of the flow at a more upstream point of the channel.

5. REFERENCES

Lau, S.; Vasanta Ram, V.; Kaniewski, M.; „Visualisation of interaction patterns of longitudinal vortices embedded in a channel flow“; In: *Book of abstracts of the 2nd International Conference on Flow Interaction cum Exhibition/Lectures on Interaction of SCIENCE & ART, pp. 23-25; Berlin*

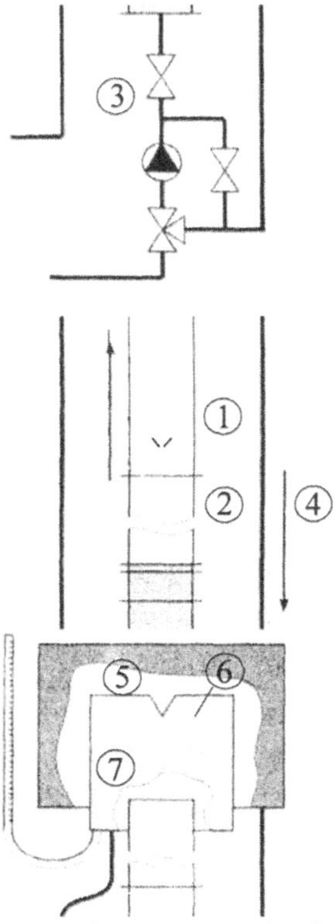

Figure 1; schematic graph of the experimental facility

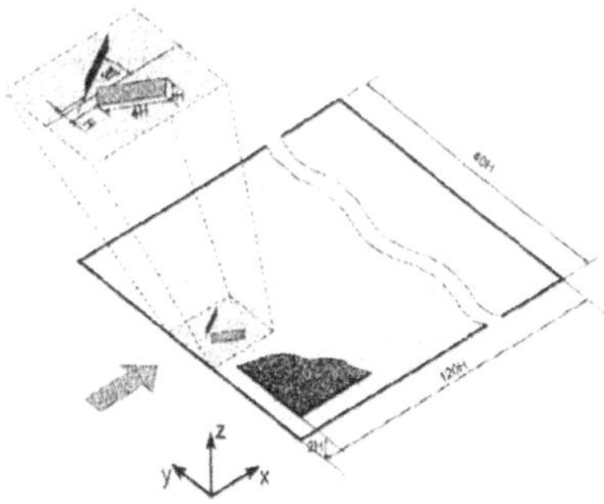

Figure 2; winglet configuration

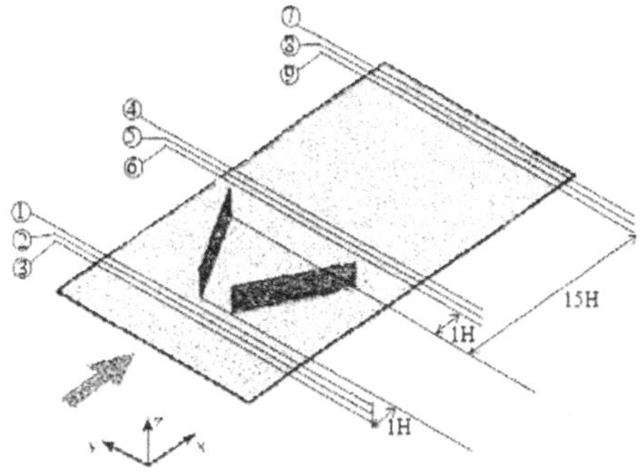

Figure 3; wire numeration

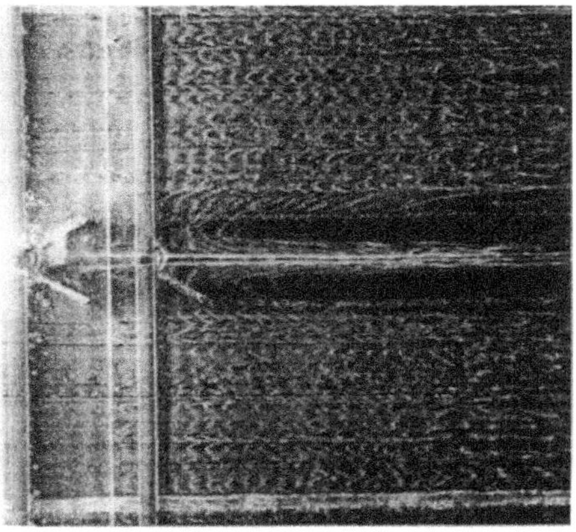

Figure 4; Re=320, wire 1, cont. run

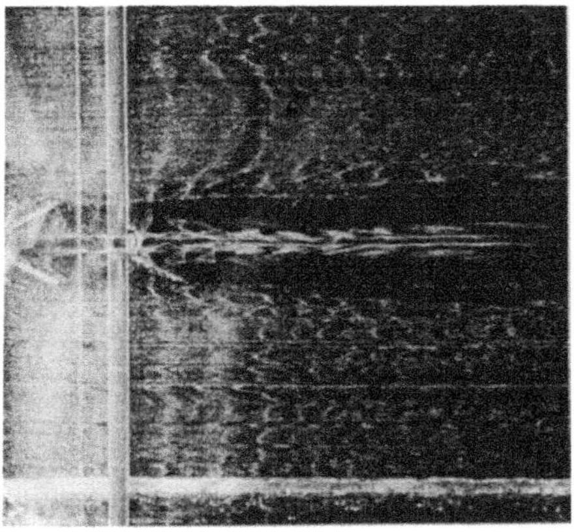

Figure 5; Re=370, wire 1, cont. run

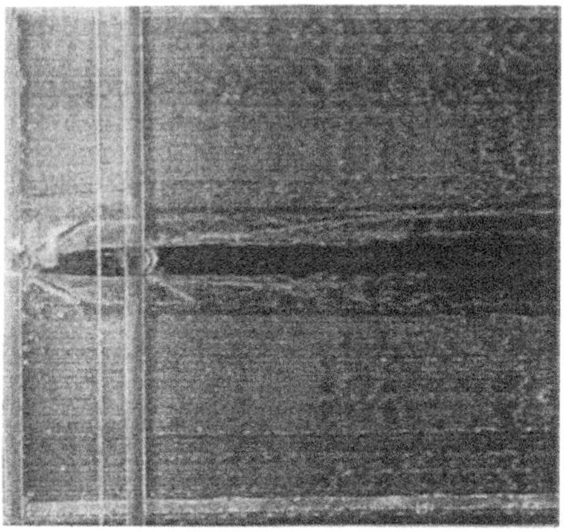

Figure 6; *Re=320, wire 2, cont. run*

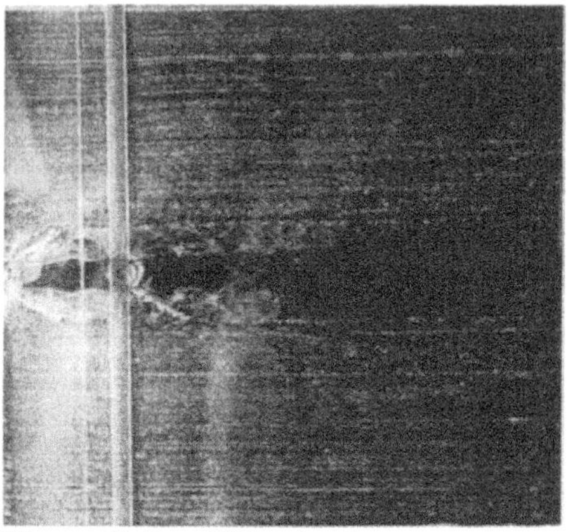

Figure 7; *Re=370, wire 2, cont. run*

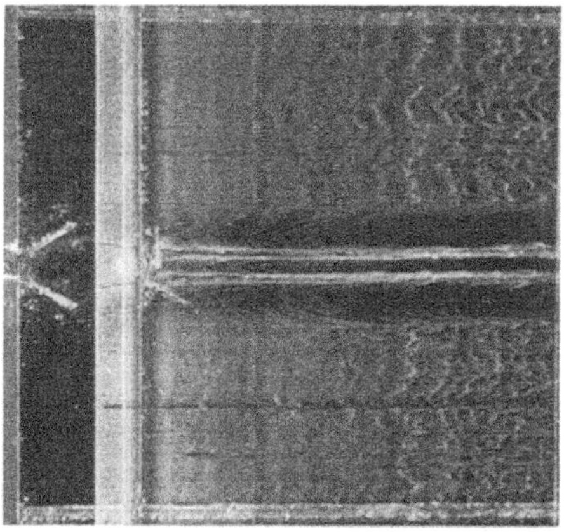

Figure 8; *Re=320, wire 4, cont. run*

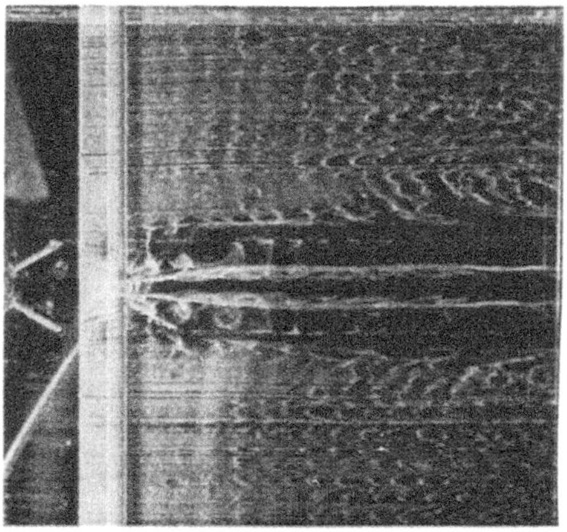

Figure 9; *Re=370, wire 4, cont. run*

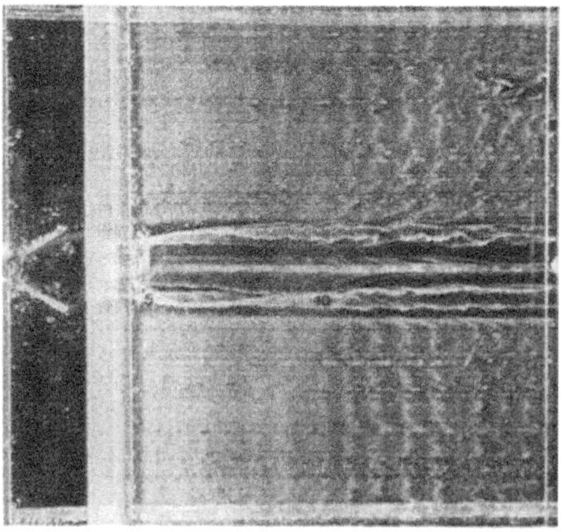

Figure 10; *Re=320, wire 5, cont. run*

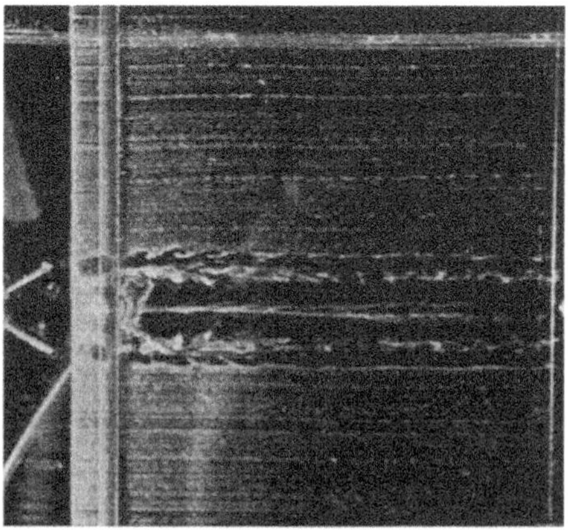

Figure 11; *Re=370, wire 5, cont. run*

Choice of representation modes and color scales for visualization in Computational Fluid Dynamics

Marie Farge

LMD, Ecole Normale Supérieure, Paris
and
CMLA, Ecole Normale Supérieure, Cachan

1 Introduction

Color visualization have become essential to analyze and communicate CFD (Computational Fluid Dynamics) results. To use them appropriately we should be conscious of their possibilities and limitations. Without such an awareness, we run the risk of being mislead by attractive images, void of information content or very difficult to interpret. This is the reason why it is necessary to notify on the visualization which representation mode and color scale have been used, otherwise it will be as meaningless as a measure without error bars, or a map without scale. We will propose some hints and rules to choose the most appropriate representation mode and color scale for a given problem. This paper will consider as example the case of two–dimensional scalar fields having zero–mean, which are often encountered in CFD.

The visual acquisition and perception process is not neutral, because it filters or distorts the information contained in the field to be analyzed. Therefore, the selection of the best representation mode and color scale should be carefully defined. It requires:

- a good knowledge of the underlying physical or mathematical problem the computation is addressing, and the relevant questions we want to answer in visualizing the field,
- an understanding of the physiology of vision and of the psychology of shape and color perception, in order to avoid, or at least be conscious of, different optical illusions or parasite phenomena affecting our visual perception, such as chromostereopsis (the shortest wavelength colors tend to focus in front of the retina and thus appear defocused [Murch]) or simultaneous contrast (a given color is perceived differently depending on the background color [Itten]),

A. Gyr et al. (eds.), Science and Art Symposium 2000, 91-100.

- an awareness of the culture dependent symbolism of forms and colors [Kandinsky], in order to reduce the risk of misinterpretation due to implicit meanings of certain colors. For instance, in Western cultures the color red is associated with heat, strength and expansion, while blue is associated with cold, weakness and retraction.

The ideas discussed in this paper are illustrated by their application to study rotating barotropic turbulent flows computed using Saint–Venant equations [Farge et al.]. The relevant fields to visualize are: *vorticity*, *divergence*, *stream function*, *velocity potential*, *pressure* and *potential vorticity*. They are two-dimensional scalar fields, which fluctuate around an average value, that we are for convenience adjusted to zero. These fields contain complex structures having very different levels of excitation and exhibit a wide range of scales (*e.g.* coherent vortices, vorticity filaments, modons, gravity waves, inertial waves...), therefore raster display should be prefered to vector display, because it gives a more detailed representation of the structures. With *raster display* an image is viewed as a set of *pixels* having different colors or different grey levels, while with *vector display* it is described as a set of curves satisfying some equations [Foley et al.]. The visualizations illustrating this paper are made of 512^2 pixels of one byte, which allows $2^8 = 256$ possible values to be displayed at 512^2 locations. We have rescaled the fields and map their range onto the 256 available integer values, between 0 and 255, in order that the zeros correspond to the value 127.

We will first discuss the choice of representation modes, then propose a normalization of color space and finally define some rules for the choice of color scales.

2 Choice of representation modes

Before analyzing a field, it is essential to state the questions we want to answer, because, to each question raised, it corresponds a representation mode more adequate than the others. Actually, each representation mode acts as an operator that reveals a given property of the field.

- To study the morphology of the most excited structures in the field, we prefer a *frontal perspective representation* with hidden surfaces removal and shadowing (cf. **Figure 1a**).

- To analyze the smoothness of the field, we propose the same perspective representation, *with a vertical cut* (cf. **Figure 1b**).

- To allow a morphological comparison between the large and small scale structures, we prefer a *perspective representation seen from the top* with shadowing (cf. **Figure 1c**).

- To have an integral view of the field, *i.e.* to enhance the large scales, we propose a *cartographic representation with a two–level color scale* (cf. **Figure 2a**), the color scale defining an equivalence between the different values of the field and several colors or grey levels.

- To have a differential view of the field, *i.e.* to enhance the small scales and gradients, we propose a *cartographic representation with a multi–level color scale* (cf. **Figure 2b**), the number of levels acting as a filter to enhance a certain scale.

- To analyze the field, both morphologically and quantitatively, for instance to compare the excitation level of different structures, we propose a *cartographic representation with a continuous color scale* (cf. **Figure 2c**). This representation is continuous in luminance, but presents three thresholds, in order to discriminate the most excited structures from the weakly excited ones. Moreover, this combination between continuous luminance and three thresholds allows us to separate the structures having negative values, which have a dark extrema, from the structures having positive value, which have a bright extrema. It also enhances the real value zero (which has become 127 after rescaling the range of variation between 0 and 255) as a discrete contour–line having the brightest luminance, since the shape of the zero contour–line conveys information on the smoothness of the field.

The choice of the representation mode depends, not only on what we are looking for, but on the actual information contained in the field to be visualized. For instance, if we choose a cartographic representation with a continuous luminance scale, varying from dark for the lowest values to bright for the highest values, which seems *a priori* to be the best default solution, the quality of the visualization critically depends on the smoothness and small scale isotropy of the field. To illustrate this, let us consider several fields of different kinds:

- the *stream function*, which is smooth, being a space integral of the velocity appears blurred (cf. **Figure 3a**), since the eye is unable to extract a shape when the luminance variation in space is too slow. Therefore, only the perspective representation (cf. **Figure 3b**) is adequate to display its smooth spatial variation,

- the *divergence*, which is not smooth (cf. **Figure 4b**), being a derivative of the velocity, also appears blurred, since it is isotropic at small scales. Therefore we again have the feeling that the picture is out of focus (cf. **Figure 4a**).

- the *vorticity*, which is not smooth, being a derivative of the velocity, does not appear blurred since it is anisotropic at small scale (cf. **Figure 5b**). The visualization is perceived as well focused, because the luminance gradients, resulting from the small scale anisotropy, are strong enough for the eye to figure out the shape of the field (cf. **Figure 5a**).

In practice, it is often necessary to use several representation modes, because each one reveals complementary informations about the underlying structure of the field. In many cases encountered in CFD, it is also interesting to use *animation*. This can be done for two purposes:

- either to follow the time evolution of a turbulent flow, in order to study its dynamics,
- or to analyze a complex localized structure, by moving the observer around it, or moving it closer to explore small details.

In some cases, the perspective representation should prefered, because the cartographic representation may be difficult to use if the value range of the field changes in time (*e.g.* as it is the case for decaying turbulent flows), which then reduces the number of levels available at each time step, which necessitates to renormalize the color scale in order to follow the flow unsteadiness.

3 Normalization of the color space

We think that the relevant approach to address the choice of color scales for CFD visualization should be pragmatic and should simplify the representation of color space as much as possible. This concern lead us to prefer the *subtractive color synthesis*, and the terminology employed by painters and printers. It is based on mixing several pigments of three different primary colors, Red-Yellow-Blue, which, added together, give the color Black (and in absence of pigment the paper remains White). We have thus discarded the *additive color synthesis*, used by computer screens. It is based on the composition of beams of three different primary colors, Red-Green-Blue, which, added together, give the color White (and in absence of light the screen remains Black). Our preference for the substractive synthesis is motivated by the fact that we are less familiar with it, as we have been trained to manipulate colors using paints and not computer screens. This comments will soon be obsolete, as our kids begin to manipulate colors with computers rather than with pencils. The subtractive synthesis also allows us to choose the color scale *a priori*, using only color pencils, without requiring a computer screen. Nowadays, due to the present lack of good software tools to manipulate colors and define color scales, we prefer to select our color scales using pencils, which again motivates our choice of the substractive synthesis.

The second simplification we propose is to *reduce of the dimensionality of color space*. According to the principle of *visual trivariance* for the perception of colors discovered by Thomas Young, the retina contains three kinds of pigments: Red, Green, Blue. Therefore the brain analyzes any visible light by filtering it through these three kinds of pigments, which explains why the color space is perceived as tridimensional. Two equivalent representations are then possible: the *cube RYB (Red-Yellow-Blue)* (cf. **Figure 6a**) and the *cone LHS (Luminance-Hue-Saturation* (cf. **Figure 6b**). The transformation matrix to exchange RYB and LHS representations has been empirically estimated by Maxwell and his wife from color tests they had performed themselves. We propose to adopt the LHS representation, because it is more natural from a perceptive point of view, since:

- *luminance* corresponds to the intensity of the perceived light,
- *hue* to the most excited ray in its spectrum of a given color,
- *saturation* to the width of its most excited spectral band.

A saturated color (*i.e.* a pure color) is monochromatic, while a desatured color (*i.e.* a pure color mixed with White or Black) has a broad band spectrum. But, in many instances, the distinction between luminance and saturation (*e.g.* the difference between dark light blue and bright obscure blue) is too subtle to be appreciated on computer screens. Therefore, we use only saturated colors, which are the most vivid colors, in order to reduce the dimensionality of the color space from three to two, considering only luminance and hue variations (cf. **Figure 6c**).

The third simplification we propose consists of discretizing and normalizing the hue scale into *twelve basic hues*, easy to recognize and from which any other hue can then be identified (cf. **Figure 7**). We have adopted the natural order of the light spectrum: Red (1), Orange (2), Yellow (3), Green (4), Blue (5), Indigo, Violet. In order to give a circular structure to the hue space (solution already proposed by Newton), we have neglected Indigo, replaced Violet by Purple (6) and added the non-spectral (*i.e.* non mono-chromatic) color Purple Red (0.5). We have also set the following rules :

- $h < 3$ for warm colors,
- $h > 3$ for cold colors,
- $h \in [2, 4]$ for colors having a weak luminance dynamics, *i.e.* which are difficult to brighten or darken without distorting their hue. This is the case for Yellow (3), which easily tends to become White or Brown. On the contrary, it is not the case for Blue (5) or Purple (6), which keep their hue when one changes their luminance (cf. **Figure 6c**).

In order to simplify the memorization and communication of colors, we have named these twelve basic hues according to the Color Naming System (cf. **Figure 7**) used for computer graphics [Berk et al.].

We have then defined a numbering scheme, from 0 to 6, which characterizes the organization of the twelve basic hues we have selected:

- the *three substractive primary colors*, Red, Yellow, Blue, are designated by the odd integers 1, 3, 5,
- the *three substractive secondary colors*, Orange, Green, Purple, by the even integers 2, 4, 6,
- *any other color*, obtained by mixing two adjacent hues h_1 and h_2, is characterized by a real number intermediate between h_1 and h_2.

To guarantee objectivity (*i.e.* the independance from the user color perception and taste) as much as possible, we have defined the basic hues in terms of the Pantone standard [Pantone], which is used worldwide by painters, printers and various industries to describe 593 different normalized colors. The Pantone standard has replaced the Munsell Color Notation [Munsell], which was used before, and it is available in Apple programming environment. We use Pantone papers, easily available in graphics art shops, to tune the computer screen, in adjusting the twelve basic hues by eye comparison with the corresponding Pantone papers. We use the Pantone markers to test our choice of color scales on paper before going to the screen, since our choice can be mislead by the additive synthesis used by the screen.

This normalization of color space, and the associated numbering scheme we have proposed, will help us to write algorithms to automatically select the color scales, independently of the user color perception and taste. We will now discuss this.

4 Choice of color scales

Johannes Itten [Itten], one of the leaders of the Bauhaus school (Weimar, 1930), has defined seven color contrasts :

- the *cold-warm contrast*, which is maximal between the warmest hue, Red (1), and the coldest hue, Green Blue (4.5), and which is minimal between neutral hues, *e.g.* the combination Purple Red (0.5)–Green Yellow (3.5),
- the *complementary contrast*, which happens between a given hue h and its complementary $\bar{h}$, diametrically opposite to h on the chromatic circle proposed by Itten (cf. **Figure 7**), such that $\bar{h} = [(h+3) \bmod 6]$,
- the *simultaneous contrast*, which is due to the physiological fact that, if an association of hues is not in complementary equilibrium, the eye tends to distort them in order to recreate this equilibrium. In consequence, a given hue is differently perceived depending on the hue of the surrounding (cf. **Figure 8**),

- the *quantity contrast*, which is function of the area covered by a given hue. The smallest quantity contrast is obtained if we respect the following area ratio proposed by Goethe [Goethe]: Yellow 3, Orange 4, Red 5, Green 6, Blue 8 and Purple 9 (cf. **Figure 9**).

- the *luminance contrast*, which exists between saturated colors having different luminances (cf. **Figure 10**). The highest luminance contrast corresponds to the combination Yellow (3)–Purple Blue (5.5), and the lowest luminance contrast to the combination Green (4)–Red Orange (1.5),

- the *hue contrast*, which is maximal between primary colors Red (1)–Yellow (3)–Blue (5) and decreases for secondary colors Orange (2)–Green (4)–Purple (6),

- the *saturation contrast*, which is strongest when we associate a saturated (vivid) color with a desaturated (whitened or blackned) color.

In practice, any association of different colors combines several kinds of contrasts. Our aim is to be able to recognize them and consciously decide to use or avoid them. For instance, to display turbulent fields we propose the cold-warm contrast to easily distinguish negative values, by associating them to cold hues, and positive values, by associating them to warm hues. We also want to *avoid simultaneous contrast*, which may cause misinterpretation, because it changes the hue of a color depending on the hue of its background. To avoid it we propose:

- to separate each color by a grey background,

- to choose n different hues in complementary equilibrium, which therefore corresponds to the hues associated with the n submits of a polygon inscribed inside the hue circle (cf. **Figure 11**). To find the ith hue in complementary equilibrium with $n-1$ other hues, we apply the formula:

$$h_{n,i+1} = [(h_i + 6i/n) \bmod 6]$$

$$\text{for } i = 1, n.$$

If we apply these rules to the cartographic representation with a continuous color scale presenting three thresholds (cf. **Figure 2c**), we choose $n = 3$. We attribute a cold hue to the values smaller than the first threshold (which corresponds to the negative values of the field), a warm hue to those larger than the second threshold (which corresponds to the positive values), and for the second threshold a hue in complementary equilibrium with the two others (which corresponds to the value zero). We thus combine the possibilities of *local analysis*, because the continuous luminance variation allows us to follow small variations (*i.e.* gravity waves), and *global* *analysis*, because the three thresholds, enhanced by the cold-warm contrast, clearly separate the most excited structures, the positive or negative coherent vortices, from the background turbulent field, the vorticity filaments (cf. **Figure 12**).

Therefore, the way we have defined our color scales to visualize turbulent scalar fields is such that all information is coded with a continuous variation of luminance, in order to avoid any loss of information if we have to publish them in black and white, and is enhanced by the superimposition of discrete hues, in order to easily distinguish one given field from the others. Each field is identified by three hues in complementary equilibrium, *i.e.* such as they correspond to the three submits of an equilateral triangle inscribed in the hue circle, (cf. **Figure 11**), with a cold hue ($h > 3$) for the negative values, a warm hue ($h < 3$) for the positive ones and any complementary hue for the value zero

To study the evolution of barotropic turbulent flows, that we have chosen as example, we need to visualize six different scalar fields. We want to characterize each of them by its color scale, which requires that each color scale should be as different as possible from the five others, to be able to identify the field at first glance. This gives the six combinations, denoted [negative-zero-positive]:

- [Blue(5)–Yellow (3)–Red(1)] for the vorticity field (cf. **Figure 12a**),
- [Purple(6)–Green(4)–Orange(2)] for the stream function field (cf. **Figure 12b**),
- [Green Blue (4.5)–Purple Red (0.5)–Orange Yellow (2.5)] for the divergence field (cf. **Figure 12c**),
- [Purple Blue (5.5)–Green Yellow (3.5)–Orange Red (1.5)] for the potential velocity field (cf. **Figure 12d**),
- [Blue(5)–Red(1)–Yellow(3)] for the pressure field (cf. **Figure 12e**),
- [Green Blue (4.5)–Orange Yellow (2.5)–Purple Red (0.5)] for the potential vorticity field (cf. **Figure 12f**).

Now, having adopted this color convention, we are able to identify at first glance each fields, which is very useful if we want to compare them. The fact that information is coded only on luminance is especially important for color–blind people (who represent 8% of the male population) as our images remain robust if they are reproduced in black and white, since we use hue only to distinguish different fields, but not to characterized their variations.

5 Conclusion

Visualization is essential to analyze and communicate the results from CFD numerical experiments. This is especially true when studying turbulent fields, due to the complexity and the variability of their spatial structure. Adequate color visualizations offer new possibilities of morphological analysis, pattern recognition and local structure decomposition, complementary to the spectral analysis which, on the contrary, is non-local.

Laboratory experiments and numerical experiments tend to use the same numerical techniques to display their results, which will promote and facilitate intercomparison between laboratory and numerical experiments. But we think that visualization will not fully play its role, unless a normalization of representation modes and color scales will be adopted, at least within a given discipline. We hope the normalization we have proposed here may bring some awareness about this need and could be considered as a first step for further implementations.

If today most people are able to easily understand a map, it is only because we are accustomed to a given representation mode and color convention. Imagine the difficulty we would face if we have to adjust to a different coding everytime we use a new map. Now imagine the non sense of having to interpret maps where the legend is not mentioned! Unfortunately, this is often the case today with computer visualizations where the color scale is not displayed on the screen. Without the adoption of a normalization, we run the risk that color visualization will be mostly used for advertisement, *e.g.* to convince granting agencies, rather that as a scientific tool useful to analyze, compare and communicate CFD results.

6 Acknowledgments

The figures illustrating this paper has been done in collaboration with Jean-François Colonna (Lactamme, Applied Mathematics, Ecole Polytechnique) and Jean-Michel Coulombier, to whom I express all my gratitude.

7 References

[Berk et al.] Berk T., Browston L., Kaufman A., "A new color-naming system for graphics languages", *Computer Graphics and Applications*, May 1982
[Farge et al.] Farge M. and Sadourny R., "Wave vortex interactions in rotating shallow water", *J. FLuid Mech.*, 1989
[Foley et al.] Foley J. D. and Van Dam A.,*Fundamentals of interactive computer graphics*, Addison-Wesley, 1982
[Goethe] Goethe J., *Traité des couleurs*, Triades, 1973

[Itten] Itten J., *Art de la couleur*, Dessain et Tolra, 1984
[Kandinsky] Kandinsky W., *Cours du Bauhaus*, Collection Médiations, Denoël, 1975
[Murch] Murch G., "Visual accomodation and convergence to multi–chromatic display terminals", *Proc. Soc. for Information Display*
[Munsell] Munsell A. H., *A color notation*, Munsell Color Inc., 1981
[Pantone] *Pantone Color Formula Guide, 18th edition*, Pantone Inc., 55 Knickerbocker Road, Moonachie, New Jersey 07074 USA, 1985

Figures from:

RONALD J. ADRIAN

Vortex packets and the structure of wall turbulence: One plate with; Figure 1(top left): PIV measurements in the x-y plane above a smooth wall of a turbulent boundary layer. Note the low momentum region (dark) indicating the footprint of a packet of hairpin vortices. Figure 2 (bottom): Complex hairpin packet that evolves from a slightly asymmetric initial disturbance. Such packets are also thought to evolve from local bumps on the wall. Figure 3 (top right): Turbulent eddies in Reynolds number = 300 channel flow. Note the similarity between the groups of eddies in this fully turbulent flow and the complex hairpin packet in Figure 2.

ANDREW POLLARD

The hidden structure and beauty in round jets. Two plates with the explanation in the text.

MARIE FARGE

Choice of representation modes and color scales for visualization in computational fluid dynamics. Five plates with 12 Figures as described in the text.

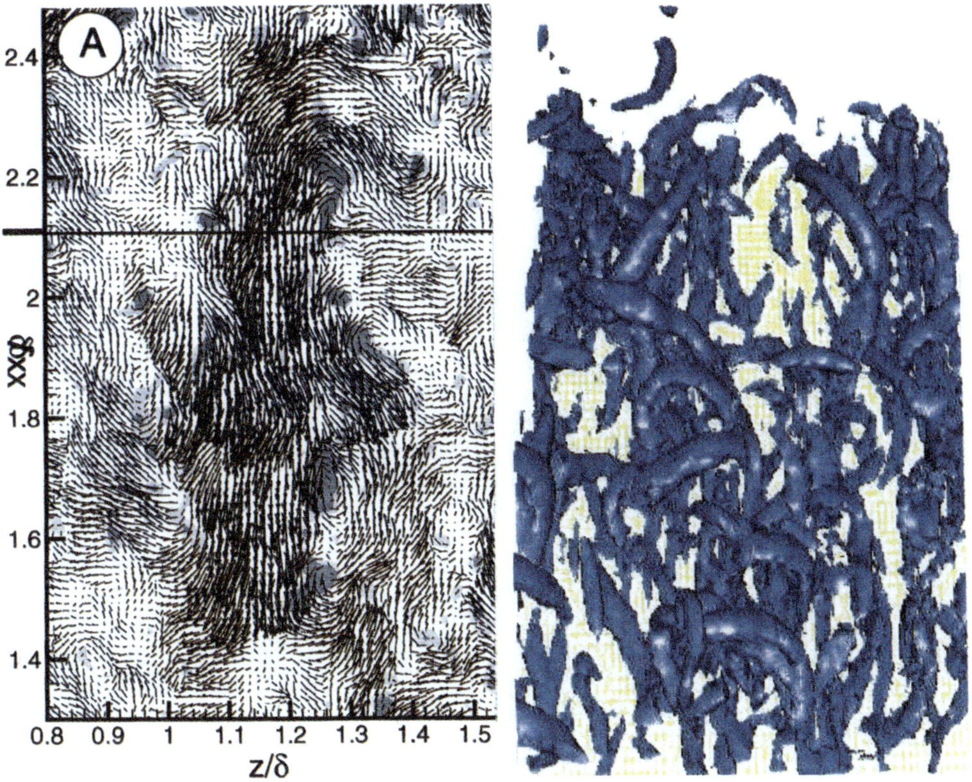
A
2.4
2.2
2
1.8
1.6
1.4
0.8
0.9
1
1.1
1.2
1.3
1.4
1.5
z/δ

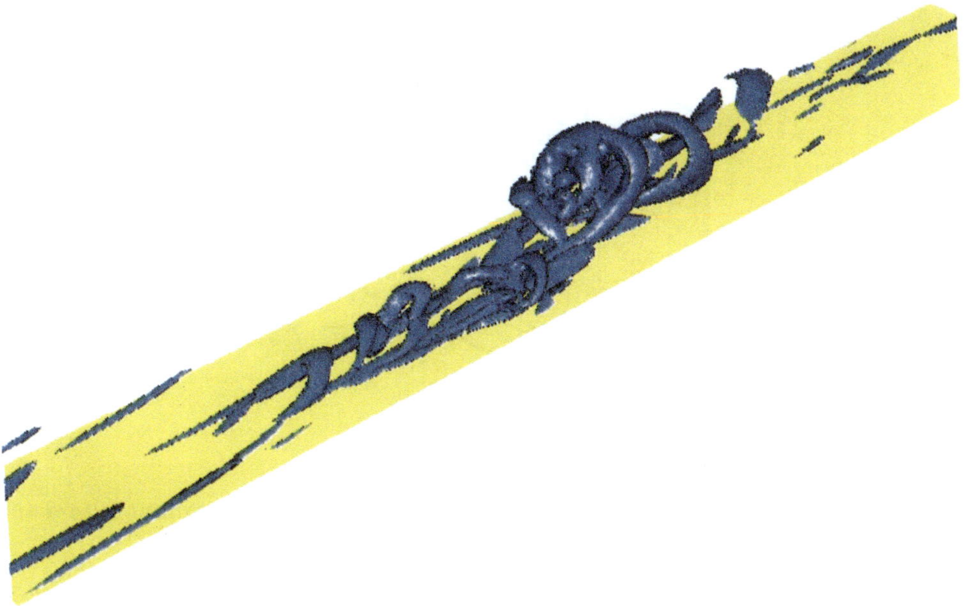

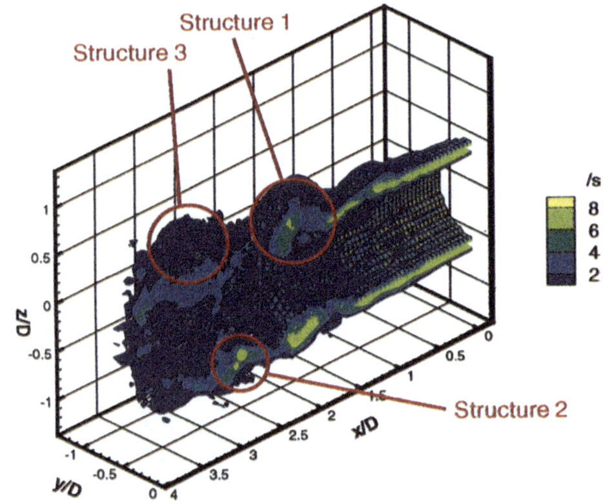

Three dimensional plot of the normalised vorticity in a round jet, sliced along the $x - z$ plane at $y = 0$.

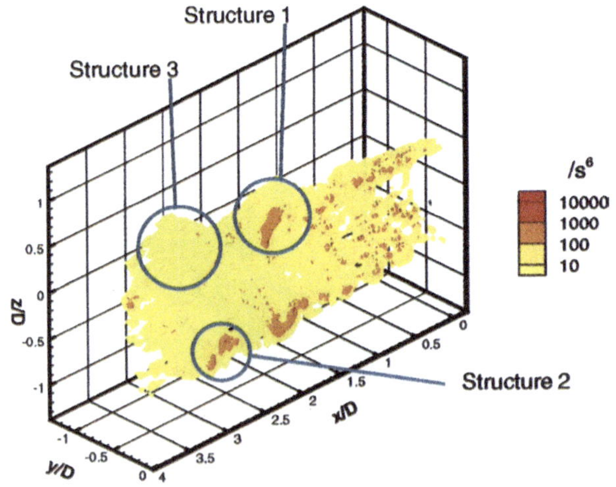

Three dimensional plot of positive values of the eigenvalue discriminant in a round jet, sliced along the $x - z$ plane at $y = 0$.

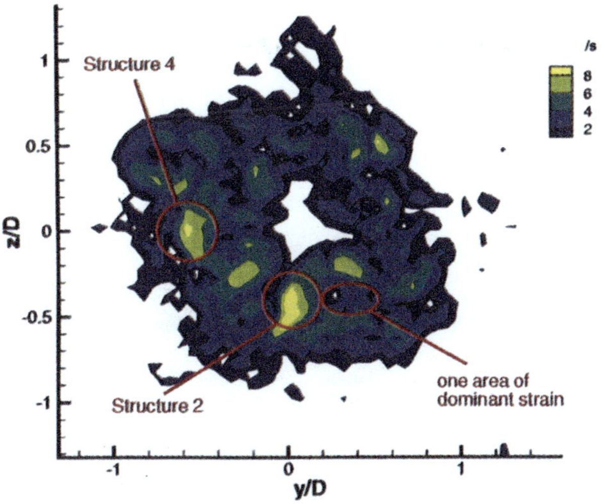

Two dimensional plot of the normalised vorticity in a round jet, along the $x/D = 3$ plane.

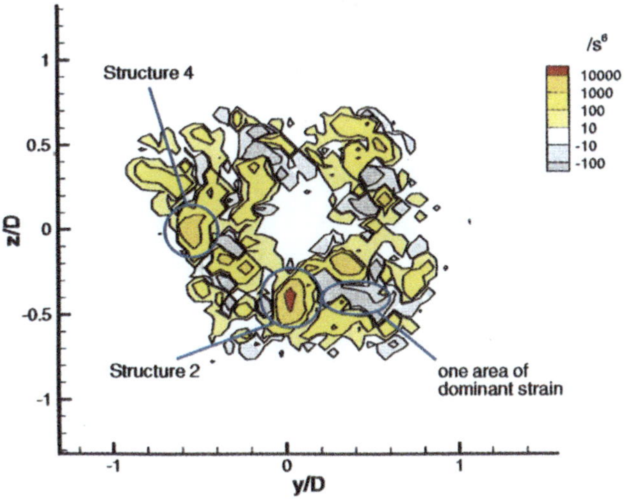

Two dimensional plot of the eigenvalue discriminant in a round jet, along the $x/D = 3$ plane.

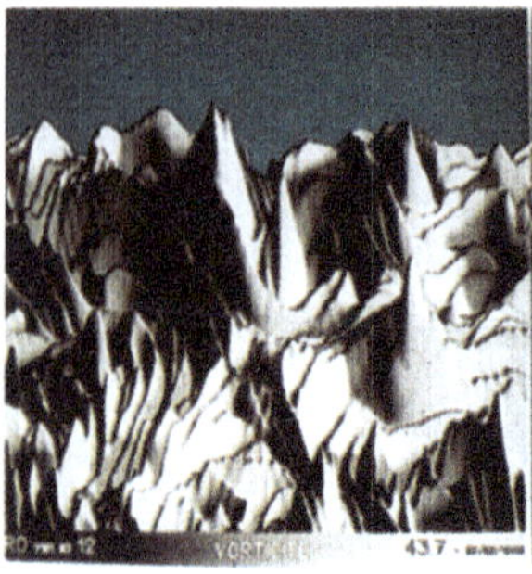

Figure 1a
Perspective representation
seen from the front

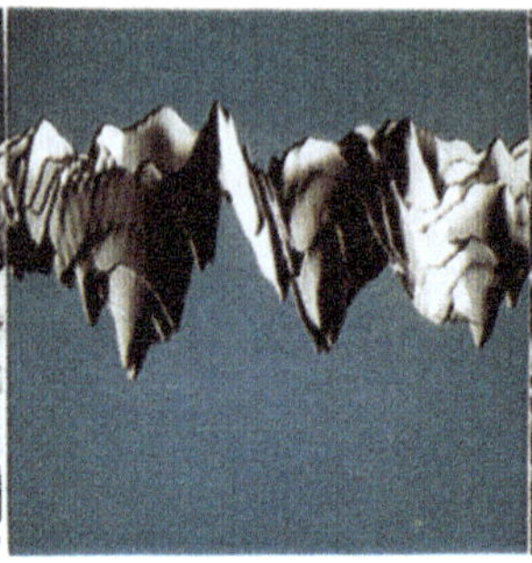

Figure 1b
Perspective representation
with a vertical cut

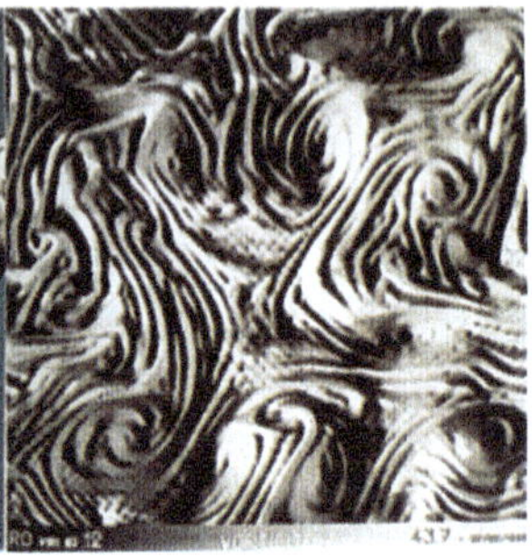

Figure 1c
Perspective representation
seen from the top

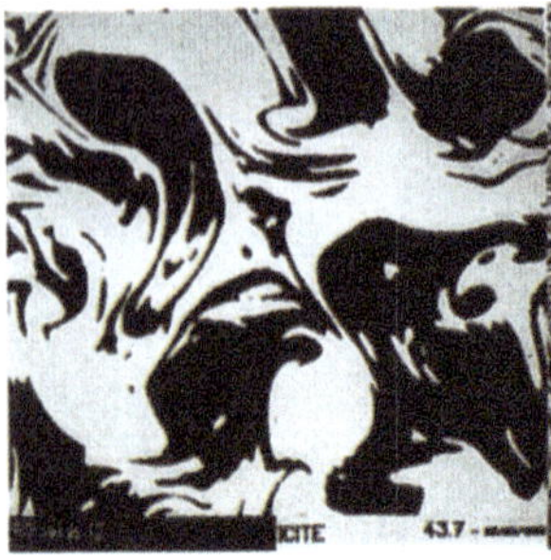

Figure 2a
Cartographic representation
with a two-level color scale

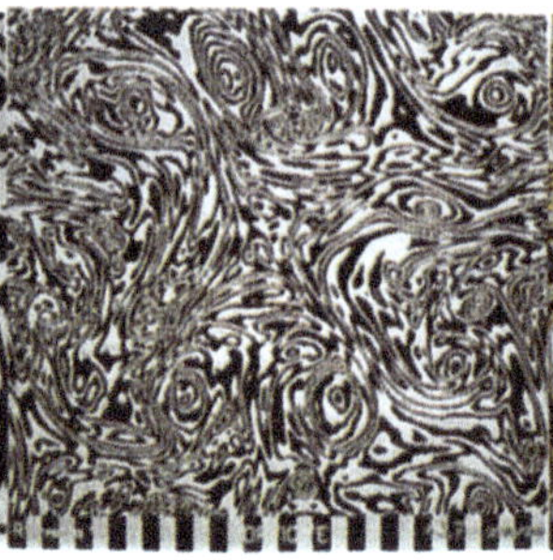

Figure 2b
Cartographic representation
with a multi-level color scale

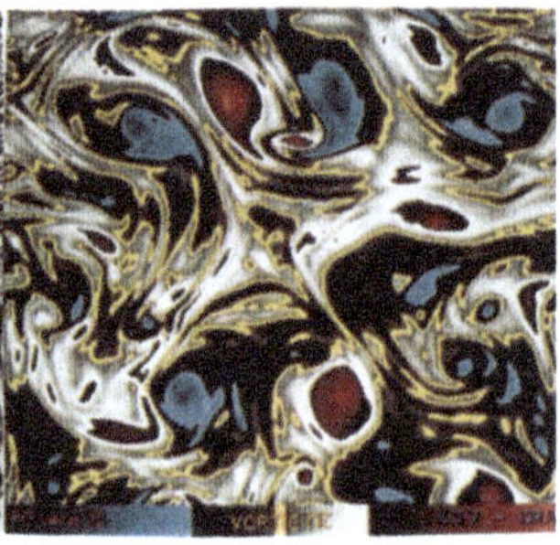

Figure 2c
Cartographic representation
with a continuous color scale

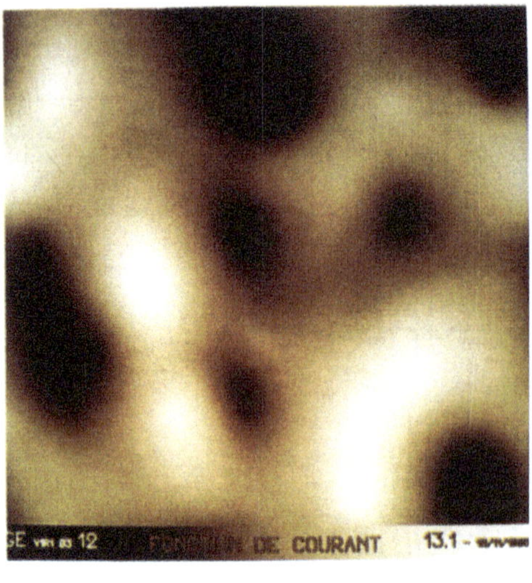

Figure 3a
Stream function (cartographic)

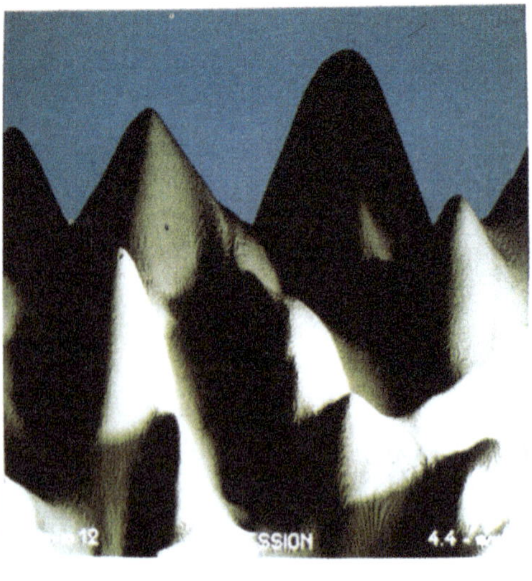

Figure 3b
Stream function (perspective)

Figure 4a
Divergence (cartographic)

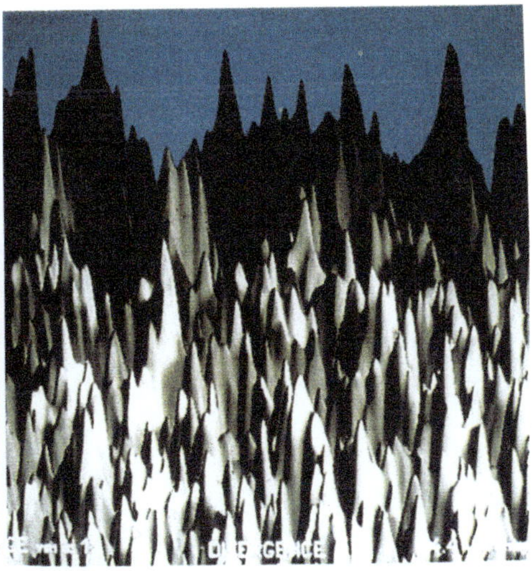

Figure 4b
Divergence (perspective)

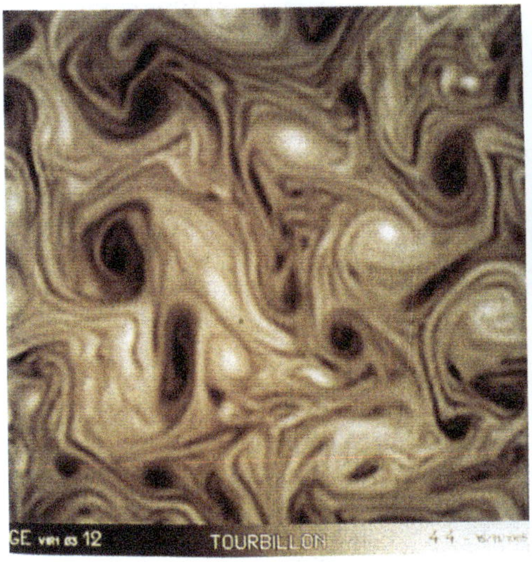

Figure 5a
Vorticity (cartographic)

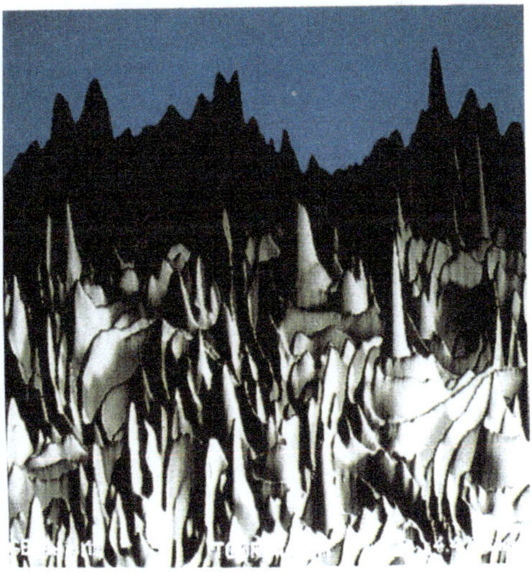

Figure 5b
Vorticity (perspective)

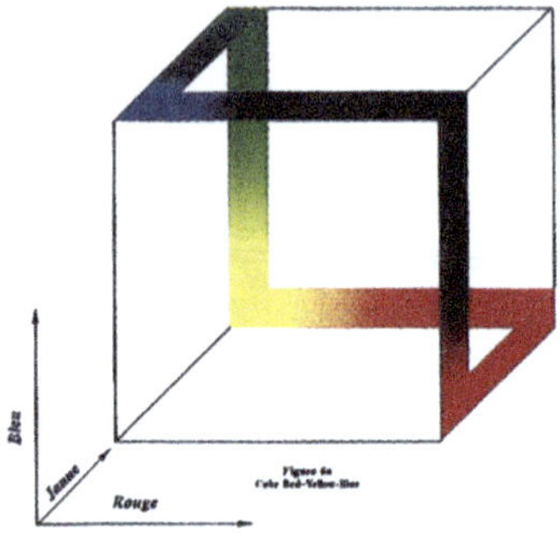

Figure 6a
Cube Red-Yellow-Blue

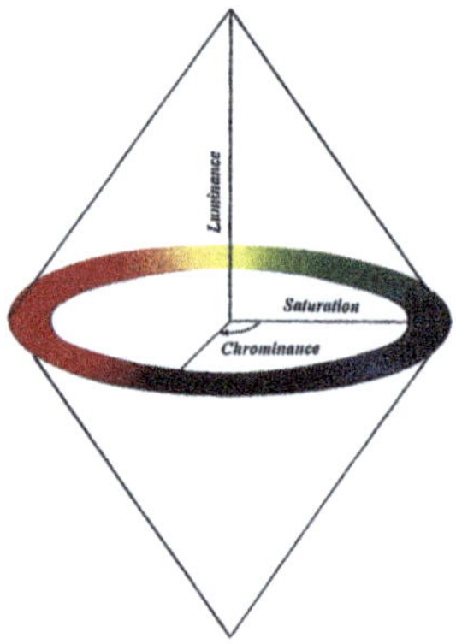

Figure 6b
Cone Luminance-Hue-Saturation

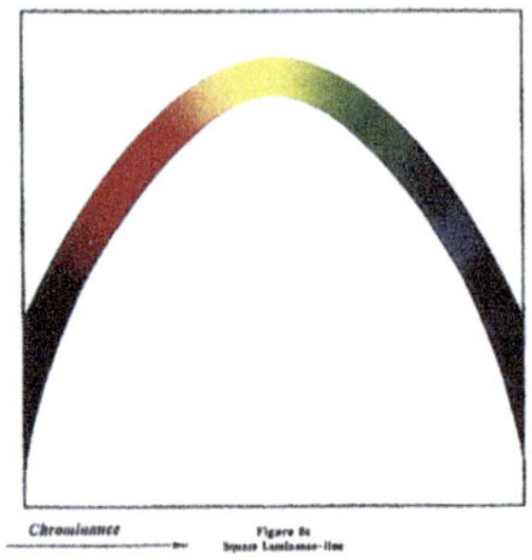

Figure 6c
Square Luminance-Hue

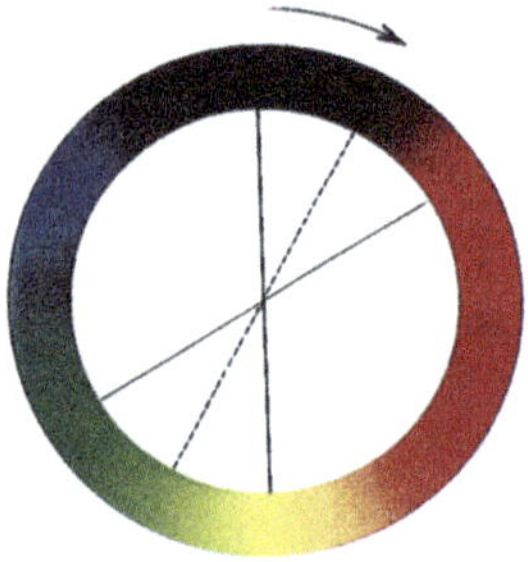

Figure 11a
Complementary equilibrium
between 2 hues

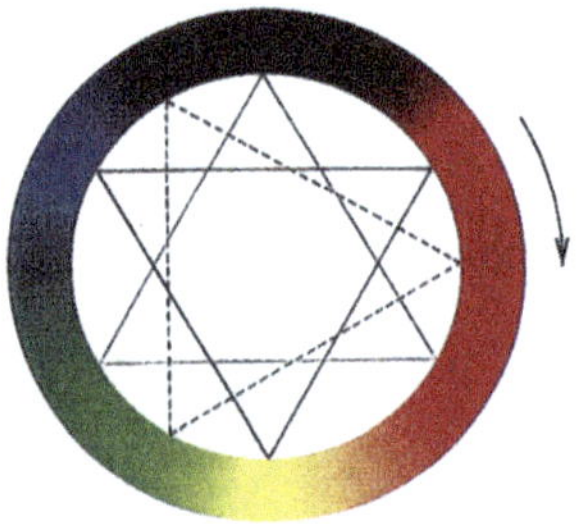

Figure 11b
Complementary equilibrium
between 3 hues

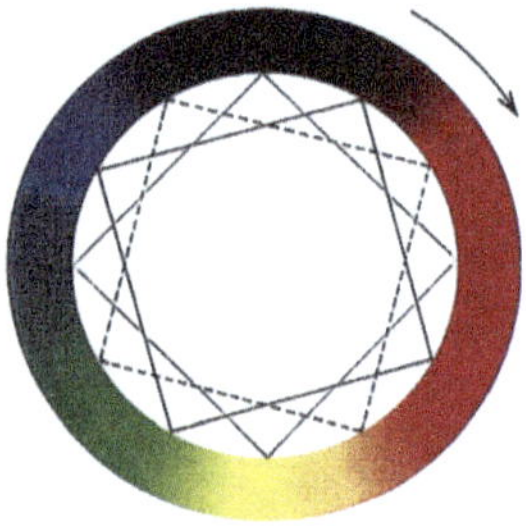

Figure 11c
Complementary equilibrium
between 4 hues

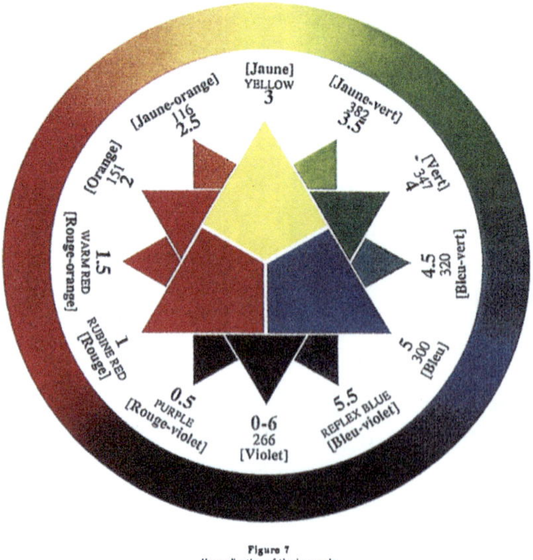

Figure 7
Normalization of the hue scale

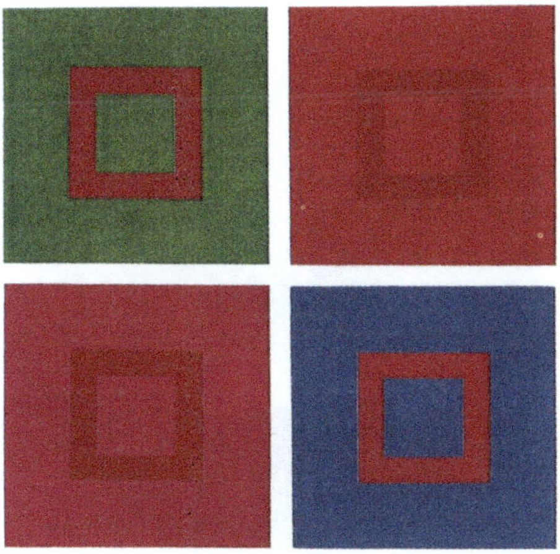

Figure 8
Simultaneous contrast

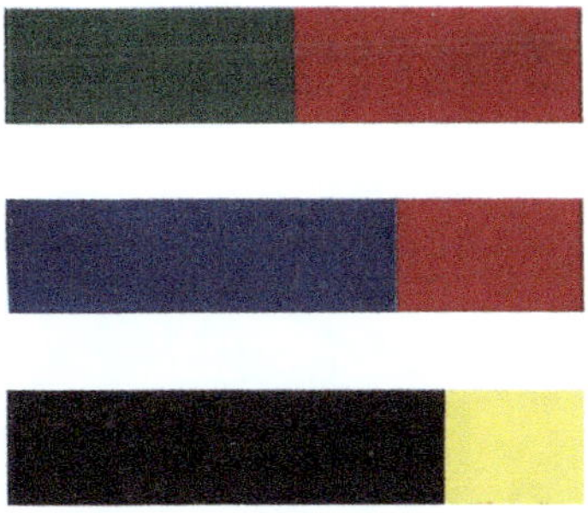

Figure 9
Quantity contrast

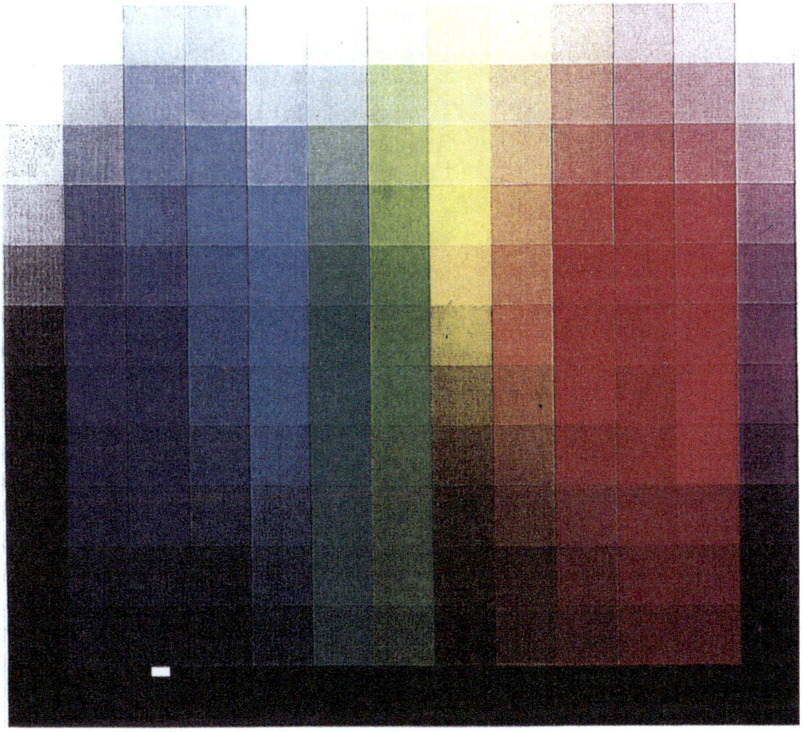

Figure 10
Luminance contrast

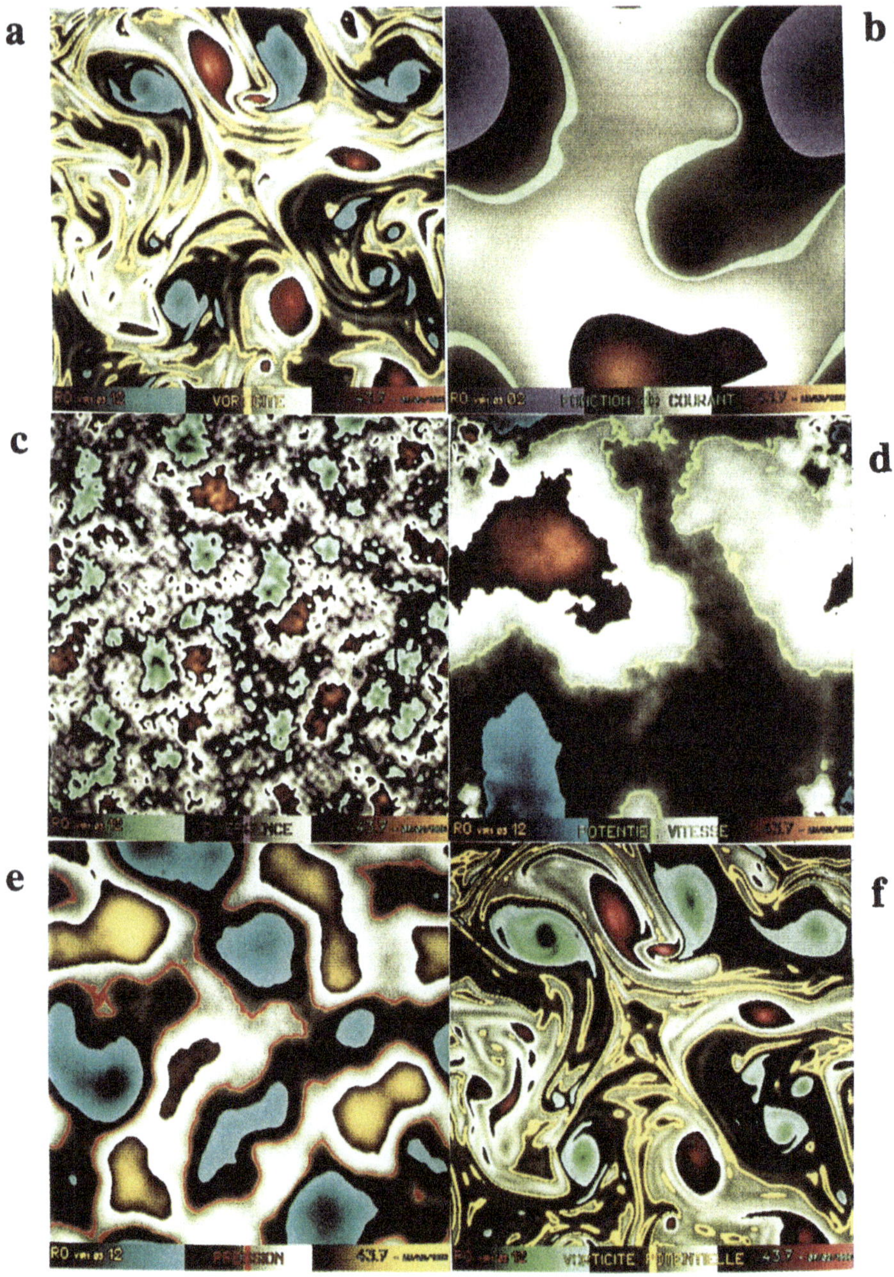

Figure 12

Extracting singularities in turbulent flow with real and complex wavelets

Maria Haase
Institut für Computeranwendungen, Universität Stuttgart,
Pfaffenwaldring 27, D-70569 Stuttgart, Germany
E-mail: mh@ica.uni-stuttgart.de

Abstract: In order to contribute to an understanding of the nature of turbulent intermittency the multifractal properties of turbulent velocity fields are explored on the basis of real and complex wavelets. The wavelet transform modulus maxima method (WTMM) is nowadays a well developed procedure to extract the singularity spectrum $D(h)$ if the signal does not contain oscillating singularities like vortex tubes. However, the presence of oscillating singularities leads to an underestimation of Hölder exponents h and thus to wrong results. In general, *two* exponents h, β are necessary to characterize the singular behaviour, where the oscillation exponent β describes the local power law divergence of the instantaneous frequency. To determine h and β we propose an improved WTMM method based on real and complex wavelets in which the maxima lines are traced directly. An appropriate partition function allows to calculate a general singularity spectrum $D(h, \beta)$.
Keywords: turbulence, wavelet transform, complex wavelets, oscillating singularities, singularity spectrum.

1. TURBULENT FLOW AND SINGULARITIES

In his celebrated K41 model Kolmogorov assumed the existence of an inertial range where turbulent flow is statistically stationary, homogeneous and isotropic. Under this assumption the velocity field would be statistically self-similar with one type of singularity only. However, experimental observations show that turbulent flow is *not* homogeneous. The mere existence of coherent structures invalidates the ergodic hypothesis and is likely to be responsible for the observed non–Gaussian behaviour of the velocity probability density function known as intermittency.

After more than a century of intensive scientific work in turbulence there is still a zoo of open questions. Among these is the problem of how small-scale structures like vortex tubes and sheets are formed and how they are statistically distributed. These structures have a quasi-singular character in the sense that they appear singular below the dissipative scale at which the smoothing effect of viscosity becomes important [Farge et al., 1999]. Therefore, parallel to theoretical investigations there is a need to develop effective and reliable methods which allow to localize and characterize singularities.

A. Gyr et al. (eds.), Science and Art Symposium 2000, 111-116.

The possible singularities of a function $f(x)$ can be subdivided into two classes, namely non-oscillating *cusps* (for example $|x|^{1/2}$), where the function or some derivative approaches infinity at a point and *spirals* or *oscillating singularities* (for example $\sin(1/|x|)$), where the frequency of the oscillation approaches infinity. Both types of singularities can either be isolated or densely distributed like in fractals.

The strength of a cusp singularity can be fully described by its Hölder exponent h, whereas for oscillating singularities one has to introduce an additional quantity, an oscillation exponent β, which describes the local power-law divergence of the instantaneous frequency [Arnéodo et al., 1997]. Turbulent flow probably contains both isolated and densely distributed cusp and spiral type singularities of varying strength.

2. LOCALIZATION AND CHARACTERIZATION OF SINGULARITIES

There are two classical tools to characterize the singularities of a function $f(x)$, the structure function method and Fourier spectra. One classical way to analyze the (multi-)fractal properties of singularities in turbulent flow, is to study the scaling behaviour of the structure function $S_q(l) = \int |v(x) - v(x+l)|^q dx \sim l^{\zeta(q)}$ of the velocity field $v(x)$, which provides information about distribution and strength of singularities. Kolmogorov's K41 theory predicts a linear scaling behaviour $\zeta(q) = q/3$, which is equivalent to a monofractal distribution of singularities of strength $h=1/3$. However, experimental observations show a deviation of higher order structure functions from the linear behaviour which is a consequence of rare events like the occurrence of coherent structures. The scaling exponents $\zeta(q)$ are closely related to the singularity spectrum $D(h) = \min_q (qh - \zeta(q) + 1)$ by a Legendre transform. $D(h)$ is the Hausdorff dimension of the set of points corresponding to the same Hölder exponent h. The drawback of the structure function method is twofold: firstly, it is restricted to $q > 0$ and secondly, it does not provide any access to information about singularities in the derivatives of the signal, i.e. it is restricted to $0 < h < 1$. Another classical tool is the Fourier transform. Since trigonometric functions have an infinite support, the Fourier transform delocalizes all information about singularities. The asymptotic decay of the power spectrum only enables to derive the strength of the worst singularity, the distribution and the type of singularities remain unknown.

In contrast, the continuous wavelet transform provides *local* information about the distribution and strength of the whole spectrum of singularities. According to the definition

$$Wf(a,b) = \frac{1}{a} \int_{-\infty}^{+\infty} f(x)\, \overline{\psi(\frac{x-b}{a})} dx \qquad (a, b \in \mathbf{R}, a > 0) \tag{1}$$

the wavelet transform (WT) decomposes a signal $f(x) \in L^2(\mathbf{R})$ hierarchically in terms of elementary components $\psi(\frac{x-b}{a})$ which are obtained from an analyzing wavelet $\psi(x)$ by dilations and translations. Here a denotes the scale, b the shift parameter and $\overline{\psi}$ the complex conjugate of ψ. The crucial point is to choose $\psi(x)$ to be well localized both in physical and Fourier space. In contrast to Fourier analysis the WT does not loose information about the position of transient phenomena and irregular structures. Together with the concept of hierarchical splitting this is the keystone for understanding why wavelets are a natural tool to analyze singularities and fractals.

A unique reconstruction of $f(x)$ is ensured if $\psi(x) \in L^1(\mathbf{R})$ satisfies a zero mean condition $\int \psi(x)dx = 0$. In order to make the entire range of singularities accessible to our analysis we further require $\psi(x)$ to be orthogonal to polynomials up to order N, i.e. $\int x^k \psi(x)dx = 0$ for $0 < k \leq N$, thus remedying one of the main disadvantages of classical approaches.

The wavelet transform of an isolated non-oscillating singularity with Hölder exponent $h(x_0) \in (n, n+1)$ scales like

$$|Wf(a,x)| \leq C a^{h(x_0)} \qquad \text{for} \quad a \to 0 \tag{2}$$

for all x within a cone of influence $|x - x_0| \leq Ka$, provided the analyzing wavelet has $N > n$ vanishing moments. An additional advantage of the WT is that all cusp singularities of $f(x)$ can easily be localized as those points towards which the maxima lines of $|Wf|$ converge as $a \to 0$ [Mallat et al., 1992].

The analysis of oscillating singularities is more complicated. The WT of an isolated spiral based on *real* wavelets leads to an accumulation of maxima lines which do not converge to the singular point. In [Arnéodo et al., 1997] it is proposed for the example of a chirp function $f(x) = |x|^h \sin(2\pi/|x|^\beta)$, $(h > 0, \beta > 0)$ to jump from one maxima line to the next always extracting the greatest value of $|Wf|$. This leads to a sequence of points (a_n, b_n) for which the WT scales like

$$|Wf\,(a_n, b_n)| \sim a_n^\alpha \qquad \text{where} \quad \alpha = h/(\beta+1) \tag{3}$$

Note that the exponent α is *smaller* than the Hölder exponent h since $\beta > 0$.

In order to simplify this procedure we construct a family of *complex progressive* analyzing wavelets $\{\psi_n(x)\}$ with an increasing number of vanishing moments. The main advantage of progressive wavelets is that the modulus of the wavelet transform is concentrated along the *ridges* [Torrésani, 1993] of the form $a = |b|^{1+\beta}$ obeying the scaling condition

$$|Wf(a,b)| \leq C a^{h/(\beta+1)} \tag{4}$$

thus allowing a direct access to h and β [Jaffard et al., 1996].

3. TWO FAMILIES OF ANALYZING WAVELETS

Confining to cusp singularities, a family of real valued wavelets $\{\psi_n(x)\}$ constructed from a Gaussian distribution $\psi_0(x)$ has proven to be very effective

$$\psi_0(x) = e^{-x^2/2}, \qquad \psi_n(x) = \frac{d}{dx}\psi_{n+1}(x) \qquad (n \epsilon \mathbf{Z}) \tag{5}$$

For $n > 0$, $\psi_n(x)$ has n vanishing moments. It can be shown that the wavelet transform based on the Gaussian family of wavelets $\{\psi_n(x)\}$ (5), fulfils a partial differential equation [Haase et al., 1998]

$$\left(a\frac{\partial^2}{\partial b^2} - \frac{\partial}{\partial a} + \frac{n}{a} \right) W^n f = 0 \tag{6}$$

where $W^n f$ denotes the wavelet transform with $\psi_n(x)$ as analyzing wavelet. Describing the lines of local maxima in a parametric form $\{a(t), b(t)\}$ a set of ordinary differential equations can be derived

$$\frac{da}{dt} = -C\frac{\partial^2}{\partial b^2}(W^n f), \quad \frac{db}{dt} = C\frac{\partial^2}{\partial a \partial b}(W^n f) \tag{7}$$

which can be integrated numerically. The main advantage of this direct method is that it allows to trace the maxima lines continously thus circumventing the usual problem of connecting maxima points [Haase et al., 1998].

This procedure can now be extended to include oscillating singularities. We propose to use the following family of complex progressive wavelets $\{\psi_n(x)\}$.

$$\psi_0(x) = (e^{i\omega_0 x} - e^{-\omega_0^2/2})\, e^{-x^2/2}, \qquad \psi_n(x) = \frac{d}{dx}\psi_{n-1}(x) \qquad (n \epsilon \mathbf{Z}) \tag{8}$$

with an increasing number of vanishing moments. For a sufficiently large frequency ω_0 ($\omega_0 > 5$ for practical purposes), ψ_0 can be replaced by the Morlet wavelet. Analogously, we derive a partial differential equation for the WT

$$\left(a\frac{\partial^2}{\partial b^2} - \frac{\partial}{\partial a} - i\omega_0\frac{\partial}{\partial b} + \frac{n}{a} \right) W^n f = 0 \tag{9}$$

and two evolution equations for the maxima lines:

$$\frac{da}{dt} = -C\frac{\partial}{\partial b}\left(\text{Real}\left(\frac{\partial W^n f}{\partial b}\overline{W^n f}\right)\right), \quad \frac{db}{dt} = C\frac{\partial}{\partial a}\left(\text{Real}\left(\frac{\partial W^n f}{\partial b}\overline{W^n f}\right)\right) \tag{10}$$

The skeleton of maxima lines can thus be continuously traced up to the desired accuracy with a reduced computational effort.

4. DETERMINATION OF SINGULARITY SPECTRUM

Since the skeleton of maxima lines carries the scaling information of singularities, Arnéodo, Bacry and Muzy developed the wavelet transform modulus maxima method (WTMM) to calculate the singularity spectrum $D(h)$ [Muzy et al., 1994], which, however, is restricted to signals with cusp singularities only.

If the function $f(x)$ contains both types of singularities, we define the partition function based on the set of maxima lines

$$Z(p,q,a) = \sum_{i \epsilon \text{max.lines}} \left(\sup_{a'<a} |W^n f(a, b_i)| \right)^q d_i(a)^p \tag{11}$$

which scales like $Z(p,q,a) \sim a^{\tau(p,q)}$. Here $d_i(a)$ denotes the distance between the i-th and the $(i+1)$-th maxima line. The two-dimensional singularity spectrum

$$D(h,\beta) = \min_{p,q}\Big(qh + p(\beta+1) - \tau(p,q)\Big) \qquad (12)$$

can be obtained by Legendre transforming $\tau(p,q)$ [Arnéodo et al., 1997].

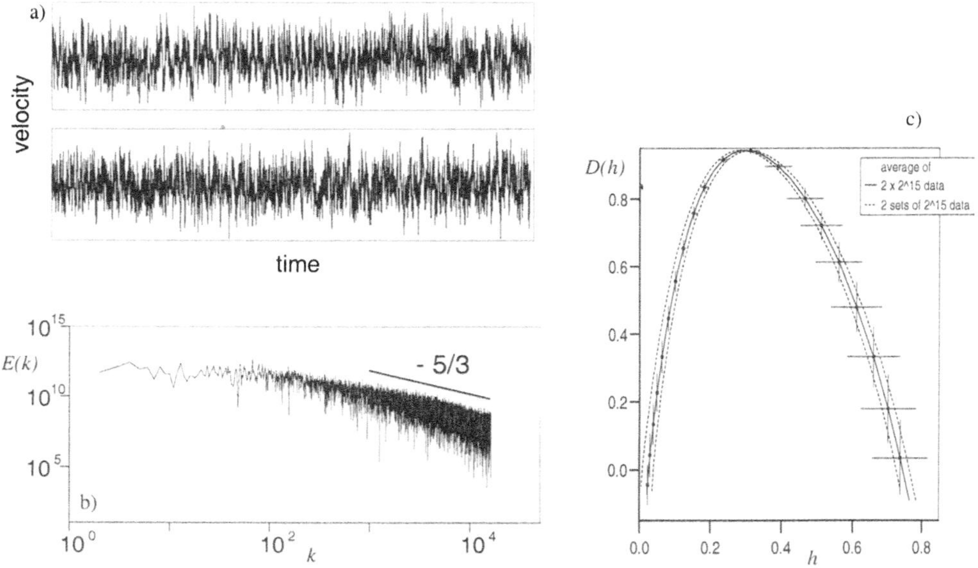

Figure 1: a) Turbulent velocity signal, b) Energy spectrum (inertial range) c) Singularity spectrum resulting from WTMM method (analyzing wavelet ψ_1).

5. WAVELET ANALYSIS OF A TURBULENT JET

In order to check the direct tracing method, we applied in a preliminary study the WTMM based on the Gaussian family of wavelets to turbulent data from an axisymmetric jet with helium at low temperature [Chabaud et al., 1994], which has been used to analyze the nature in turbulent intermittency. Figure 1a shows two data sets, each containing 2^{15} samples of the velocity field. As can be seen from the energy spectrum (Fig. 1b), the data are taken from the inertial range. For each of the two data sets the singularity spectrum is calculated using ψ_1 as analyzing wavelet (Fig. 1c).

It can be seen that the Hölder exponents fluctuate considerably in a wide range between 0 and 0.8. The most frequent exponents h are found for the maximum of the spectrum close to the classical Kolmogorov value of 1/3. Although these results are in good agreement with previous observations [Muzy et al., 1994], one should keep in mind that it is assumed that the signal does not contain spiral singularities.

6. CONCLUSIONS AND FUTURE WORK

We have presented some recent work on the detection and characterization of cusp and oscillating singularities in turbulent flow using wavelet analysis. The aim is to explore the range and the statistical distribution of cusps and spirals in order to get more insight into the phenomenon of intermittency and to compare the predictions of existing turbulence models with experimental observations.

The WTMM method based on real and complex wavelets provides a unified multifractal description since it can be applied both to functions and measures. To our knowledge it is the only method which allows to extract the whole range of cusp and oscillating singularities and which gives at the same time access to singularities in the derivatives of the function. In addition, the WTMM method seems to be numerically more stable than direct methods since only averaged quantities are involved and since the restriction to maxima insures that small errors are less taken into account. Although much work has been done on the theoretical side on the multifractal formalism, it still remains much to do on the numerical side, especially in the field of oscillating singularities.

REFERENCES

Arnéodo, A.; Bacry, E.; Jaffard, S.; Muzy, J.F.; 'Oscillating singularities on Cantor sets: a grand-canonical multifractal formalism'; J.Statist.Phys. Vol. 87, Nos 1/2, pp. 179–209 (1997).

Chabaud, B.; Naert, A.; Peinke, J.; Chillà, F.; Castaing, B.; Hébral, B.; 'Transition toward developed turbulence', Phys.Rev.Lett.73, pp. 3227–3230 (1994).

Farge, M.; Kevlahan, N.K.-R.; Perrier, V.; Schneider, K.; 'Turbulence analysis, modelling and computing using wavelets'; In: *Wavelets in Physics*, Ed. van den Berg, J.C.; Cambridge University Press, pp. 117–200 (1999).

Haase, M.; Lehle, B.; 'Tracing the skeleton of wavelet transform maxima lines for the characterization of fractal distributions'; In: *Fractals and beyond*, Ed. Novak, M.M., World Scientific; Singapore; pp. 241–250 (1998).

Jaffard, S.; Meyer, Y.; 'Wavelet methods for pointwise regularity and local oscillations of functions', Memoirs of the American Mathematical Society Vol. 123, No 587 (1996).

Mallat, S.; Hwang, W.L.; 'Singularity detection and processing with wavelets', IEEE Trans.Inf.Theory 38; pp. 617–643 (1992).

Torrésani, B.; 'Some time-frequency aspects of the continuous wavelet transform'; In: *Progress in wavelet analysis and applications*, Eds. Meyer, Y.; Roques, S.; Editions Frontières (1993).

Vortism a Basic Art

Georg Emanuel Koppenwallner und Ludwig Plotter,
D-37085 Göttingen/ Germany
Himmelsstieg 1 / Fax – 0049 551 7989512

Abstract: Basic art is an effort to fill the gap between science and art. Vortism uses vortices as iconographic symbols. These symbols can be combined to create synoptic imaginations for painting. They can also explain in a more image-like fashion flow phenomena in bionics or technics. This can be applied to invent new machines and ideas. The adaption of vortism in suction devices is presented. The terms frontal-flow, frontal-vortex and Coanda-vortex are introduced. The swimming of fluidic-beings thunas and sharks is presented in a vortistic manner. The relation between science and art is shown from both sides. A turntable experiment is proposed that is inspired by the basic art of vortism. This experiment shall use frontal- vortex generators.

Keywords: Vortism, Coanda, Basic Art, Painting, Invention

1.WHY ART - WHY BASIC ART

Out of a technical point of view art is the human ability to create material things. Out of a more mental angle of view art is the human ability to create imaginations. We share and cultivate these imaginations until they influence our lives. Human evolution, later human history, human technical and social development shows a lot of these live governing cultivations. These are: Thougts, things, rituals, religious believes, social orders, technical know how, science etc. . But our often narrow minded reception of the world does not longer fit to our global possibilities
The global society works still mainly on the death line [=*Todeslinie*] Beuys (1976). Technical applications are evolving into mass production, mass consumption causing mass destruction and mass nature changes. Joseph Beuys postulated the idea of a social plastic [=*Soziale Plastik*] to de-imaginaze the old system by the aid of all individuals. Everybody is creating the social plastic. That sounds difficult and is difficult. But art is the tool or more mental the idea to manage this. Basic art shall help to shape this social plastic.
Art is more than creating things and having thougts. Art is also a source of emotion.
We can neither understand our world in details nor in totality. But we must act and survive in this world. Art is a human way to do this. Art does not create theories that become more and more restricted or even false. Artworks remain as true as the emotions behind.
We observe today an increasing drift between highly specialized knowledge and more complex and universal points of view. This is a big problem.

A. Gyr et al. (eds.), Science and Art Symposium 2000, 117-126.

A basic art shall not be a concurrence to isolated sciences. It shall bridge the regions excluded for sciences on the one side and for the multimedial blindness on the other side.
Art is already a method to be applied to the structure of houses, planes, submarines or cars. We may call this work architecture, industrial design, bionics or ingeneering. A piece of art is always included. All these applied arts must be usefull, practical and of economical, scientific or other practical value. Pure art can be more related to humor, nonsense and other dangers of strict orders and rigid systems. Basic art shall form the fundamentals to more synoptic views of our world. In computional simulations we need diagnostic fields to create prognostic fields. We need arts to find or invent new images that are forerunners of scientific, technical and political ideas

2. VORTISM A BASIC ART

Vortism uses vortices as iconographic elements. The most common examples are the monopolar vortex (source-sink flow), the bipolar vortex (convection) and the tripolar vortex (a common instability mode of monopolar vortices especially in rotating systems). Vortices of varying scales are a common phenomena in flows about a certain Reynolds-number. This phenomena of complex vortex interaction is termed turbulence, a 'parametrization' of non-understanding. In technical situations it is often usefull to control or to remove such flows. This is of growing importance in a future of restricted energy consumption. But to understand turbulent flows interacting with their surrounding is just one feature of ‚hopeless‘ many influences on beings that are living or machines that are working with or in fluids. Basic art tries to regard all these influences.

2.1 Vortism in suction devices

Human skill creates artificial atmospheres in houses, production halls, ships, planes etc.. To remove polluted gas is a daily event also in kitchens. Above the oven is often a cooker hood. The common way to increase suction efficiency there is to increase flow rate.

A more sophisticated method is the use of jet flows towards the suction areas. These jets can be free, swirling or wall jets, Recknagel Sprenger (RS) (1999). A use of vortistic iconography shall show the situation in suction devices. A thermal jet ascending from a heat source hits a horizontal boundary and spreads as a kind of density current under this plane. If the flow is more heavy compared to the surrounding we just have to turn this drawing, Fig. 1 a), b). The plane can contain a suction area and is comparable to a cooker hood.

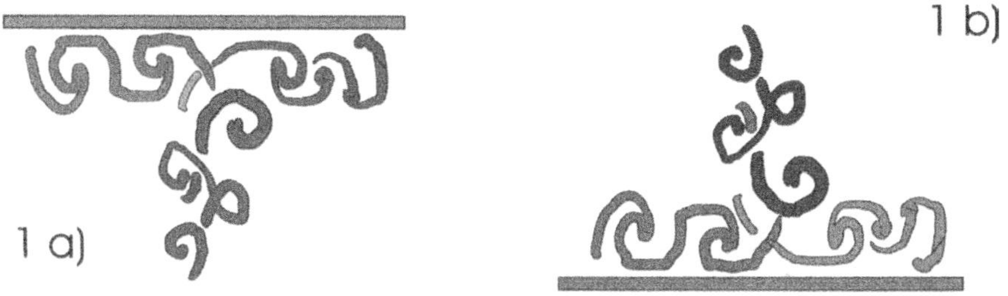

***Figure 1**. Density currents and thermal flows a) $\rho < \rho_{umg}$, b) $\rho > \rho_{umg}$*

The suction flow velocity **u** is rapidly decreasing in a distance **x** form the opening, RS (1999) p. 1643.

$$u(x) \sim \frac{1}{x^2}, \qquad \frac{u(x)}{u(x=0)} = C * \frac{A}{10x^2 + A}$$

A is the area of an rectangular or circular suction mound. It is **C** = 1 for a free opening and **C** = 1,33 for a mound with a flange. One diameter in distance the suction velocity is reduced to about 10%.
To control density currents around suction openings at undersides the use of frontal flows increases suction efficiency

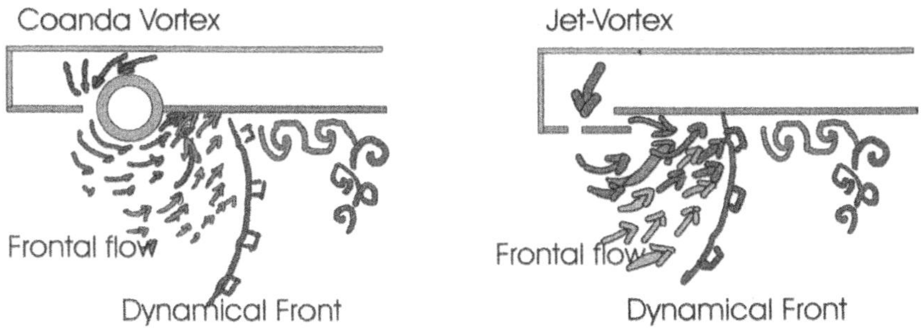

***Fig. 2**. **Frontal-vortex generators**, a) Coanda-vortex b) Jet-vortex (Strahlwirbel) type*

The flow 2 a) is a circular shear flow near a surface. The flow 2 b) is a vortex flow near a surface. Both flows create a dynamical front near the surface. I propose for these flows the term: **Frontal flow**.

Fig. 2 shows two different types of frontal vortex generators. I also propose the term **Coanda-vortex** for a vortex without core, Fig. 2 a). So it is possible to say:
Frontal-vortices stop and reverse escaping density-flows and thermals near boundaries. These kinds of frontal-vortex generators are applied in cooker hoods and other suction devices. Fig. 2 c) shows the principle of a frontal vortex cooker hood. At the left is a schlierenfoto of the frontal region.

c)

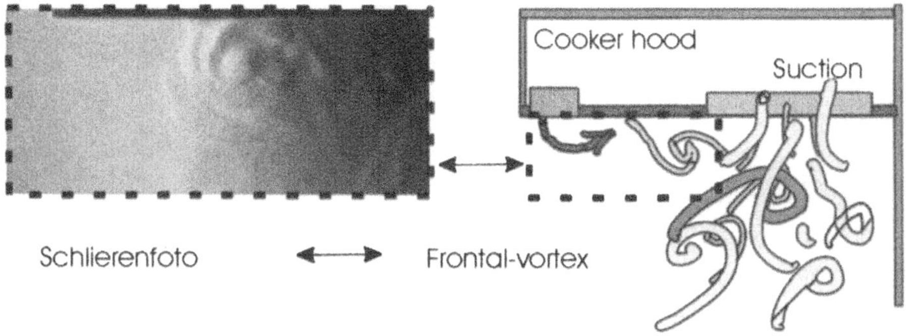

Fig. 2 c) *Principle of a frontal vortex cooker hood, left Schlierenfoto of the flow*

The coanda effect is named after H. Coanda (1932) who described the jet to wall and jet to jet attachment in a patent .
The parameter range is described in terms of a **Re**-number. Fig. 3 b) shows a typical experimental setup. Newman (1961) defines a Re-number with the exit-velocity **u**, the radius of the coanda-tube **R**, the kinematic viscosity ν and the channel width **h**.
Köster and Löhr (1964) use only the tube radius **R** .
The Coanda-effect is not observed for **R/d** < 3.

$$\mathbf{Re} = \sqrt{\frac{u^2 Rh}{2\nu^2}} \quad \text{(Newman 1961)}, \qquad \mathbf{Re} = \frac{uR}{\nu} \quad \text{(Köster und Löhr 1964)}$$

Fernholz (1966) states that for **R/h** > 14 the channel width h has no influence.
Under **R/h** < 7 the influence of the channel and its boundary layer is important.

$$\mathbf{Re}_{\mathbf{suc.}} = \sqrt{\frac{u^2 Rd}{2\nu^2}} \quad \text{(suction devices)}$$

For the use in suction devices I propose the use of the slot width **d** because of the strong varying channel widths and the relative small ratio **R/d** ≥ 4 .

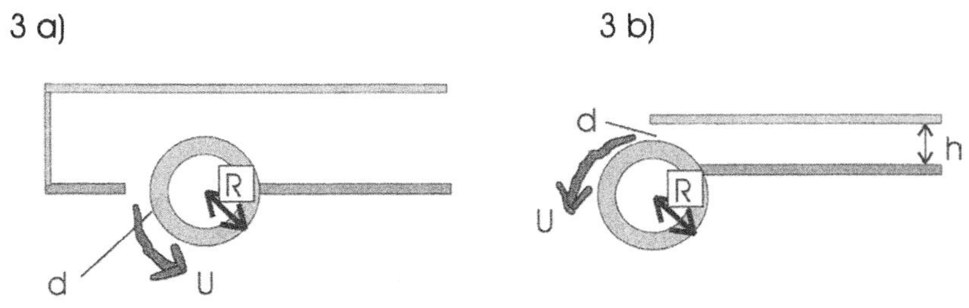

Fig. 3. Characteristic lengths of a) coanda-generator in suction devices, b) typical experimental setup

The coanda generators in suction devices work on the lower limit of the coanda effect in a range of **R/d** ≥ 4, $\mathbf{Re}_{suc.} \sim 1000$ - 10000.
These frontal flows can be also structured in a **U**-shaped form. Fig. 4 shows the case of a laboratory suction device with a moveable lid. This flow increases the inflow velocity near the boundaries. Flows shown in Fig. 1 b) are reversed and easier removed.

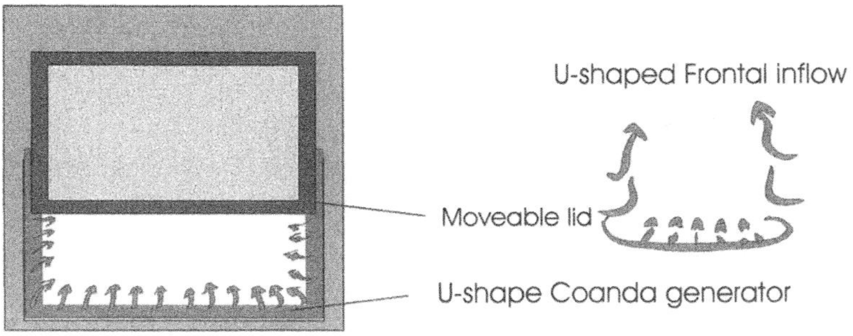

Fig. 4. U-shaped frontal vortex for a laboratory suction device

2.2. Vortism and fluidic beings

In the following part I will develop a 'vortistic imagination' of 'the flow tracks' of two typical fluidic beings, sharks and bonitos - a small thuna kind. This imagination was developed during painting studies of these animals. It cannot replace scientific research on these marine animals but shall create an interaction between art and sciences as marine research, fluid mechanics, biology etc.. The sharks belong to the elasmobranchs. The thunas are member of the teleosts.

U-shape or horseshoe vortex is also a common feature for divers and other fluidic beings Make following experiment in clear water with a foot paddle. Move the paddle near the surface. You see two strong line vortices visible by the air entrainment. As a diver you would have a better sight of this phenomen

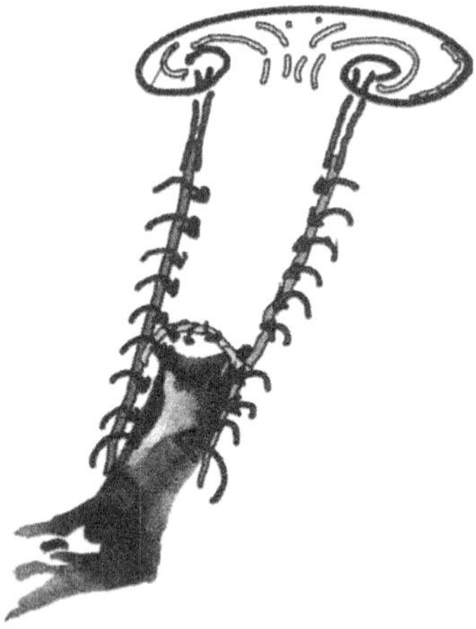

Fig. 5 Paddle-vortex

This vortices are generated by the flow around the fin edges. My idea concerning the propulsion of this is as follows. The long side vortices of the horseshoe vaste energy but also accelerate the fluid to propulse the 'fluidic being'. Fluidic beings that depend on economic long distance swimming have to minimize these losses. These fish are for example a) bonitos, macrels of the *teleost* group or b) sharks of the *charcharhinidae* group. The fins of group a) (bone fish) are of lunatic shape. The fins of group b) many members of the *elasmobranch* group, sharks, are of more or less asymmetric form. This tail shape is also common for sturgeons. Sharks of the *lamnidae* group developed nearly lunatic tails, Lindberg (1974) .

The front view of these animals reveals also differences. The fish with lunatic tails have an elliptical front cross section, Fig. 6. The front section of sharks is of a triangular form, Fig. 7. The breast fins of normal fish can be pressed to the body. The breast fins of shark are more rigid and cannot be pressed to the body. Sharks swim therefor with permanent lift by this wing like fins. The flow pattern around a moving shark will be different from the pattern of a moving thuna of comparable size or of comparable characteristic Reynolds - or Froude - number .

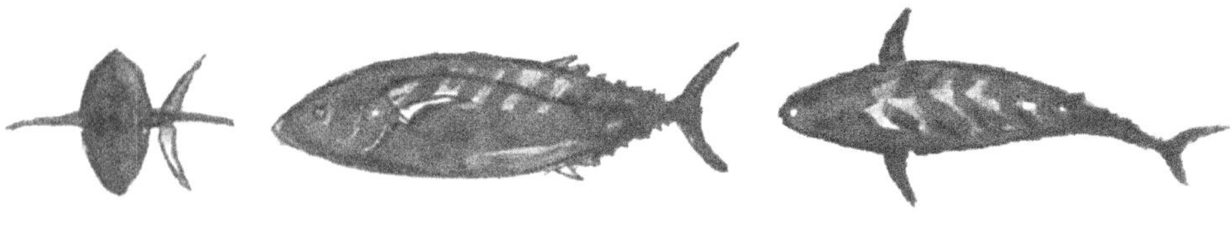

Fig. 6 *Fin shape and cross section for bonitos or thunas*

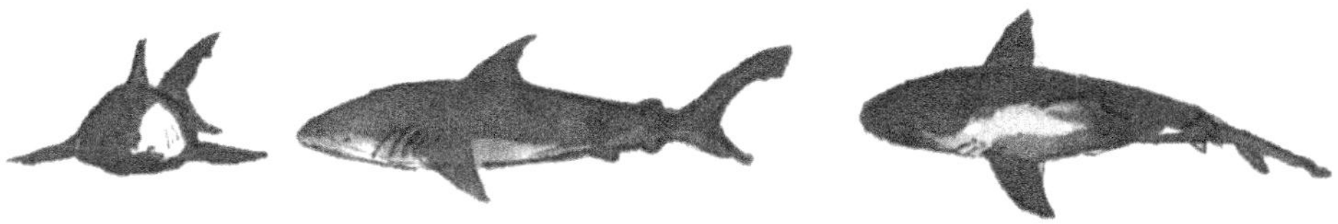

Fig. 7 Fin shape and cross sections for a 'typical' shark

The flow around a thuna is divided by sharp spoilers before and on the middle of the fin. These spoilers seem to divide the flow in a upper and a lower part. Fig. 8 a shows a sketch of a bonito tail. A possible flow field around the tail is sketched in fig. 8 b. The fin stroke is to the left and is symbolized by the small fish. Left/right is understood looking in the direction of motion.

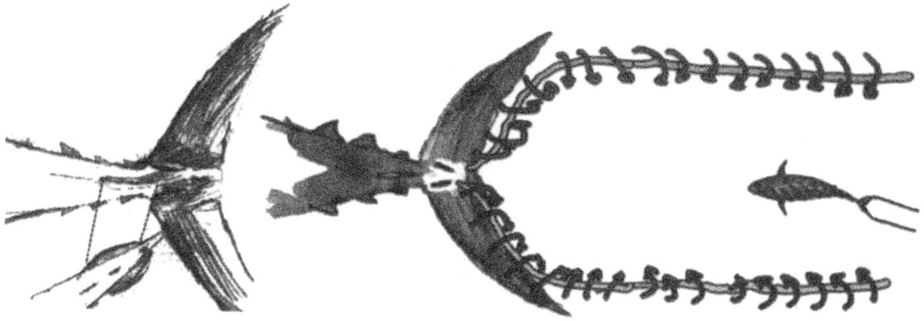

Fig. 8 *Bonito fin a) side and top view b) flow structure*

Fig. 9 shows the 'flow track' of a bonito. The flow around the left/right body side of the fish is accelerated by a tail stroke to the left/right and concentrates in two line vortices. The fluid of the upper and lower body side is accelerated separately, Fig. 9 a.
In a front view the flow fields of these strokes are of a bipolar kind, Fig. 9 b. The line vortices of the left/right strokes can be joined by alternatively joining the upper and lower line vortices. The bonito has a vortex track of two line vortices of opposite sense of rotation, Fig. 9 c.

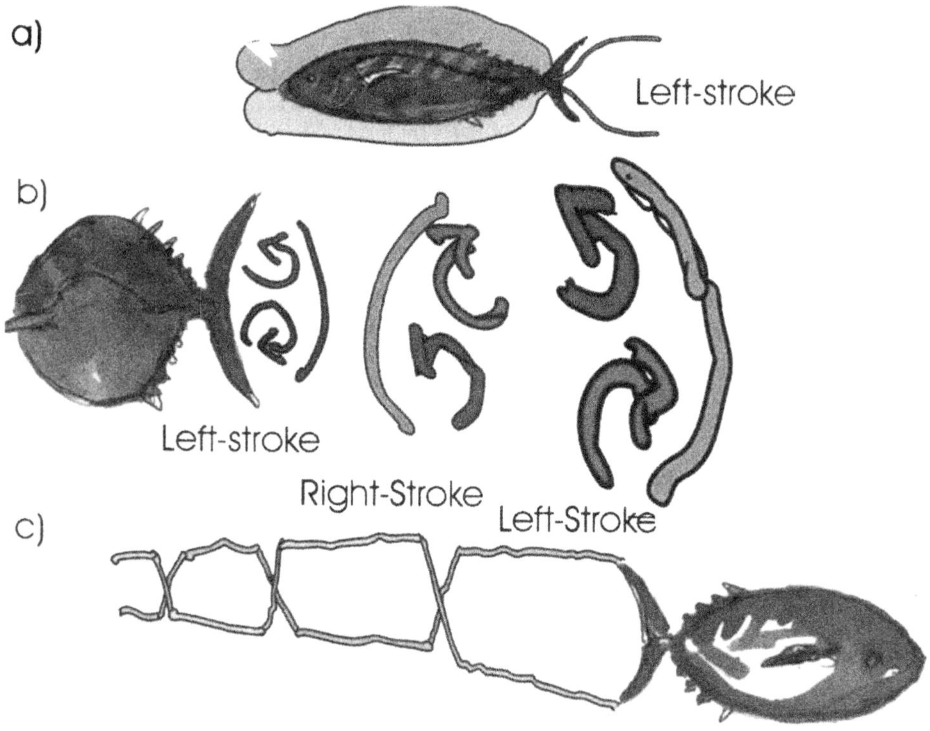

Fig. 9 *Flow structure of a swimming bonito, a) seperated flow around upper and lower body sides, b) bipolar flow structure, c) propulsion vortex lines.*

The flow track of a shark is of different kind. The shark uses only one line vortex with each stroke. Fig. 10 proposes a flow structure with one line vortex originating from the propulsion fin, Fig. 10 a), b). This vortex can be connected to the trailing vortices of the breast fins, Fig. 10 a),c) . The flow track of a shark is shown in Fig. 10 d). This asumption bases on the asymmetric form of the sharks fin and the typical 'cut' in the tail fins upper part.

These 'vortistic' imaginations lead to the idea of a single and a double vortex line propulsion method for fluidic beings. The being as a whole creates a typical vortex system. Each vortex system gives different ways to reduce drag. This system is also just part of the beings total 'concept'. Elasmobranchs (sharks) are different in the muscle structure compared to teleosts fishes. The shark also uses a special skin. The term 'shark skin' became synonymous with a low friction surface. The bonito itself is a low resistance body. It is compared to profiles, Hertel (1963). My guess is that the interaction of body and fin vortices with propulsion vortices is important and characteristical for the movement of many fluidic beings. The interaction of fin vortices with tail vortices is also observed in submarine dynamics, Burcher and Rydill (1994).

The fore planes interact with the after planes and influence the neutral stability point of the ship.

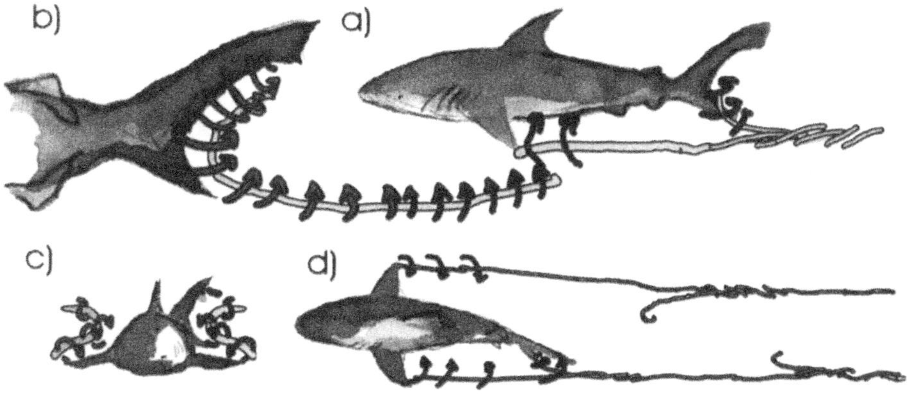

Fig. 10 *Flow structure of a 'typical' shark*

A more scientific view on the movement of aquatic fluidic beings is described in the book 'Fish locomotion', Blake (1983). The movement of a 'ground' shark, a dogfish, is described as an anguilliform propulsion mode. The whole body is moved in a propulsive wave. This mode is named after the eel, *anguilla.*
The bonito moves in a thuniform mode. The fish moves mainly the tail. The body anterior of the tail shows nearly no lateral movement.

3. RÉSUMÉ

What is the difference between any scientific, for example a bionical, biological or fluid-mechanical point of view compared to an inventive, imaginative or more artistical -a painters- point of view towards the same things, the nature, our mind and its restrictions. The clue to this question is hidden in the restrictions. The artist is less restricted in the way he works, obvious **disfunction**, nonsense and many artistic degrees of freedom are 'allowed'. But the perception of the artworks is dominated by the restriction of the public. The scientific way is different. A scientific publication must fullfill a certain degree of scientific rules and a high degree of **function**. The perception of the scientific communitiy is less restricted. The assumption is that a scientist can either understand, reject or develop further new scientific ideas. But these new ideas must accept permanent proofs. This disfunction / function criterion between art and science is described by Bierens (1999). But this criterion shows that he looks from the scientific side. Let us apply this criterion to this paper. The vortism of suction devices has at least industrial function, is patented or applied for patent. The vortism of fluidic beings is an unproofed speculation generated by fish paintings and drawings. The pictoral art of vortism cannot be presented in a semi-scientific basic-art publication like this. But we can make an artistical transfer for science and propose the use of circular or linear frontal-vortex generators, compare Fig. 2 a), b) in a barotropic turn-table experiment at the bottom (or the vertical walls) to simulate frontal dynamics in a rotating system. A appropriate turn-table with a high Ekmann layer for such an experiment is described in Chabert d'Hieres et al. (1991). It should be possible to block and steer the coastal current of this experiment. It should also be possible to create a source sink vortex with frontal dynamics near the bottom.

An aim of basic art is to **find functions and pictures** interacting with the results of such experiments. And this is comparable to the relation of science and art.

4. REFERENCES

[Beuys J., 1976] Ich durchsuche Feldcharakter; In: *,Soziale Plastik'; p. 121;* Editors Harlan et al.; Achberger Verlagsanstalt, Achberg ; ISBN 3-88103-011-5

[Bierens, C., 1999] Sculpture of a Particle Accelerator; Katalog Formule 2 Exhibition; p. 110 -115; Künstlerhaus Bethanien; Vice Versa Verlag; Berlin; ISBN 3 932 754 05 0

[Blake, R.W., 1983] Fish locomotion; Cambridge Univ. Press.; Cambridge; ISBN 0 521 24303 3

[Burcher R., Rydill L., 1994] Concepts in submarine design; Cambridge University Press;p. 180; ISBN 0 521 41681 7

[Chabert d'Hieres G., Didelle H., Obaton D., 1991] A laboratory Study of Surface boundary Currents: Application to the Algerian Current; Journ.of Geophys. Res.; Vol 96; p. 12.539 - 12.548

[Coanda H., 1932] Procédé de propulsion dans un fluide; Brevet d'Invention Gr. 6 Cl . 2; No. 762688, République Francaise.

[Fernholz H.H., 1966] Zur Umlenkung von Freistrahlen an konvex gekrümmten Wänden, TU Berlin, DLR Bericht 66-21

[Hertel, H., 1963] Struktur Form Bewegung; Mainz

[Köster H., Löhr R., 1964] Untersuchungen der Umlenkung eines ebenen Strahles durch einen Kreiszylinder (Coanda-Effekt); Institut für Aerodynamik der DFL Braunschweig; DFL Bericht 64/30

[Lindberg, G.U., 1974] Fishes of the world; John Wiley & Sons, Chichester, ISBN 0 470 53565 2

[Newman B.G., 1961] The Deflection of Plane Jets by Adjacent Boundaries – Coanda Effect ; In: Boundary Layer and Flow Control; Ed. G.V. Lachmann; Pergamon Press; London

[Recknagel Sprenger Schramek, 1999] Taschenbuch für Heizung + Klima Technik 2000; Ed. E.R. Schramek; pp. 1439 – 1442, 1638 – 1656; Oldenbourg V.; München; ISBN 3-486-26215-7

Gallery of physical phenomena

S. Alekseenko, D. Markovich
Institute of Thermophysics, Lavrentyev Ave., 1, Novosibirsk,630090, Russia
e-mail: aleks@itp.nsc.ru

Abstract: The paper presents the gallery of physical phenomena found and/or described in the Institute of Thermophysics of Russian Academy of Sciences (Novosibirsk, Russia). The majority of results is related to hydrodynamical or thermophysical processes. The main method of describing processes is their visualisation. For each concrete phenomenon the photograph is shown and brief description is given.
Keywords: Visualisation; waves; jets; vortices; coherent structures.

1. INTRODUCTION

The visualisation of physical phenomena plays an important role in hydrodynamics and thermophysics. It provides a possibility of understanding the essence of the physical phenomenon and effective representing the results of both experimental measurements and numerical simulation. The visualisation is particularly important for three-dimensional nonstationary flows and complex processes of heat transfer. The "Album of Fluid Motion" by Van Dyke, 1982 and regular publications in "Physics of Fluids" under the title "Gallery of Fluid Motion" may be cited as the most impressive examples of flow visualisation.

In this work, the different methods of visualisations are used as an effective means for representation and description of various physical phenomena from the field of hydrodynamics and thermophysics. These phenomena were found and/or described in the Institute of Thermophysics. The results presented may be classified as follows: waves, jets, vortex and coherent structures.

2. WAVES ON FALLING LIQUID FILM

The remarkable peculiarity of falling liquid thin films is in their instability which results in formation of running surface waves even at the Reynolds numbers Re of the order of unit [Alekseenko et al., 1994]. Figure 1 shows the typical wave pattern at Re=30. As follows from the experiment, film waves are usually three-dimensional, nonstationary and strongly nonlinear. Among great variety of disturbances, one can clearly separate the structural elements - solitary waves of horseshoe shape. Such a wave was first cal-

A. Gyr et al. (eds.), Science and Art Symposium 2000, 127-138.

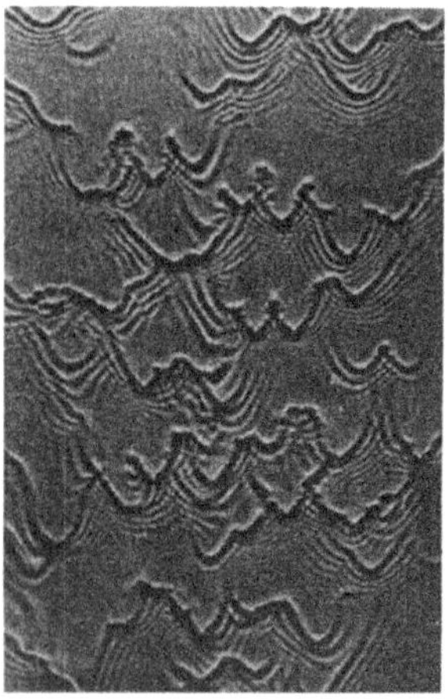

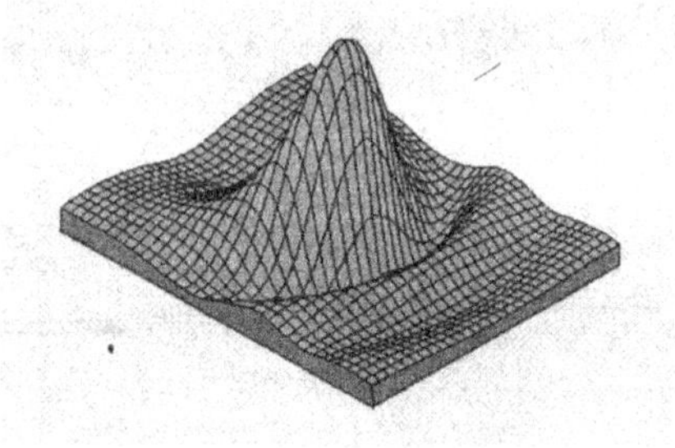

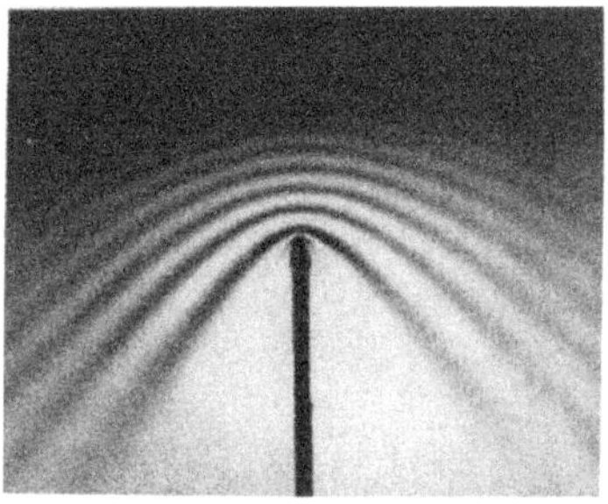

Figure 1; *Waves on falling liquid film.*

Figure 2; *Solitary wave in thin liquid layer*

Figure 3; *Standing waves in liquid films*

culated by O. Tsvelodub (see p. 216 in [Alekseenko et al., 1994]) on the basis of the model equation (Fig. 2).

Besides the running film waves, one can observe standing three-dimensional waves which may be generated by a point obstacle. Figure 3 demonstrates the example of standing waves arising due to touching the smooth surface of falling ethanol film with a needle tip [Alekseenko et al., 1994].

3. SPIRAL WAVES IN DISPERSIVE MEDIA

If the obstacle touches the free surface of rotating fluid, a more complicated pattern of standing waves appears in comparison with Fig. 3. In [Alekseenko and Cherep, 1994], a fairly general case is theoretically considered where a point source moves along the circle in dispersive media. The arising pattern of standing waves is completely defined by the parameter $\gamma = c_g / c_p$, where c_g and c_p are the group and phase velocities correspondingly. Figure 4, a shows the case of $\gamma = 1.5$ (capillary waves on deep water). We may point on that the wave crests have a spiral shape, and wave pattern exists only in definite area bounded by the broken line. The experiment, presented in Fig. 4, b, confirms the theoretical conclusions. Here, "+" is the centre of the circumference along which the point source moves. The same method is applied to description of spiral structure of galaxies $(\gamma = 1)$ and gravity waves generated by the ship during its motion along the circle $(\gamma = 0.5)$.

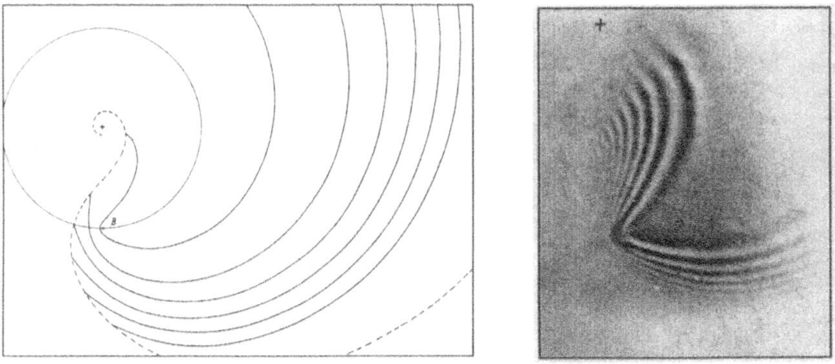

a b

Figure 4; Spiral waves in dispersive media

4. RAYLEIGH - BENARD CONVECTION

The instability, arising under heating the liquid layer from below, is known as the Rayleigh - Benard convection and is defined by a single dimensionless parameter - Rayleigh number Ra. Figure 5 shows the evolution of spatial structure of the Rayleigh - Benard convection in ethanol layer of thickness H between two solid plates with temperature difference ΔT [Berdnikov et al., 1990]. The flow visualisation is made with the aluminium powder. For $Ra = 3.7 \cdot 10^3$, $(\Delta T = 4.85^\circ C, \quad H = 2.0 mm)$ the pronounced two-dimensional cellular structure is formed (Fig. 5, a). With increasing Ra up to $1.98 \cdot 10^4$ $(\Delta T = 1.87^\circ C, \quad H = 5.08 mm)$ the bimodal form is observed when transverse rolls are superimposed on the main flow (Fig. 5, b). This is a consequence of the "crossroll instability" which is typical for stratified boundary layer near the horizontal plane. Finally, Figure 5, c demonstrates the turbulent regime of convection ($Ra = 3.92 \cdot 10^5$, $\Delta T = 5.68^\circ C, \quad H = 9.0 mm$)

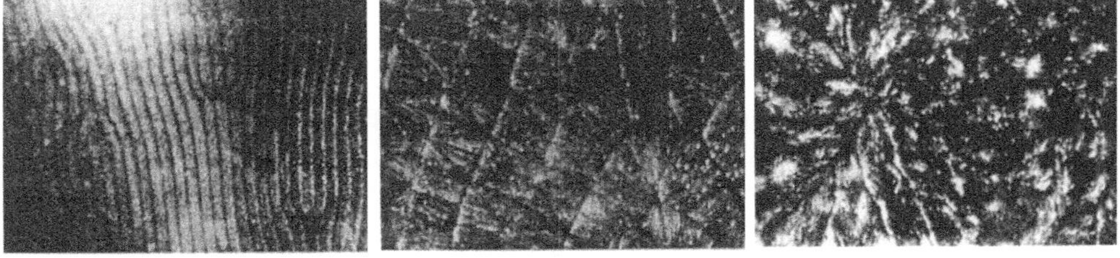

a b c

***Figure 5;** The evolution of spatial structure of Raleigh-Bernard convection with the growth of Ra number.*

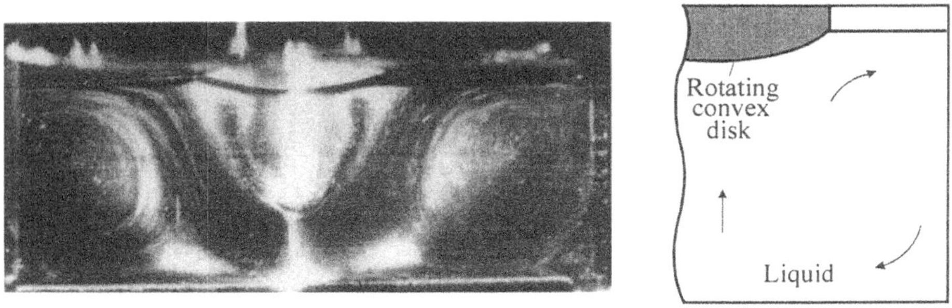

Figure 6; *Flow structure in Chokhralski method. a - visualisation, b - scheme*

4. CHOKHRALSKI'S METHOD

This method is a main way of growing the silicon crystals. A crystal quality depends largely on hydrodynamics which is characterised by a great variety of flow regimes. One of the regimes found recently is shown in Fig. 6. Here, the flow pattern is presented for isothermal conditions [Berdnikov et al., 1989]. The crystal is modelled by the rotating disk with convex surface at $\mathrm{Re} = 2570$. Near the side wall of a crucible the toroidal vortex is clearly seen. In the upward swirl flow under the disk, one can observe a vortex breakdown with a complex inner structure.

5. COANDA EFFECT

When the submerged plane jet propagates in a bounded space, for example, along an adjacent wall the rarefaction arises between the wall and jet due to different conditions of ejection. As a result, the pressure drop appears across the jet which leads to distortion and stable attachment of a jet to the adjacent wall. This phenomenon is referred to as the Coanda effect by the name of Romanian engineer who first described and applied it to some technologies. Figure 7 illustrates the Coanda effect for the case of submerged water jet issuing from the plane nozzle with width of 10 *mm* at $\mathrm{Re} = 46000$ [Alekseenko and Markovich, 1995].

6. COHERENT STRUCTURES IN SUBMERGED IMPINGING JET

The use of impinging jets is the most effective way of enhancement of heat and mass transfer between the liquid and solid wall. A remarkable peculiarity of the impinging jet is in existing large-scale coherent structures (vortices) which influence significantly on hydrodynamics and heat and mass transfer. The coherent structures arise due to instability of free shear layer in the axisymmetric submerged jet (Kelvin-Helmholtz instability), and then they penetrate into the near-wall radial jet. This process is clearly seen in

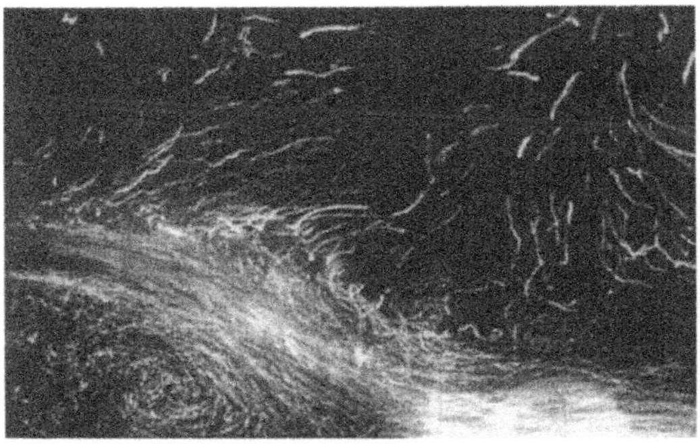

Figure 7; Coanda effect

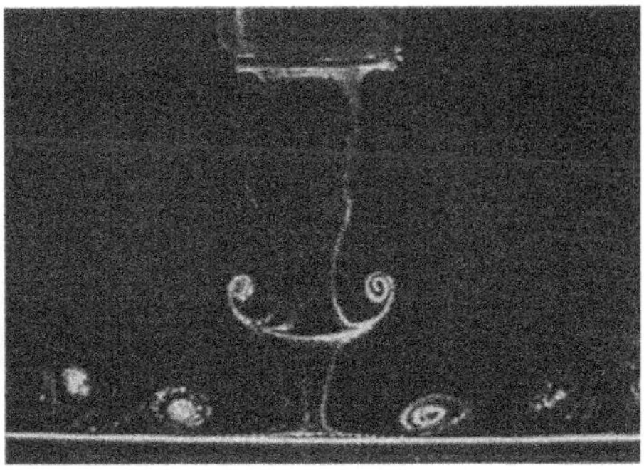

Figure 8; Coherent structures in impinging jet.

Fig. 8, where submerged jet issues from the round nozzle with diameter of 10 mm in water chamber [Alekseenko et al., 1999].Reynolds number is 1000, nozzle-to-wall distance equals 3, flow is visualised with hydrogen bubbles. It is shown that the coherent structures may be generated by means of periodic forcing. The periodic forcing as well as adding small air bubbles may be used for effective controlling turbulent flow.

7. CUMMULATIVE EFFECTS IN TWO-PHASE FLOW

In [Pribaturin and Lezhnin, 1994], the cumulative mechanism of destruction of the gas plug in a liquid from a shock wave was first found. This phenomenon is demonstrated in Fig. 9. The gas plug rises in the vertical tube of 25 mm diameter filled with water (Fig. 9, a). From the top down, a shock wave with intensity of 0.5 bar falls. Due to the convex shape of the plug front, a cumulative funnel begins to form (Fig. 9, b). Then a cumulative jet appears (Fig. 9, c) and finally destruction of a plug occurs (Fig. 9, d).

8. VORTEX FILAMENTS

A vortex filament is a clearly defined example of coherent structures. It represents a longitudinal vortex with the near-axis concentration of a vorticity. The dynamics of a vortex filament has a very complicated character. As a result, the formation of the various structures and phenomena is observed such as follows: spiral vortices, waves on vortices, vortex breakdown, vortex rings. Figure 10 shows so-called double helix - two entangled helical vortex filaments of the same sign which are generated in the vertical hydraulic vortex chamber with the cross-section area of 188×188 mm [Alekseenko and Shtork, 1992, Alekseenko et al., 1999]. The chamber has tangential inlet, narrow exit orifice, and a bottom with two inclined slopes. The flow visualization is provided with air bubbles. The light helical lines consist of air bubbles which visualize the vortex axes. As a whole, this vortex structure is immobile. The number of turns in a double helix de-

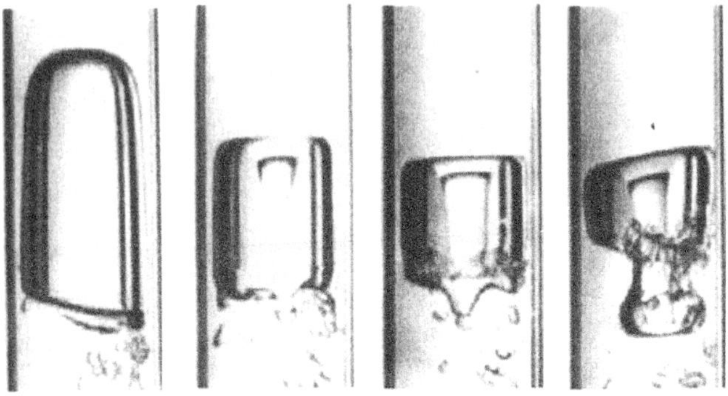

a b c d

Figure 9; *Cumulative destruction of the gas plug in a liquid from a shock wave.*

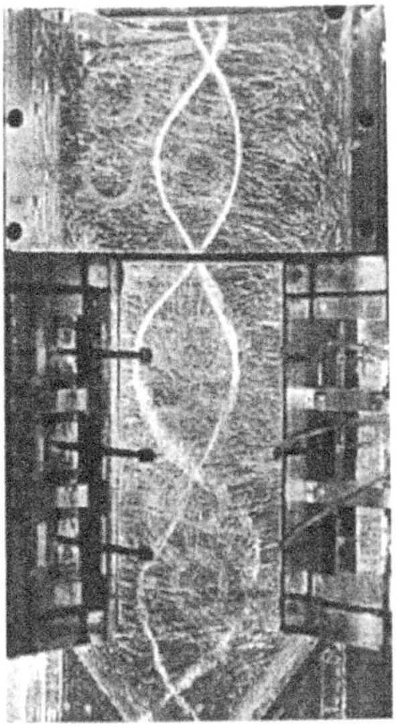

Figure 10; *Double helix - two entangled helical vortex filaments*

pends on the inclination angle of slopes at the bottom. The model [Alekseenko et al., 1999], based on the assumption of helical symmetry, confirms and explains the experimental data. To some approximation, the equations of motion of a vortex filament may be reduced to the non-linear Schrodinger equation which admits of soliton solutions. Such a soliton was first calculated by H. Hasimoto in 1972. Hasimoto soliton (or spiral soliton) represents a solitary spiral wave which rotates and propagates along the vortex filament. This phenomenon is shown in Fig. 11 where the light line is a vortex axis (but not vortex core which is thicker) [Alekseenko and Shtork, 1992].

A vortex breakdown is one of the most important phenomena in vortex dynamics that has yet to be described. It manifests itself in a sudden deviation of the vortex filament from the flow axis and formation of a zone of reverse flow. Figure 12 demonstrates the spiral type vortex breakdown in the case when a vortex filament penetrates through the orifice of 110 mm diameter into the upper unbounded space [Alekseenko and Shtork, 1992]. The spiral structure rotates as a whole together with a swirl flow.

9. CELLUAL STRUCTURE OF DETONATION FRONT

During detonation of gaseous mixtures the front of detonation wave is known to be unstable that leads to formation of characteristic cellular structure. Such a structure was first described in experiment [Voitsekhovskii et al, 1963] on detonation combustion of a mixture: $2CO + O_2 + 5\%H_2$. The trace on smoky wall of a plane channel reflects the existence of a cellular structure of detonation wave which has propagated through the channel (Fig. 13, a). The similar pattern was computed in [Borissov and Sharypov, 1989] on the basis of nonlinear model of evolution of detonation front (Fig. 13, b).

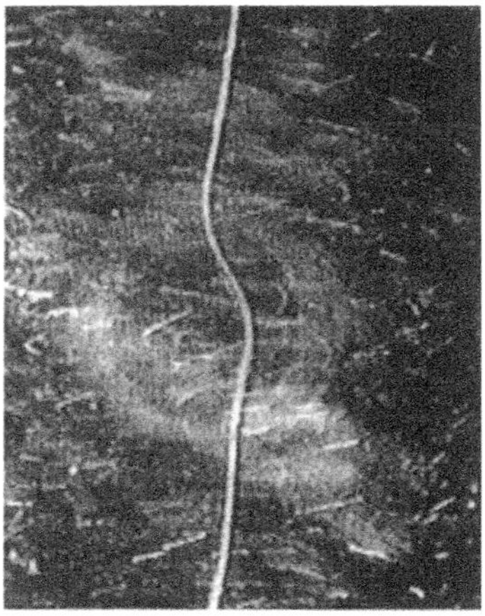

Figure 11; Hasimoto's soliton (spiral solitary wave on vortex filament).

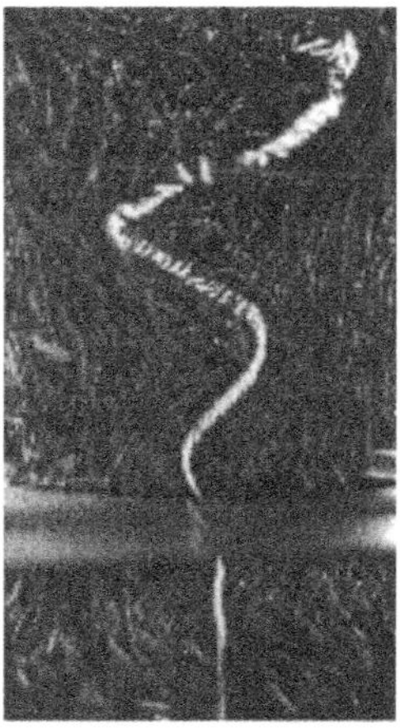

Figure 12; Vortex breakdown of spiral type.

10. SOLIDIFICATION OF LIQUID METAL DROPLET ON A SUBSTRATE

Spreading microdroplets of a liquid metal over the cold solid surface is accompanied by the processes of heat and mass transfer and phase transition [Predtechensky et al., 1999]. It is shown that the spreading of droplets has a oscillating character. Simultaneous propagation of crystallization front into the liquid bulk of a droplet leads to the formation of peculiar surface perturbations in solidified droplet as it is demonstrated in Fig. 14. The experimental conditions are described in the table:

Position	Droplet diameter, mkm	Velocity, m/s	Droplet temperature, °C	Substrate temperature, °C	Substrate substance
a	105	5.4	236	50	Gold
b	45	4.5	220	40	Gold
c	40	2.6	190	50	Gold
d	90	1.9	250	50	Poliamide, covered with gold

Droplet substance is tin-lead solder (63% Sn - 37% Pb).

11. TURBULENT JET

Figure 15 shows the shadowgrams of the same supersonic turbulent jet at different exposition times [Novopashin, 1985]: (a) - 10^{-2} s, (b) - 10^{-8} s. Here, a jet of the nitrogen is-

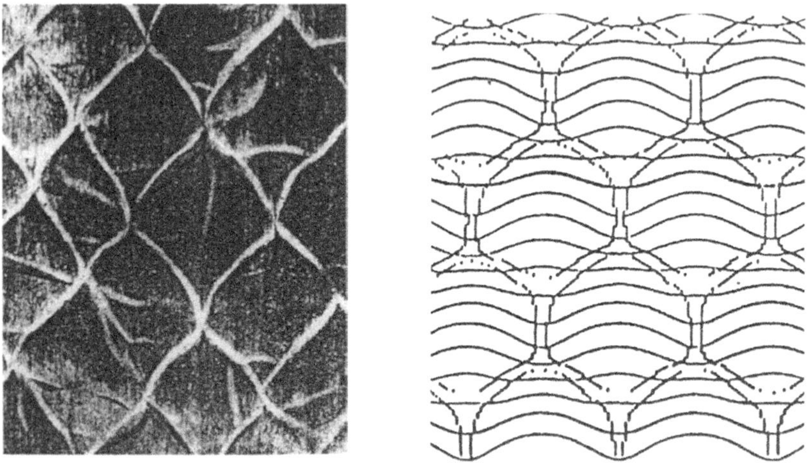

a b

Figure 13; *Cellular structure of detonation front.*

sues from the nozzle of 3 mm diameter into the air at pressure of 10^5 Pa. Pressure before the nozzle is 10^6 Pa. In Fig. 15,a, we can see a usual pattern of supersonic jet with density jumps. However, only short exposition time allows us to see a turbulent structure through the density pulsations.

12. INTERNAL WAVES IN STRATIFIED FLUID

The translatory motion or oscillations of a solid body in stratified medium result in a generation of so-called internal waves. The visualisation of the internal waves is considered to be a nontrivial task. Recently the method of color Hilbert visualization (new version of the shadow method) was developed. Its principle is in bichromatic Hilbert filtering, for example, in red and green intervals [Dubnistchev, 1988]. By this means, the spatial gradients of a phase field of optic density are effectively separated. Figure 16, a presents the coloured Hilbert shadowgram of internal waves arising due to translatory motion of a cylinder from left to right in stratified fluid. The diameter of viewing field is 60 mm. In Fig. 16, b a shadowgram is presented for cross-like internal waves which are generated by horizontal low-frequency oscillations of a cylinder in the medium near the thermodynamical critical point. This phenomenon was first described in [Leonenko, 1995].

13. RANQUE EFFECT

The Ranque effect is that the strong swirling of gas flow leads to the temperature separation of a gas. A cold gas issues from the central orifice of the vortex tube (Ranque-

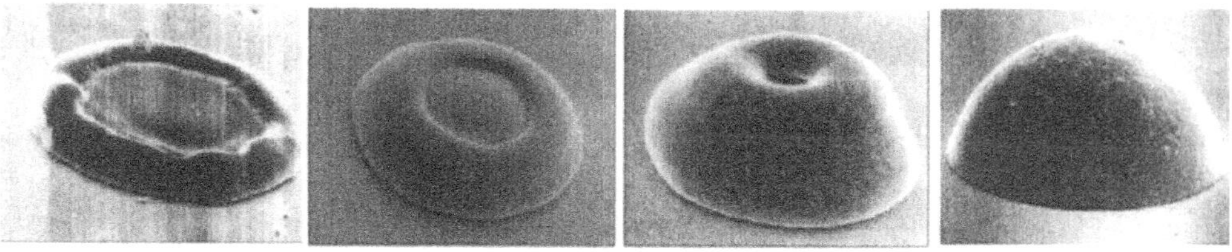

a b c d

Figure 14; *Solidification of liquid metal droplet on a substrate*

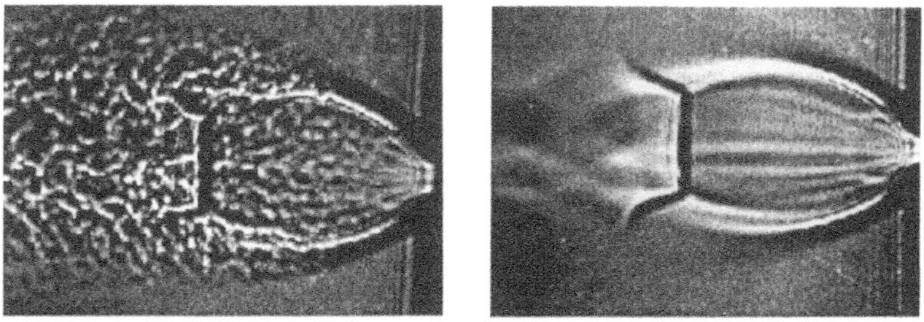

a b

Figure 15; *Turbulent supersonic jet.*

Hilsch tube) near the tangential inlet, but a hot gas leaves through the periphery ring slot at another end of a tube. The temperature difference attains 100°C. Up to this point an adequate description of the Ranque effect is absent. Figure 17 shows the two-vortex flow structure in the air vortex tube which was first observed in [Arbuzov et al., 1997]. The method of Hilbert visualization was applied. The vortex tube has square cross-section of 34×34 mm. Here, on the contrary to Fig. 11, the vortex structure rotates. Besides, at some distance from the "cold" exit the reconnection of vortices and subsequent their breakdown occur.

14. MARANGONI EFFECT

The method of infrared thermography is used to study the temperature distribution on a free surface of the liquid film falling down a vertical wall with local heat source of 6.5×13 mm (shown with black frame in Fig. 18) [Kabov et al., 1996]. As is seen, surface temperature has nonuniform distribution that leads to an appearance of the surface shear stress (so-called Marangoni effect or thermocapillary instability). As a result, a standing wave arises which has a complex triple cellular structure caused by the secondary instabilities.

15. SUPERSONIC RING JET

Recently a novel way of deposition of SiH films for solar cells was proposed on the basis of a flow plasma-chemical method [Sharafutdinov et al., 1994]. This method provides for the formation of a supersonic ring jet of argon and monosilan. The photograph of a jet is shown in Fig. 19 [Sharafutdinov et al., 1996]. A flow visualisation occurs due to own irradiation of the plasma mixture of gases under their activation with an electronic beam. The diameter of a ring nozzle is 20 mm, a width of a slot is 2 mm, and Mach number is 5.

16. FOUR-NOZZLE ARRANGEMENT OF A SPACECRAFT

Laboratory modelling jets of engines of orientation and control is a necessary stage of creating spacecrafts. An effective approach to modelling such jets is related to use of visualisation with an electronic beam. Figure 20 shows a flow structure in cross-sections behind the four-nozzle arrangement of a spacecraft model [Gerasimov et al., 1980]. The supersonic jets of nitrogen issue from the nozzles of 1 mm diameter into a vacuum. The diameter of the inner body is 60 mm. Figure 20, a represents a flow pattern in the plane near the nozzles. Other pictures are obtained at some distances from the nozzles plane. An observed black area in the patterns is the result of interaction of supersonic jets with the central body, and coloured cross arises due to mutual interaction of neighbouring jets.

17. WAVELET ANALYSIS OF A VORTEX FLOW

A container with rotating wall is a convenient subject for studying the vortex breakdown and unstable phenomena. Correspondingly, a wavelet analysis is one of the most acceptable methods for investigating complicated and transition processes. Figure 21 presents the results of wavelet analysis of pulsations of vertical velocity component in rectangular chamber with rotating upper cover (disk) [Meledin and Naumov, 1999]. Here, the cross-section of a chamber is 100×100 mm; Reynolds number equals 9000; the chamber is filled with a water. A coloured area presents wavelet transform of velocity pulsations. A projection of this field on the horizontal axis represents a basic signal (velocity pulsations), and a projection of the field on the vertical axis gives the energy spectrum. In particular, the existence of a low-frequency precession of the vortex follows from these data.

18. VORTEX COMBUSTOR

The visualisation methods play an important role when it needs to present the data on numerical simulation of the complex subjects such as combustors. Figure 22 shows the results of numerical simulation of three-dimensional, two-phase, turbulent flow in the pulverised-coal combustor of a vortex type [Krasinsky, 1999]. The fields of velocity vector and concentration of disperse phase are shown. Such a representation alleviates significantly the analysis of combustion process and allows optimising the furnace geometry and regime parameters.

19. SPHERICAL STRATIFICATION OF A GLOW DISCHARGE

Figure 23 presents a phenomenon of spherical stratification of a glow discharge in a high-molecular gas [Nerushev et al., 1997]. Previously, only plane stratas in tubes with glow discharge were observed. In this experiment, a vacuum chamber is used with typical size of 0.5 m and pressure range from 10^0 to 10^4 Pa. In the centre of a chamber, a small positive electrode is placed, and the steel chamber walls are the negative electrode. The voltage drop on the discharge was about 400 V. The colour and number of stratas depend on the kind of a gas and value of a pressure. The distortion of a spherical shape, shown in Fig. 23, b, occurs in the case of a strong anisotropy of external boundary conditions. The observed phenomena are explained by the ionization instability.

REFERENCES

[Alekseenko and Shtork, 1992] Alekseenko, S.V.; Shtork, S.I.; "Swirling flow large-scale structures in a combustor model"; *Russian J. Eng. Thermophysics;* V. 2; pp. 231-266.

[Alekseenko and Cherep, 1994] Alekseenko, S.V.; Cherep, A.A.; "Spiral waves in dispersive media"; *Acta Mechanica;* V. 105; pp. 143-160.

[Alekseenko and Markovich, 1995] Alekseenko, S.V.; Markovich, D.M.; "Interaction of plane turbulent jets"; *Russian J. Eng. Thermophysics;* V. 5; N 2; pp. 125-144.

[Alekseenko et al., 1994] Alekseenko, S.V.; Nakoryakov, V.E.; Pokusaev, B.G.; "Wave Flow of Liquid Films"; Begell House; NY; 313 p.

[Alekseenko et al., 1999a] Alekseenko, S.V.; Bilsky, A.V.; Markovich, D.M.; Semenov, V.I.; "Evolution of instabilities in an axisymmetric impinging jet"; *Proc. of 4th Int. Symp. on Engineering Turbulence Modelling and Measurements*; Corsica; France; pp. 637-645.

[Alekseenko et al., 1999b] Alekseenko, S.V.; Kuibin, P.A.; Okulov, V.L., Shtork S.I. "Helical vortices in swirl flow"; *J. Fluid Mech.;* V. 382; pp. 195-243.

[Arbuzov et al., 1997] Arbuzov, V.A.; Dubnistchev, Yu.N.; et al. "Observation of large-scale structures in vortex tube and Ranque effect"; *Pis'ma Zh. Tekhn. Fiz.*; V. 23; N 23; pp. 84-90. (in Russian).

[Berdnikov et al., 1989] Berdnikov, V.S.; Borisov, V.L.; Markov, V.A.; Panchenko, V.I.; "Simulation of the hydrodynamical front and annular cross section"; *Heat transfer. Soviet Research;* V. 21; N 6; pp. 828-845.

[Berdnikov et al., 1990] Berdnikov, V.S.; Getling, A.V.; Markov, V.A.; "Wave number selection in Rayleigh-Benard convection: Experimental evidence for the existence of an inherent optimal seale"; *Exp. Heat Transfer;* V. 3; N 3; pp. 269-288.
[Borissov and Sharypov, 1989] Borissov, A.A.; Sharypov, O.V.; "Simulation of structure of unstable detonation front"; *Izv. SO RAN SSSR, Ser. tekhn. Nauk*; V. 2; pp. 50-55. (in Russian).
[Dubnistchev, 1988] Dubnischev, Yu.N.; "Optical methods for diagnosis of the flows"; *Optoelectronics, Instrumentation and Data Processing;* N 6, pp. 74-83.
[Gerasimov et al., 1980] Gerasimov, Yu.I.; Polopezhentsev, S.A.; Yarygin, V.N.; "Studying gasodynamical structure of combined jets"; *Proc. VI USSR Conf. Dynamics of Rarefied Gases;* Novosibirsk, Part. 2; pp. 132-137 (in Russian).
[Kabov et al., 1996] Kabov, O.A.; Marchuk, I.V.; Chupin, V.M.; "Thermal imaging study of the liquid film flowing on vertical surface with local heat source"; *Russian J. Eng. Thermophysics;* V. 6; N 2, pp.105-138.
[Krasinsky, 1999] Krasinsky, D.V.; "Numerical simulation of two-phase turbulent reacting flows in pulverized-coal combustor of vortex type"; PhD thesis; Novosibirsk; Inst. of Thermophysics.
[Leonenko, 1995] Leonenko; Yu.G.; "Effect of gravity on processes in medium at near-critical parameters of a state"; *J. Exper. Theor. Physics;* N 3, pp. 765-773 (in Russian).
[Meledin and Naumov, 1999] Meledin, V.G.; Naumov, I.V.; Private communication; Novosibirsk; Inst. of Thermophysics.
[Nerushev et al., 1997] Nerushev, O.A.; Novopashin, S.A.; Radchenko, V.V.; Sukhinin, G.I.; "Spherical striation in a glow discharge"; *JETP Lett.;* V. 66; pp. 711-714.
[Novopashin, 1985] Novopashin, S.A.; "Density pulsation and laminar-turbulent transition in the mixing layer of supersonic jets"; In book: *Laminar-turbulent transition,* IUTAM Symp., Novosibirsk; Springer-Verlag; Berlin; Heidenlberg; pp. 529-533.
[Predtechensky et al., 1999] Predtechensky, M.R.; Varlamov, Yu.D.; Vl'yankin, S.N.; "Solidification of liquid metal droplets on a substrate - I. Experiment"; *Int. J. Heat Mass Transfer* (to be published).
[Pribaturin and Lezhnin, 1994] Pribaturin, N.A.; Lezhnin, S.I.; "Nonstationary wave processes in pipes with two-phase media"; *Pressure Vessels and Piping;* V. 240; pp. 45-50.
[Sharafutdinov et al., 1994] Sharafutdinov, R.G.; Skrynnikov, A.V.; Polisan, A.A.; "Method of deposition of hydrogenized silicon films"; *Russian Patent* N 2100477, 18.10.1994; (in Russian).
[Sharafutdinov et al., 1996] Sharafutdinov, R.G.; Skrynnikov, A.V. et al.; "High-rate deposition of a SiH films using a flow plasma-chemical method with electron beam"; *J. Appl. Phys.*; V. 79; N 9; pp. 7274-7277.
[Van Dyke, 1982] Van Dyke, M.; "An Album of Fluid Motion"; The Parabolic Press; CA; USA.
[Voitsekhovskii et al., 1963] Voitsekhovskii, B.V.; Mitrophanov, V.V.; Topchiyan, M.E.; "Structure of Detonation in Gases"; SO AN SSSR; Novosibirsk; (in Russian).

The role of color in the optical visualization of flows with variable fluid density

W. Merzkirch, N. Ramesh
Lehrstuhl für Strömungslehre, Universität Essen
D-45117 Essen, Germany
e-mail: wolfgang.merzkirch@uni-essen.de

Abstract: The generation of colors in optical flow visualization records supports the comprehensibility of the technical and scientific information, but it does not add to improving the situation for quantitative measurements. Examples of color-schlieren records and „white-light" interferograms are shown in the presentation.

1. INTRODUCTION

Flow visualization is an established tool for analyzing fluid flows. Since both gases and liquids are normally transparent, it requires special techniques for making their motion visible. In a rough approximation one can distinguish between two fundamental principles: flow visualization by means of light scattering from tracer particles with which the fluid is seeded, and visualization by means of light transmittance through the flow that must exhibit changes of the fluid density or the refractive index. Methods relying on the latter principle are often classified as optical visualization (Merzkirch 1981, 1987). The information provided by these visualization methods is either qualitative, thus allowing for interpreting the mechanical and physical processes involved in the development of a flow, or quantitative, so that measurements of certain properties of the flow field (velocity, density, etc.) can be performed.

We focus here on the so-called optical visualization methods with which the properties of the transmitted light wave, altered by the refractive index variation, are converted into a signal perceptible by the eye. Variations of light intensity are such a visible signal and, therefore, visualized flow patterns are often displayed in form of a black-and-white photograph. This applies particularly to quantitative recordings, e.g. interferograms, which are generated with monochromatic light so that the information after possible binarization of the recorded data is condensed to either bright or dark. In addition of the differentiation between bright and dark, the eye can also distinguish between colors. Use has been made of the component color in a number of optical visualization methods, but rarely for quantitative purposes, and most often for supporting the qualitative direct visual perception and the resulting interpretation of the flow pattern, thereby also providing an aesthetic aspect of the visualization process. The component

A. Gyr et al. (eds.), Science and Art Symposium 2000, 139-142.

color is either implemented into the optical system by a color-coding element (e.g. color schlieren system), or it is inherently included in the optical principle due to its dependence on the light wave length (e.g. interferometry).

2. COLOR SCHLIEREN METHOD

In a Töpler schlieren system a light beam is transmitted through the transparent object field (fluid flow) and then focused into a focal point. Here, a knife edge reduces the overall light intensity and, acting as a spatial filter, it increases or decreases the local light intensity in a recording plane depending on the local refractive light deflection in the flow under investigation. Numerous modifications of the schlieren method have been described in the literature, and there are not many which were reported during the past decades that were not yet mentioned in the review of Schardin (1942). An important modification are the color schlieren techniques. In their simplest form they use a three-color-filter in place of Töpler's knife edge such that the undisturbed light passes through a specific color filter, e.g. blue, and the positively or negatively deflected light appears, e.g., red or yellow, respectively (Fig. 1).

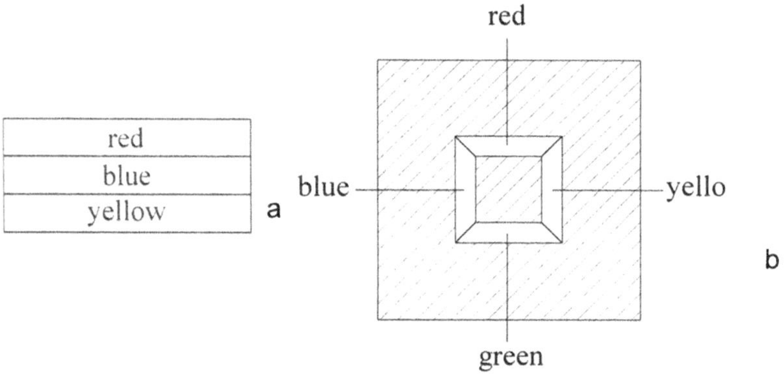

Fig. 1; *Examples of filters for producing a color schlieren record. Filter (a) allows to identify light deflection in positive and negative vertical direction, filter (b) in the vertical and horizontal directions, simultaneously (see Settles 1985).*

Color schlieren systems have been reviewed by Settles (1985). When using color filters like those sketched in Fig. 1 the choice of colors is arbitrary and a criterion for their selection is a good color contrast. If one uses a color-coding element based on light dispersion, e.g. made of certain crystals, one is bound to the natural sequence of colors („rainbow schlieren", Howes 1984). Color schlieren systems are very sensitive, inexpensive, informative, and rarely used for quantitative purposes, though it is not too complicated to convert colors into digital data for the further processing of the experimental results.

3. INTERFEROMETRY

The majority of interferometers used for fluid mechanical measurements are based on the dual beam principle. The visible signal in the form of an interferometric fringe system is a measure of the difference in optical path length, $\Delta\ell(x,y)$, between the two interfering light waves. Here, x and y are the coordinates of the recording plane where the fringe system appears, while the light is transmitted through the flow in z-direction. The signal carries information on the density distribution in the flow, and it is either dependent on the absolute value of the fluid density („reference beam interferometer“) or on the first derivative of the density („shearing or schlieren interferometer“). Bright interference fringes appear where $\Delta\ell(x,y)$ is equal to an integer number (including zero) times the light wave length λ. These numbers designate the „fringe order“.

Interferometry requires the interfering light rays to be optically coherent. This requirement is met best with the use of monochromatic, coherent light from a laser. Then, a great number of undistinguishable fringes can be produced in the field of view, and their order is not evident in the fringe pattern (Fig. 2). Interferometry is also possible with a light source that emits within a limited range of wave lengths, but at the expense of reducing the total number of fringes available. Only the fringes of low order have a hight contrast, while the fringes of higher order (on both sides of the zeroth order) become more and more diffuse. In addition, the fringe pattern becomes increasingly colored with increasing order due to the dependence of locations of the bright and dark fringes on the wave length λ. Thus, the loss of optical coherence, associated with the appearance of colors, reduces the potential for quantitative measurements.

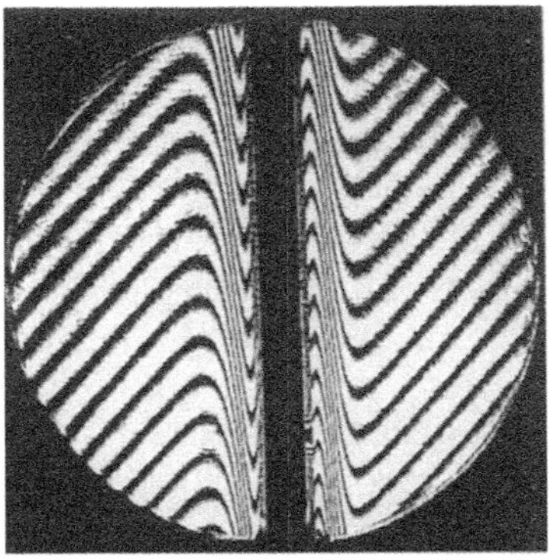

Fig. 2; *Shearing interferogram of the natural convective flow along a vertical heated flat plate, taken in monochromatic (laser) light.*

If the spectral composition of the light source is such that two wave lengths are dominating, then a spatial beat frequency can be superimposed to the fringe pattern; i.e., starting from the zeroth order, the fringe clarity decreases, then increases again and reaches a second maximum of contrast, and this pattern is repeated several times and at regular intervals (beat frequency) until the overall contrast becomes so low that no fringes are distinguishable at the higher orders. Only the zeroth order is a white fringe ($\Delta\ell$ being zero for all wave lengths). This allows to identify the fringe orders, in contrast to using monochromatic light, and it can be helpful for one particular flow situation, a shock wave where the fluid experiences a step-like change, thus causing a jump of the interference fringes across the shock front. Only the „white-light" illumination with colored fringes allows to identify the fringes of the same order on both sides of the shock and to determine the discontinuous density increase.

„White-light" interferograms appear most impressive when the interferometer is adjusted in the „infinite fringe width" mode. It can be shown that, with this adjustment, a shearing interferogram is equivalent, both in pattern and sensitivity, to a schlieren record.

4. CONCLUSIONS

The appearance of colors in optical flow visualization supports the comprehensibility of the technical or scientific information provided by the flow patterns, and it may add an aesthetic component. A great number of examples are shown in the oral presentation, including visualization of compressible flows, convective heat transfer problems, and flows in a density-stratified fluid. In these examples, colored patterns are generated in the physical experiment. As an alternative way, color can also be added to a black-and white pattern by digital image processing.

REFERENCES

[Howes, W.L. 1984]; „Rainbow schlieren and its applications"; 23, 2449-2460.

[Merzkirch, W. 1981]; „Density sensitive flow visualization"; In: *Methods of Experimental Physics,* Vol. 18: Fluid Dynamics, ed. R.J. Emrich, pp. 345-403, Academic Press, New York.

[Merzkirch, W. 1987]; „Flow Visualization, 2nd edition"; *Academic Press*, New York.

[Schardin, H. 1942]; „Die Schlierenverfahren und ihre Anwendungen"; *Ergebnisse der Exakten Naturwissenschaften 20*, 303-439.

[Settles, G.S. 1985]; „Color-coding schlieren techniques for the optical study of heat and fluid flow"; *Int. J. Heat Fluid Flow 6*, 3-15.

Direct Navier-Stokes simulations of unsteady vortex breakdown in a swirling jet

Peiji Chen and Eckart Meiburg
Department of Aerospace and Mechanical Engineering
University of Southern California
Los Angeles, CA 90089-1191, USA
eckart@spock.usc.edu

Abstract: Three-dimensional, fully unsteady Navier-Stokes simulations are carried out in order to study the mechanisms governing vortex breakdown in swirling jets. For sufficiently large swirl numbers, the initial breakdown is followed by the formation of an axisymmetric, quasisteady velocity field. In the wake region of the primary breakdown bubble, a finite region of absolute instability is observed to exist that gives rise to the formation of a temporally periodic, helical instability, thereby resulting in the familiar 'spiral vortex breakdown structure.' This finding hence suggests that the transition from axisymmetric to helical breakdown is a manifestation of a global mode instability.

Keywords: Swirling jets, vortex breakdown, three-dimensional Navier-Stokes simulations, absolute instability, global mode

1. INTRODUCTION

Vortex breakdown represents one of the few classical phenomenona in the field of fluid dynamics whose mysteries are still only partially resolved, inspite of the continuous and detailed attention it has received by experimental, theoretical, and computational researchers over many years. Our enduring fascination with vortex breakdown can certainly be traced to its fundamental nature, as well as its central importance in a number of technical applications, ranging from issues in aircraft control to combustion stabilization. At the same time, however, the flow patterns observed when vortex breakdown occurs often have a strong aesthetic appeal, which has added to the attraction of trying to understand their governing physical principles. As can be recognized in the many beautiful flow visualization pictures produced over the years (e.g., Werlé [1960], Sarpkaya [1971], Maxworthy [1972], Billant, Chomaz and Huerre [1998], Petitjeans and Wesfreid [1997], Althaus, Brücker and Weimer [1995], and many more), vortex breakdown usually represents the sudden appearance of a free stagnation point within the flow, away from solid walls and accompanied by the formation of a downstream recirculation region. To the untrained observer, it hence may appear as if the flow senses an obstacle where there is none. As a consequence, the usually slender structures observed in flow visualizations up-

A. Gyr et al. (eds.), Science and Art Symposium 2000, 143-147.

stream of the vortex breakdown point abruptly transition into bubble-like or helical structures, so that the overall appearance can at times resemble that of a flower at the tip of a long stem. At fairly low flow rates, the flow can maintain helical or axisymmetry to a high degree of fidelity even downstream of the initial breakdown, which can add significantly to the beauty of flow visualizations. At larger Reynolds numbers, vortex breakdown often signals the transition from laminar, orderly flow to a turbulent state characterized by strong fluctuations and disorder. Topical reviews are provided by Hall [1972], Leibovich [1978], and Escudier [1988].

Over the years, various fundamental concepts have provided some partial insight into the physical mechanisms underlying the vortex breakdown phenomenon. Hall [1972] stresses the similarity to a separating boundary layer in order to analyze the flow upstream of the vortex breakdown event. Squire [1960] investigates the role of waves propagating upstream within the vortex core. Benjamin [1962] proposes that vortex breakdown represents the transition from a supercritical to a subcritical flow, featuring stationary finite amplitude waves. Maxworthy, Hopfinger and Redekopp [1985] investigate the role of traveling, kinked solitary waves in triggering vortex breakdown. Leibovich [1978] and coauthors address vortex breakdown from the perspective of a stability problem. In recent years, this point of view has gained renewed relevance through the works of Delbende, Chomaz and Huerre [1996], as well as Lim and Redekopp [1998]. Their findings that convectively unstable jet- or wake-like velocity profiles can be made absolutely unstable by the addition of moderate amounts of swirl suggests that vortex breakdown may be related to the appearance of an absolute instability. The numerical simulations to be reported in the present paper partially aim at evaluating this relationship.

2. GOVERNING EQUATIONS AND NUMERICAL TECHNIQUE

The simulations to be reported below are based on the three-dimensional Navier-Stokes equations in cylindrical coordinates (r, θ, z). Rewriting these equations in terms of the quantities $q_\theta = v_\theta, q_r = rv_r$, and $q_z = v_z$, where v_θ, v_r, and v_z are the velocities in the azimuthal, radial, and axial directions, respectively, the accurate numerical treatment of the centerline is greatly facilitated, as described by Verzicco and Orlandi [1996]. At the upstream boundary, axisymmetric profiles for all three velocity components are prescribed in dimensionless form as

$$\begin{aligned}
0 \leq r \leq \infty &: \quad q_r(r) = 0 \ , \\
0 \leq r \leq 1 &: \quad q_\theta(r) = Sr(2 - r^2) \ , \quad q_z(r) = \alpha + (1 - \alpha)r^2(6 - 8r + 3r^2) \ , \\
1 \leq r \leq \infty &: \quad q_\theta(r) = S/r \ , \quad q_z(r) = 1 \ .
\end{aligned}$$

At the radial boundary, typically ten jet radii away from the centerline, slip walls are assumed, and at the downstream boundary, usually twenty jet radii from the inflow, radiation boundary conditions are applied to each of the variables [Orlanski

1976, Verzicco and Orlandi 1996]. As dimensionless parameters we have a Reynolds number Re, a swirl number S, and a coflow ratio α

$$Re = \frac{q_{z,cl}\delta}{\nu}, \quad S = \frac{\delta \frac{1}{2}\frac{\partial v_\theta}{\partial r}(r=0)}{q_{z,cl}}, \quad \alpha = \frac{q_{z,cl}}{q_{z,\infty}}.$$

where $q_{z,cl}$ is the axial centerline velocity at the inflow, $q_{z,\infty}$ indicates the axial coflow velocity, δ denotes the vortex core radius, and ν represents the kinematic viscosity.

The governing equations are solved on a staggered grid by means of second order accurate central finite differences in space. The temporal discretization employs a fractional step method [Kim and Moin 1985, Verzicco and Orlandi 1996] in conjunction with a third order Runge-Kutta scheme. A coordinate transformation in the radial direction allows us to concentrate the grid points near the centerline, where most of the interesting dynamical interactions occur. The numerical code was validated by comparison with the earlier axisymmetric results of Grabowski and Berger [1976].

3. RESULTS

Both axisymmetric as well as fully three-dimensional, unsteady simulations were carried out for a number of different parameter combinations. Figure 1 displays the streakline patterns for $Re = 200$, $\alpha = 1$, and $S = 1.095$ for times t=50, 500, 650, and 750. A detailed analysis of the numerical results shows that up until $t \approx 600$ the flow is nearly identical to the corresponding axisymmetric case. In particular, it attains a quasisteady state that lasts for approximately 400 dimensionless time units. Beyond that time, however, an instability is seen to grow that leads to the formation of a pronounced helical structure in the wake of the initial vortex breakdown bubble. Once the growth of this instability has saturated, the entire flow field behaves strongly periodic in time. By tracking the evolution of temporal changes in the axial velocity component near the centerline, it can be shown that the evolution of this instability is accompanied by waves propagating in the upstream direction. Since such waves are not observed for swirl numbers too small to result in breakdown, this observation lends credence to the applicability of the absolute/convective instability framework. In order to further explore this issue, we analyzed in more detail the quasisteady, axisymmetric velocity field that gave rise to the helical instability. By matching the profiles at closely spaced intervals in the streamwise direction with the Batchelor vortex profile investigated by Delbende et al. [1996], we were able to characterize each streamwise location as locally stable, convectively unstable, or absolutely unstable. The temporally periodic, helical instability was seen to evolve when a finite region of absolute instability existed in the wake of the primary breakdown bubble. This observation strongly suggests that the growth of this instability can be characterized as a global mode (Huerre and Monkewitz [1990]).

Since the experimental literature knows several examples of streaklines providing somewhat misleading information, we reexamined the above results on the basis of data for vortex lines and contours of both vorticity magnitude as well as passive scalar concentration. These showed the same qualitative structure and thereby confirmed the existence of the helical instability in the wake of the breakdown. In particular, the relevant plots indicated that the streaklines qualitatively indicate the shapes of the vortex lines, and that they mark regions of high values of vorticity magnitude as well as high passive scalar concentration.

Further interesting observations were made for the case $Re = 200$, $\alpha = 1$, and $S = 1.2$. The initial stages are very similar to the above case, i.e., axisymmetric breakdown followed by the evolution of a quasisteady state and the subsequent growth of a helical instability. However, several hundred dimensionless time units after the formation of the primary vortex helix, a second helical structure of the same axial wavelength emerges and grows to similar strength. Subsequently, the two helical structures peacefully coexist with each other, while continuously corotating around the centerline. This finding resembles corresponding experimental observations of multiple helical structures by Billant et al. [1998].

The above analysis is currently being extended to larger families of base flows, ranging from jets with and without coflow to wakes, in order to evaluate the general applicability of the above findings. Detailed results will be reported elsewhere.

ACKNOWLEDGMENTS

The authors gratefully acknowledge helpful discussions with Professors Maxworthy, Redekopp and Huerre. This work is funded by the National Science Foundation. Supercomputing time is provided by the San Diego Supercomputer Center, which is funded by NSF.

REFERENCES

Althaus, W., Brücker, C., and Weimer, M. 1995 "Breakdown of slender vortices" In: Fluid vortices, ed. S.I. Green, Kluwer Acad. Pub.

Benjamin, T.B. 1962 "Theory of the vortex breakdown phenomenon" J. Fluid Mech. 14, 593.

Billant, P., Chomaz, J.-M., and Huerre, P. 1998 "Experimental study of vortex breakdown in swirling jets" J. Fluid Mech. 376, 183.

Delbende, I., Chomaz, J.-M., and Huerre, P. 1996 "Absolute/convective instabilities in the Batchelor vortex: a numerical study of the linear impulse response" J. Fluid Mech. 355, 229.

Escudier, M.P. 1988 "Vortex breakdown: observations and explanations" Prog. Aerosp. Sci. 25, 189.

Grabowski, W.J. and Berger, S.A. 1976 "Solutions of the Navier-Stokes equations for vortex breakdown" J. Fluid Mech. 75, 525.

Hall, M.G. 1972 "Vortex breakdown" Ann. Rev. Fluid Mech. 4, 195.
Huerre, P. and Monkewitz, P.A. 1990 "Local and global instabilities in spatially developing flows" Ann. Rev. Fluid Mech. 22, 473.
Kim, J. and Moin, P. 1985 "Application of a fractional-step method to incompressible Navier-Stokes equations" J. Comp. Phys. 59, 308.
Leibovich, S. 1978 "The structure of vortex breakdown" Ann. Rev. Fluid Mech. 10, 221.
Lim, D.W. and Redekopp, L.G. 1998 "Absolute instability conditions for variable density, swirling jet flows" Eur. J. Mech., B/Fluids 17, 165.
Maxworthy, T. 1972 "On the structure of concentrated, columnar vortices" Astr. Acta 17, 363.
Maxworthy, T., Hopfinger, E.J., and Redekopp, L.G. 1985 "Wave motions on vortex cores" J. Fluid Mech. 151, 141.
Orlanski, I. 1976 "A simple boundary condition for unbounded hyperbolic flows" J. Comp. Phys. 21, 251.
Petitjeans, P. and Wesfreid, J.E. 1997 "Vortex stretching and filaments" Appl. Sci. Res. 55, 279.
Sarpkaya, T. 1971 "On stationary and traveling vortex breakdowns" J. Fluid Mech. 45, 545.
Squire, H.B. 1960 Dep. of Aero. Rep. 102, Imperial College, London.
Verzicco, R. and Orlandi, P. 1996 "A finite difference scheme for three-dimensional incompressible flows in cylindrical coordinates" J. Comp. Phys. 123, 402.
Werlé, H. 1960 Rech. Aeronaut. 74, 23.

The transition within the subcritical regime of flow over a circular cylinder

K.W. Lo[+] *and N.W.M. Ko*[++]
[+] *Maunsell Environmental Management Consultants Ltd., Grand Central Plaza Tower 2, Shatin, Hong Kong, China*
kwlo@maunsell.com.hk
[++] *Department of Mechanical Engineering, The University of Hong Kong, Hong Kong, China*
nwmko@hkucc.hku.hk

Abstract: Within the subcritical regime of a circular cylinder, different groups of secondary vortices are found. The one-group secondary vortices are related to the early stage of transition. With a splitter plate, the transition in the subcritical regime is suppressed. The two-group secondary vortices are dominant in the later stage of transition.
Keywords: Transition, subcritical regime, splitter plate

1. INTRODUCTION

Within the Reynolds number range $7 \times 10^3 < Re < 2 \times 10^4$, the flow over a smooth circular cylinder can be classified into two sub-regimes, the lower and upper subcritical regimes. Kourta et al. [Kourta, Boisson, Chassaing and Ha Minh, 1987] observed that the division is at $Re \approx 1.6 \times 10^4$, while Norberg [Norberg, 1987] suggested the division at 7.3×10^3. These results imply a transition between these two sub-regimes. With the introduction of splitter plate at the back of the cylinder, the large scale primary vortices are suppressed or inhibited [Apelt and West, 1975; Anderson and Szewczyk, 1997]. Thus, this paper attempts to explore this transition under the effect of suppressed primary vortices and the interaction between the small scale secondary vortices and the large scale primary vortex sheet.

2. APPARATUS

The experiments were carried out in a low speed wind tunnel of test section 0.4m x 0.4m. The freestream turbulence intensity was 0.2% at the freestream velocity of 20m/s. The diameter d of the cylinder was 50mm and the aspect ratio was 4. The blockage ratio of the cylinder was 12.5%. Splitter plate of length $l_s/d = 3.0$ was placed at the back of the cylinder and along the centreline to suppress the formation of the

A. Gyr et al. (eds.), Science and Art Symposium 2000, 149-158.

large scale primary vortices [Apelt and West, 1975; Anderson and Szewczyk, 1997]. The Reynolds number range, based on the diameter, was $4 \times 10^3 < Re < 3 \times 10^4$, covering the transition between the lower and upper subcritical regime. A new conditional sampling scheme was developed to reveal the interaction of the secondary vortices and the primary vortex sheet [Lo and Ko, 1999]. In this scheme, the trough of the primary vortex sheet time history was located and the secondary vortices in the trough region were sampled according to three threshold levels, $1 < \sigma_i < 2$, $2 < \sigma_i < 3$ and $3 < \sigma_i$, where σ_i is the threshold level of the rms value u_i' of the secondary vortices. The primary vortex sheet was then sampled again. Since the formation and presence of the secondary vortices on the primary vortex sheet are not stationary, cancellation may occur during sampling. Therefore, instead of sampling the steamwise velocity u_i of the secondary vortices, their steamwise energy u_i^2 was adopted. The triple decomposition of the fluctuating velocity field yields [Hussain, 1983],

$$u = u_v + u_i + u_r ,$$

where subscripts v, i, r denote the primary vortex sheet, secondary vortex and random components respectively. At the minimum velocity $(u_v)_{min}$, the time of the jth sample of the primary vortex sheet is t_v^j. The maxima of the secondary vortices within the time period of the primary vortex sheet were located in the determination of the threshold level σ_i^j. ΔT_v^j is the period of the successive maximum velocities of the primary vortex sheet. As there are three threshold levels the function

$$g(\sigma_i^j) = 1 \text{ if } \sigma_{min} < \sigma_i < \sigma_{max}$$
$$=0 \text{ if } \sigma_{min} > \sigma_i \text{ or } \sigma_i > \sigma_{max}.$$

Thus, the valid samples of the secondary vortices within the period of $(t_v^j - \Delta T_v^j/2) < t < (t_v^j + \Delta T_v^j/2)$ are acquired. The recovered steamwise energy of secondary vortices of n samples of primary vortex sheet, based on the alignment with t_v^j is

$$<u_i^2> = \sum (u_i^j)^2 g(\sigma_i^j) / n.$$

The recovered steamwise velocity of the primary vortex sheet, based on the alignment t_v^j is

$$<u_v^2> = \sum (u_v^j) g(\sigma_i^j) / n.$$

The total number of ensemble averaging of each threshold level was greater than 2000, about 40 – 70% of total samples.

As will be shown later, some of the patches of secondary vortices within a cycle of the primary vortex sheet may contain beside one group, two and three groups. A group is defined as the vortices appear to have the increase, the maximum and decrease in velocity. Therefore, separate ensemble averaging was carried out at the categories of one-, two- and three-groups. The definitions of the recovered time history are shown in Figure 1.

3. RESULTS AND DISCUSSION

Figure 2 shows the time histories of the 1/3 octave filtered streamwise velocities at the centre frequency f_v and f_i of the primary vortex sheet and secondary vortices at different Reynolds numbers. The measurement locations were at the early state of the developing primary vortex sheet before its roll up and just after the formation of the secondary vortices. In the lower subcritical regime of Re = 4.7×10^3, the secondary vortices at $x_s/d = 0$, $y/d = 0.64$ occur randomly throughout the time history of the primary sheet (Figure 2a). In addition, the primary vortex sheet has the well known low frequency modulation. However, there is no coupling between the primary vortex sheet and the secondary vortices. At Re = 1.08×10^4, the velocity fluctuations at $x_s/d = 0$, $y/d = 0.56$ show frequent busts of intensive activities of the secondary vortices (Figure 2b). These high energy groups occur near the troughs of the velocity of the primary vortex sheet, suggesting coupling between the primary vortex sheet and the secondary vortices. These bursts of intensive activities as packets of shear layer fluctuations are also observed at $x/d = 1.0$ $y/d \approx 0.8$, within the formation region of the primary vortex, by Prasad & Williamson [Prasad and Williamson, 1997] at the Reynolds number range of 2.5×10^3 to 8×10^3. They are only found for parallel shedding and are more frequent at higher Reynolds numbers. At this Reynolds number of 1.08×10^4, Figure 2b also shows some individual packets of secondary vortices within a cycle, dividing into two groups. In the upper subcritical regime, at Re = 3.1×10^4, at $x_s/d = 0$, $y/d = 0.56$, there are more frequent occurrences of the secondary vortices (Figure 2c). Coupling is still found and the modulation of the velocity of the primary vortex sheet seems to mainly occur on the negative side. At this Reynolds number of 3.1×10^4, more secondary vortices in two and three groups are found.

Figure 3 shows the distribution of the secondary vortices within the three threshold levels. In the following discussion, the high, medium and low energy secondary vortices are referred to those recovered by the three respective threshold levels of $\sigma_i > 3$, $2 < \sigma_i < 3$ and $1 < \sigma_i < 2$. Within the present Reynolds number range, the medium energy secondary vortices are only about 20% of the total number. The low energy vortices ($1 < \sigma_i < 2$) are more frequent below the crossover Reynolds number of 7.5×10^3, while that the high energy ones ($\sigma_i > 3$) are more frequent at higher Reynolds numbers. At Re > 2×10^4, the number of the high energy ones is nearly constant.

Figure 4 shows the effect of threshold level on the maximum recovered energy $\langle u_i^2 \rangle_{max}/u_i^2$ of the secondary vortices. The low ($1 < \sigma_i < 2$) and medium ($2 < \sigma_i < 3$) energy vortices vary only slightly with Reynolds number. The high-energy ($\sigma_i > 3$) vortices increase in energy within the lower subcritical regime, reaching a maximum at

Re = 7 × 10^3 and decrease with the increase in Reynolds number until Re = 2 x 10^4 in the upper subcritical regime. This phenomenon confirms the dominant role of these high-energy vortices in the early stage of transition.

Figure 5 shows the distributions of the three groups of the high-energy ($\sigma_i > 3$) secondary vortices without and with splitter plate. Without splitter plate, the number of one-group secondary vortices is maximum in the lower subcritical regime at Re < 7 × 10^3. Then a lower peak is found at Re = 7 × 10^3. At higher Reynolds number, the number of one-group secondary vortices decreases. However, the two-group vortices increase in number and reaches a maximum at Re = 1.5 × 10^4. The three-group vortices also increase rapidly to a local maximum at Re ≈ 1.5 × 10^4 and become dominant in the upper subcritical regime. Thus, the increasing dominance of the high-energy secondary vortices during the transition is related to the one-group vortices in the early stage of transition and two-and three-groups in the later stage.

The phase shift $\Delta\tau_i$ of the recovered peak and its bandwidth $\Delta\tau_b$ (-3db from the peak) of the high energy secondary vortices are shown in Figure 6. The phase shift $\Delta\tau_i$ decreases with the increase in Reynolds number in the lower subcritical regime. In the upper subcritical regime at Re > 2 × 10^4, the phase shift is near zero and the bandwidth is higher. In the transition subregime, the bandwidth is lower than those in the lower and upper subcritical regime, also indicating more regular nature of the secondary vortices.

Figure 7 shows the ratio $<u_v>_{+1}/<u_v>_{+2}$ of the primary vortex sheet with the maximum recovered energy of the secondary vortices, including those of the three groups of medium-and high-energy vortices. As the energy of the low- energy groups is two orders of magnitude lower, it is presented as a single group. At $<u_i^2>_{max}/u_i^2 < 1$, the waveform of the primary vortex sheet is not affected, indicating that the low-energy secondary vortices do not affect the vortex sheet. At $<u_i^2>_{max}/u_i^2 > 1$, the waveform is affected and the ratio is less than unity. The higher the energy of the secondary vortices, the more significant is the effect on the development of the primary vortex sheet. The higher the number of group of secondary vortices, the greater is the effect on the primary vortex sheet.

With splitter plate, the growth of the primary vortices is suppressed and the formation length is longer [Unal and Rockwell, 1988; Cimbala and Garg, 1991]. In the present study, the 3d long splitter plate suppresses significantly the primary vortex shedding. Figure 3 shows the numbers of secondary vortices of the three energy levels with splitter plate. Instead of the crossover point of the high and low energy secondary vortices at Re = 7.5 × 10^3, because of the splitter plate, the crossover Reynolds number is higher at 2.1 × 10^4. Although the medium energy secondary vortices have slightly high percentage, their number does not vary significantly with Reynolds number. These results of splitter plate also indicate the importance of the high and low energy secondary vortices on the transition.

Figure 4 also shows that the high-energy secondary vortices with splitter plate reach the maximum at $Re \approx 1.2 \times 10^4$ and decrease with the increase in Reynolds number until $Re = 1.6 \times 10^4$. This Reynolds number range is narrower than that without splitter plate ($7 \times 10^3 < Re < 2 \times 10^4$), suggesting the suppression of the transition by the splitter plate in the subcritical regime. At $1 \times 10^4 < Re < 1.3 \times 10^4$, the $\langle u_i^2 \rangle_{max}/u_i^2$ of the high-energy vortices with splitter plate are nearly the same as those without splitter plate. It implies that with suppressed primary vortex sheet and its strain, the high-energy secondary vortices still develop to their peak energy in this stage of transition. The primary vortex sheet and its forcing effect play no significant effect in this part and its effect is significant in other parts of the regime.

The three different groups of secondary vortices are also found and the numbers of the high-energy vortices are also shown in Figure 5. Although the one-group vortices are still the most numerous in the lower subcritical regime, their number decreases rapidly in the transition regime. The two-group vortices become the most frequent at $Re = 7 \times 10^3$, which in turn are dominated by the three-group ones. This illustrates that with the reduced forcing effect of the primary vortex sheet, the dominant roles play by the one-group vortices before the transition, the two-group vortices in the early stage of the transition and the three-group vortices in the later stage of transition and in the upper subcritical regime. With the splitter plate, the effect is shifted more towards lower Reynolds number than that without splitter plate.

The phase angles $\Delta\tau_i/\Delta T_v$ and the bandwidth $\Delta\tau_b/\Delta T_v$, of the high energy secondary vortices with splitter plate are shown in Figure 6. During the transition, the phase angle and bandwidth are significantly higher than those without splitter plate, implying earlier formation and smaller coupling of the secondary vortices.

The recovered velocity ratios $\langle u_v \rangle_{+1}/\langle u_v \rangle_{+2}$ of the high-energy secondary vortices with splitter plate are also shown in Figure 7. For $\langle u_i^2 \rangle_{max}/u_i^2 > 2$, the $\langle u_v \rangle_{+1}/\langle u_v \rangle_{+2} < 1$ suggests the effect on the primary vortex sheet by the secondary vortices. Although the splitter plate is to suppress the primary vortices [Apelt and West, 1975], the present ratio in the formation region is the same as that of no splitter plate.

4. CONCLUSION

The high-energy secondary vortices play an important role in the transition from the lower subcritical to upper subcritical regime. The one-group secondary vortices are dominant in the lower subcritical regime, while the two-group secondary vortices are dominant in the early stage of the transition. In the later stage of transition and in the upper subcritical regime, the three-group secondary vortices are dominant. With the splitter plate, the transition in the subcritical regime is suppressed. Unlike that without splitter plate, the one-group secondary vortices are dominated before the transition, while the two-group and three-group secondary vortices are dominated in the early and later stage of transition, respectively.

5. REFERENCES

[Kourta, Boisson, Chassaing and Ha Minh, 1987] Kourta, A; Boisson, H.C.; Chassaing, P.; Ha Minh, H.; "Nonlinear interaction and the transition to turbulence in the wake of a circular cylinder", In: *J.Fluid Mech.*, **181**, pp. 141-161.

[Norberg, 1987] Norberg, C; "Effects of Reynolds number and a low intensity freestream turbulence on the flow around a circular cylinder", In: Ph.D. Thesis, Chalmers University of Technology, Sweden.

[Apelt and West, 1975] Apelt, C.J.; West, G.S.; "The effects of wake splitter plates on the flow past a circular cylinder in the range $10^4 < Re < 5 \times 10^4$. Part 2", In: *J. Fluid Mech.*, **71,** pp. 145-160.

[Anderson and Szewczyk, 1997] Anderson, E.A.; Szewczyk, A.A.; "Effects of a splitter plate on the near wake of a circular cylinder in 2 and 3 dimensional flow configurations", In: *Exp. In Fluid*, **23**, pp. 161-174.

[Lo and Ko, 1999] Lo, K.W.; Ko, N.W.M.; "Vortex interaction in the formation region of a grooved circular cylinder", In: *Fluid Dyn. Res.,* **24**, pp. 161-183.

[Hussain, 1983] Hussain, A.K.M.F.; "Coherent Structures – reality and myth.", In: *Phys Fluids*, **26**, pp. 2816 – 2850.

[Prasad and Williamson, 1997] Prasad, A.; Williamson, C.H.K.; "The instability of the shear layer separating from a bluff body", In: *J. Fluid Mech.*, **333**, pp. 513 –525.

[Unal and Rockwell, 1988] Unal, M.F.; Rockwell, D.; "On vortex formation from a cylinder. Part 2: Control by splitter plate interference", In: *J. Fluid Mech.*, **190**, pp. 513 –525.

[Cimbala and Garg, 1991] Cimbala, J.M.; Garg, S.; "Flow in the wake of a freely rotatable cylinder with splitter plate." In: *AIAA J.*, **29**, pp. 1001 –1003.

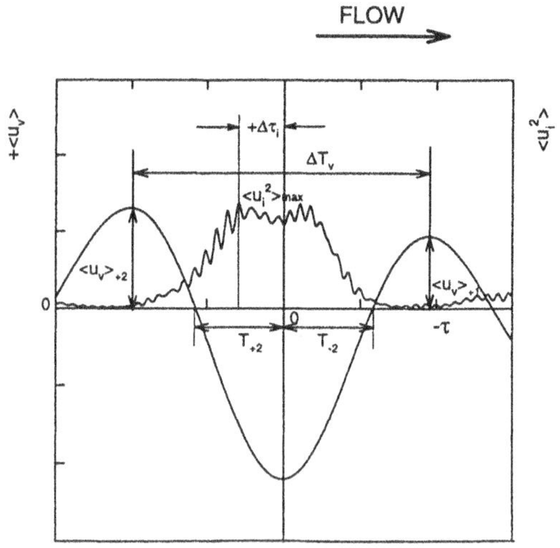

Figure 1; Definitions of recovered time histories.

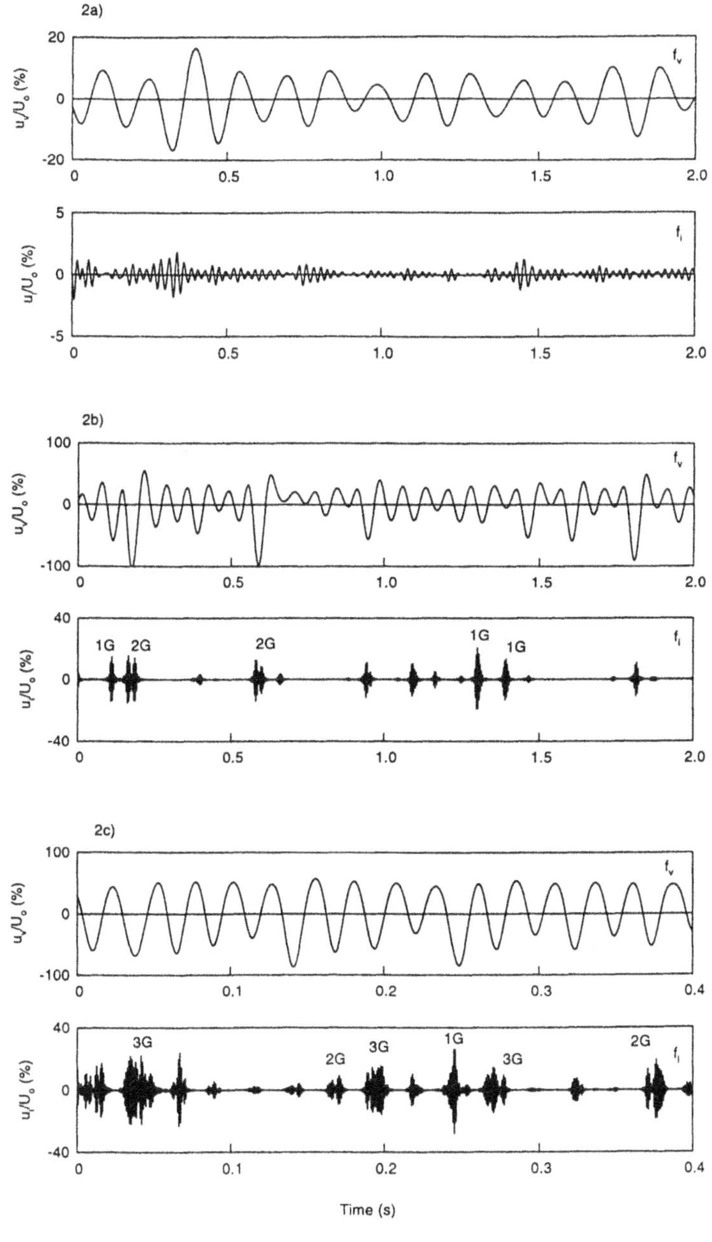

Figure 2; Time histories of velocity of primary vortex sheet and secondary vortices. (a) Re= 4.7 x 10^3; (b) Re = 1.08 x 10^4; (c) Re = 3.1 x 10^4.

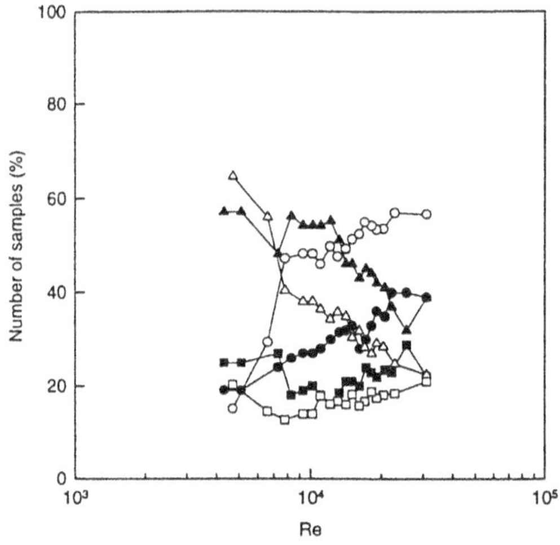

Figure 3; Variations of numbers of samples at different threshold levels.
Δ, $1 < \sigma_i < 2$; □, $2 < \sigma_i < 3$; o, $\sigma_i > 3$.
Open symbol: no splitter plate; solid: splitter plate.

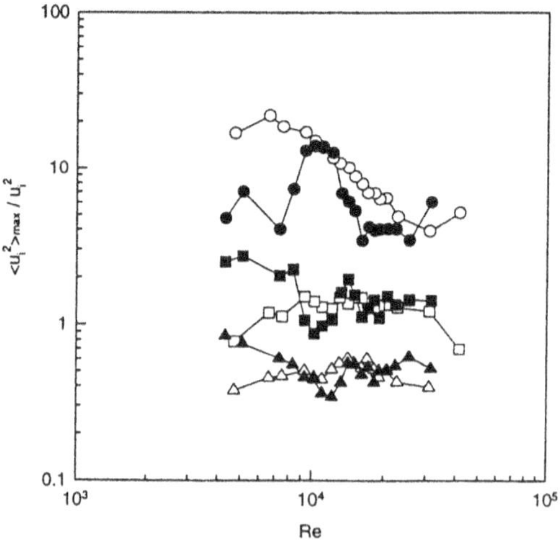

Figure 4; Variations of maximum recovered energies of secondary vortices at different threshold levels.
Δ, $1 < \sigma_i < 2$; □, $2 < \sigma_i < 3$; o, $\sigma_i > 3$.
Open symbol: no splitter plate; solid: splitter plate.

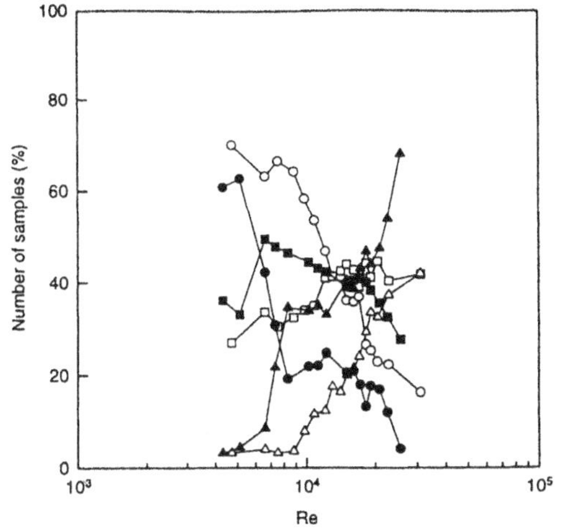

Figure 5; Variations of numbers of samples with Reynolds number ($\sigma_i > 3$).
o, One group; □, two groups; Δ, three groups.
Open symbol: no splitter plate; solid: splitter plate.

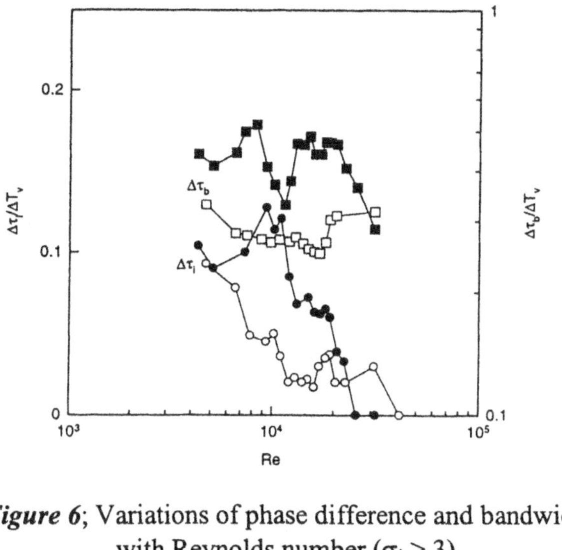

Figure 6; Variations of phase difference and bandwidth
with Reynolds number ($\sigma_i > 3$).
o, $\Delta\tau_i$; □, $\Delta\tau_b$.
Open symbol: no splitter plate; solid: splitter plate.

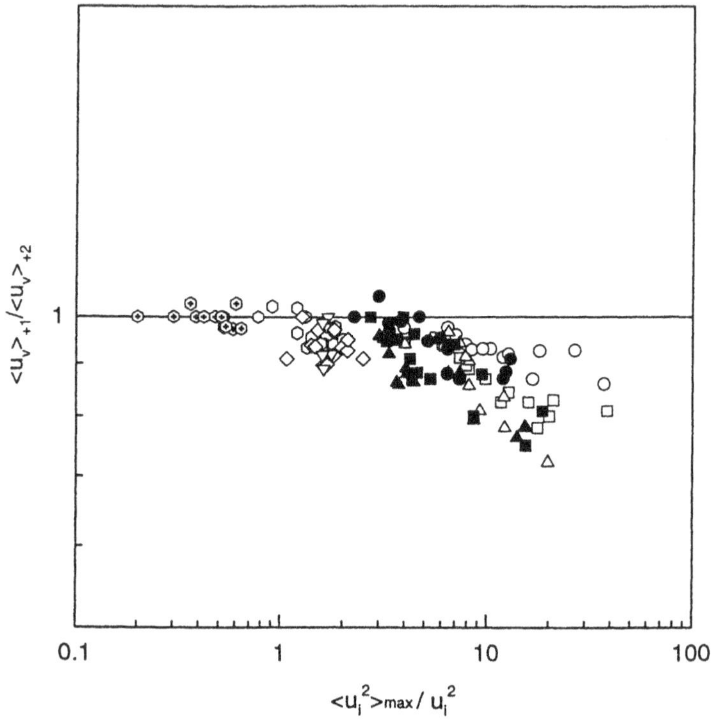

Figure 7; Variations of recovered velocity ratios of primary vortex sheet.
$1 < \sigma_i < 2$: ⊛ , one-group.
$2 < \sigma_i < 3$: ⬡ , one-group; ◊, two-group; ∇, three-group.
$\sigma_i > 3$: o , one-group; □, two-group; Δ, three-group.
Open symbol: no splitter plate; solid: splitter plate.

Microfluidic System for Biological Agent Detection

Yih-Far Chen*, Joon Mo Yang**, Jen-Jr. Gau*, Chih-Ming Ho*, Yu-Chong Tai**
**Mechanical & Aerospace Engineering, University of California, Los Angeles, USA*
***Nanogen, Inc., San Diego, California, USA*
****Electrical Engineering, California Institute of Technology, Pasadena, USA*
Email: chihming@ucla.edu, http://ho.seas.ucla.edu

Abstract: A MEMS-based microfluidic system capable of biological agent collection, treatment and detection is presented in this paper. The system incorporates microfluidic devices and an electrochemical DNA sensor to carry out airborne particle collection, sample preparation and identification with specificity. A silicon-nitride membrane filter was fabricated for airflow filtration, the design rule for pressure dissipation through μ-filter has been established through aerodynamic experiments and numerical simulations. An automated compact fluid handling system is built with a flexible configuration to accomplish sample preparation. Nucleic acid extraction has been successfully practiced with different protocols. By applying the synthesized oligonucleotide probes, the microfluidic system together with electrochemical DNA sensor has been characterized with cultured *Escherichia coli*. The test results show that positive identification of a specific biological agent can be achieved with the proposed microfluidic system.
Keywords: Microfluidics, MEMS, and Biological agent detection.

1. INTRODUCTION

Currently employed biochemical and immunological tests are limited by throughput and processing speed, an evolving trend is toward miniaturization and automation to accelerate the developments of molecular diagnostics. Since the ion exchange of bio/chemical reaction takes place in aqueous form, the desired features including parallel screening tests, walk-away operation system and economical reagent consumption can be realized with automated microfluidic system. Microfluidics provide the systematic integration instructions to carry out the operation for complex bio/chemical processes in micro scales.

Originating from the integrated circuit technique, the microfabrication technique can precisely etch reaction chambers and connecting channels on the substrate to create flow networks. An automated fluidic system can be built by integrating controllable liquid delivery devices to mimic bench top operations. Incorporating microfabrication technique and microfluidics, it is feasible to design a microfluidic system to accomplish

A. Gyr et al. (eds.), Science and Art Symposium 2000, 159-168.

the procedures of analytical assays. This provides a powerful tool in life science research such as high throughput screening for drug discovery, toxicological assays, and genetic diagnostics. As a multidiscipline demonstration of microfluidics, a compact MEMS-based fluidic system for pathogen detection and identification with electrochemical DNA sensor is reported in this paper. The designed system consists of three modules with the function of airborne particle collection, sample preparation and DNA identification.

Pathogenic bacteria are the major concern among biological agents and can be easily found in the aqueous nutrient medium. Ultrafiltration is a necessary step for granule and bacteria removal in water supply treatments. Airborne particles possibly attached with pathogens also serve as the transport medium and are taken into consideration herein. A microfilter is fabricated for airborne particle filtration from the airflow; high capture efficiency and low friction loss are required for practical operation. Wind tunnel experiments were conducted to investigate the pressure drops and power consumption of the μ-filter and establish the design rule for the parameters of opening ratio, hole-size and thickness.

MEMS technique can fabricate micro actuators that can be incorporated as fluid delivery devices. The design and fabrication of thermopneumatic microfluidic devices, μ-valves and μ-pumps, have been reported in a series of papers from the presented research (Yang *et al.*, 1997; Grosjean et al. 1999; Grosjean and Tai, 1999). These devices are of similar fabrication process and thus are appropriate for monolithic system integration. With the microfluidic devices under development, we used miniature solenoid valves and peristaltic pumps to do the sample preparations and to work out the system configuration for micromachined devices.

DNA-based diagnosis can identify unknown biological agents with high specificity based on the genetic variation of the species. By synthesizing the oligonucleotide probes complimentary to the target gene sequence, the sample can be examined with hybridization technique. With the probes modified with fluorescent or electrical labeling, the occurrence of hybrid formation can be detected with an optic system or electrochemical sensor. Due to the great variety of species, the sample preparation process is target dependent and consists of many fundamental operations of laboratory technique. It is the major task of microfluidic system design to efficiently carry out these operations in micro scale.

2. AIRBORNE PARTICLE COLLECTION MODULE

Airborne pathogens and bio-warfare agents are typically 10 μm or less; the micromachining technique is preferable for this particle threshold to fabricate perforated membrane filter. An $8\times8mm^2$ μ-filter made of silicon-nitride and reinforced with Parylene C coating was fabricated by Yang *et al.* (1998). High efficiency filtration with low power consumption is required for the membrane filter to be used for airborne particle collection. Power dissipation was measured in a small wind tunnel as show in

Figure 1. The μ-filter was fabricated with circular, hexagon and rectangular shapes of through hole as shown in Figure 2. With the configuration of 1~3 μm thickness and 4~12 μm hole size, the opening ratio ranges from 4% to 45%.

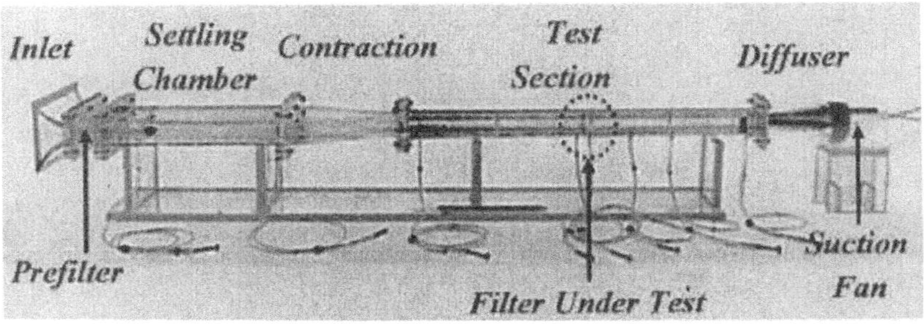

Figure 1; *Wind tunnel setup for power dissipation measurements.*

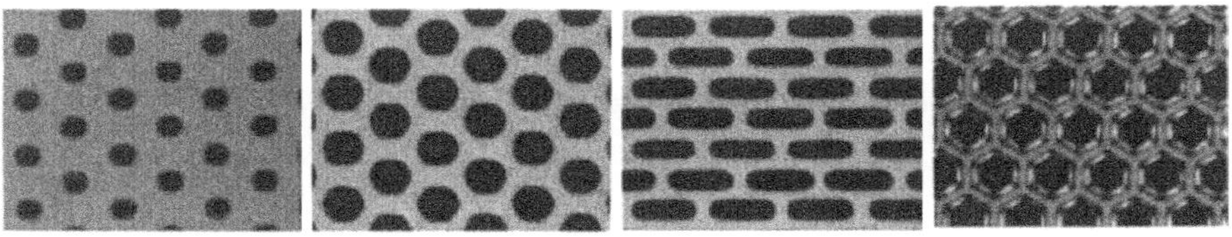

Figure 2; *Micromachined membrane filters with hole sizes of 4~12 μm and thickness of 1~3 μm. The opening ratio ranges from 4% to 45%.*

With thorough investigation on the geometric factors of μ-filter, a semi-empirical design rule is established by Yang *et al.* (1999), Figure 3 shows the consistency among the experimental data and numerical simulation results. The test results show that low power consumption can be achieved with the fabricated μ-filter, for example, 0.03 Watts for 20 l/min airflow through 1 cm^2 screening area. It is also shown that the gaseous flow in filters strongly depends on the opening factors, moreover, the thickness, sidewall profile and size distribution of the through holes should also be taken into consideration for accurate power dissipation estimation.

The surface-slip effect of airflow in a micro-channel reported by Chiu (1998) has only minor effects when testing the μ-filter. The reason is simply because not enough distance for the microfluidic phenomenon to develop through the thin filter. In the case of the micro-channel flow, the surface-slip effect is experienced due to the fact that the channel length is long compared to the height. Parylene C and Teflon were used as surface-coating material in the airflow tests, nevertheless, the surface coating does not affect the pressure drop in the tests.

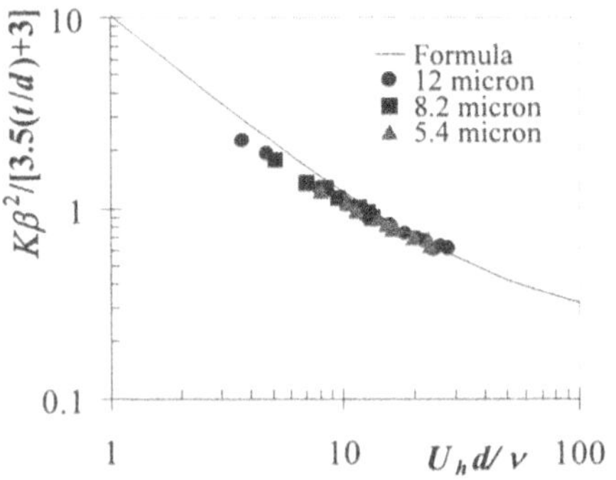

t	: filter thickness
d	: hole size
β	: opening ratio of filtering area
ΔP	: pressure drop through filter
U_{in}	: inflow velocity
U_h	: airflow velocity through filtering area
ρ	: density of air

$$\text{Design Rule: } K = \frac{2\Delta P}{\rho U_{in}^2} = \beta^{-2}\left[3.5\frac{t}{d}+3\right]\left[10\frac{\nu}{U_h d}+0.22\right]$$

Figure 3; *The design rule for microfilter power dissipation estimation*

In the laboratory test conditions, particle clogging on the filter is occasionally observed during the experiments. Continuing the downstream biochemistry analysis, particles of the size of interested, 5~10 μm, will be transported into liquid reservoir. The particles collected on the μ-filter are manipulated with electrostatic field generated by electrodes deposited on top of the filter. Lee *et al.* (1999) realized the dielectrophoretical particle transport using a 3-phase travelling voltage, the preliminary tests with borosilicate glass particles of 8 μm diameter showed that the efficiency of particle transport is around 10%~25%. No obstruction of particle movements by the surface tension of air/liquid interface was observed when particles entered the liquid reservoir.

3. SAMPLE PREPARATION MODULE WITH MICROFLUIDIC DEVICES

Sample preparation involves procedures required to treat the sample and extract the necessary components for detection, it is extremely varied in a broad variety of bio/chemical reagents and materials used when conducting the bioassay. As one of the least automated steps in the analytical processes, laboratory automation is acquiring more attention in the field of instrument development. Also benefiting from the automation, enormous screening tests can be accomplished in parallel; this urges the attempts for miniaturization to conserve the precious reagents. Microfluidics is the kernel of this concept - the ability to carry out the procedures of weighting, delivery and process control and to perform the complex operations such as mixing and separation for bio/chemical reactions with micro-volume liquids.

Many MEMS-based fluidic devices have been proposed in the literature, the driving mechanisms of which fall into two categories: electric fields and mechanical actuation. The former method is based on electrophoretic effects to do the electrokinetic valving and pumping. By applying an electric field, charged molecules can be moved along channel or be retained in reservoirs. This method does not require moving structures in the system and has been successfully practiced on the Capillary Electrophoresis (CE) chips. The mechanical actuation method does the liquid pumping or flow-passage blocking by driving micro actuators within the cavity chambers. A self-priming micro pump has been developed with piezo-membrane by Institute of Microtechnology Mainz, German.

Based on the rapid phase change property of specific working fluid (Fluorinert^{TM}), a thermopneumatic actuation method was proposed by Yang *et al.* (1997). When current is applied to the working fluid, a flexible silicone membrane is activated with the pneumatic pressure of gaseous Fluorinert^{TM}. A thermopneumatic μ-valve was made by Yang *et al.* (1997); Grosejeans and Tai (1999) integrated three thermopneumatic valves into a bi-directional, self-priming thermopneumatic peristaltic μ-pump. The preliminary test shows that this μ-pump is capable of 6.2 μl/min flow rate with thermopneumatic actuation and up to 120 μl/min with pneumatic actuation. The thermopneumatic actuation is adopted in the microfluidic system design based on the premise of system integration.

In order to conduct the sample preparation protocol in micro scale, the operations of fluidic devices are programmed to attain the effects of each individual step. It is prerequisite to characterize the performance of automated sample preparation processes for system integration. In anticipation of the MEMS-based fluidic devices, the automation of sample preparation is practiced beforehand with miniature controllable fluidic parts.

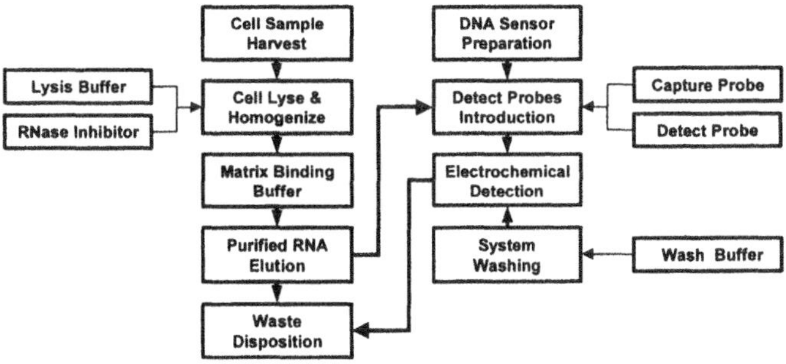

Figure 4; *Operation diagram of RNA isolation and purification from gram-negative bacteria detection.*

Figure 4 shows the diagram indicating the system operations of sample preparation together with electrochemical sensor detection. For DNA-based tests, the intracellular component of nucleic acids is required for specificity identification. A certified r*RNA* extraction protocol was chosen to practice the automation operation, cultured *E. coli* was used as test biological agent. Cells were harvested from culture medium by centrifuge and went through the following steps: Cell samples are first lysed and then homogenized. Ethanol is added to the lysate to provide ideal binding condition to the matrix of commercial available columns. The lysate is then loaded onto the column for r*RNA* binding, then the debris is washed away. Then purified r*RNA* is eluted from binding matrix with deionized water. Timing and temperature control are also specified in the protocol.

Several modifications have been made to accomplish the sample prep protocol with programmable fluidic devices, such as: the dead-end filtration and cross-flow filtration were used for cell harvesting instead of high *g* centrifugation, the repetitive injection is used as the substitute for vigorous agitation. The commercial column is limited to single use only and the continuous operation is an important feature for the system, the protocol is modified to include the affinity chromatography for r*RNA* purification. The modified protocol was practiced manually to verify its validity. The effect of eliminating specific or low temperature control is also examined. It is found that nucleic acids will degrade under the effects of denaturing enzyme when temperature control is removed.

A microfluidic system was built with miniature fluidic devices to test the modified sample preparation protocol. Figure 5 shows a booksize system assembled with peristaltic pumps, solenoid valves and 8-channel computer I/O interface for fluidic device control. The peristaltic pump can deliver liquid with flow rate of 0.34~4000 μl/min. The wetted surface of miniature fluidic devices and the tubing materials are pre-selected as bio-inert or low binding to prohibit cross-contamination from successive test runs, an additional washing is also included in the design.

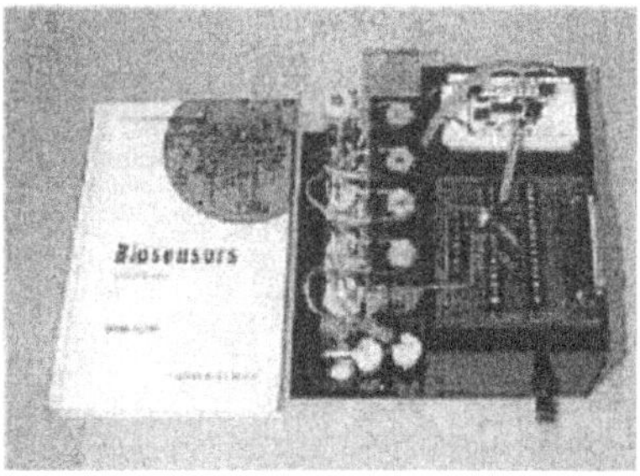

Figure 5; *The automated sample preparation module based on modified protocol.*

With the modified protocol, overnight cultured *E. coli* is fed into the packaged system. Cells are concentrated from culture medium with a 0.2 μm porous membrane. By adding the lysis buffer, cells are lysed and the intracellular materials are released. Bio/chemical reagents are successively added into lysate to complete the required bio/chemical reactions in designated order.

The system performance is examined by running bench top operation in parallel, the outputs were tested with gel electrophoresis. Gel electrophoresis is a traditional technique of molecular biology to separate nucleic acids based on their size. The efficiency of r*RNA* extraction by the automated microfluidic system is shown in Figure 6 in comparison to the standard operation based on original protocol. Size effect can be found in the extraction efficiency, the extraction efficiency ranges from 55%~70% of 31s r*RNA* to 35%~40% of 16s r*RNA*. Low efficiency in 16s r*RNA* might be due to incomplete cell lysis and the elimination of temperature control.

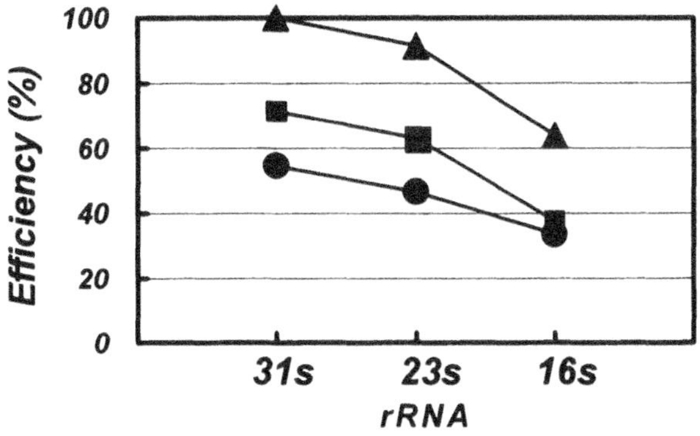

Figure 6; *rRNA obtained from cell lysate by the sample preparation module.*
Δ: *modified protocol with manual operation,* □,O *:automated operation.*

4. DNA IDENTIFICATION MODULE

The specificity detection is made possible by introducing synthesized oligonucleotide primers as capture probes and detection probes into the fluidic system, hybrids will form among the probes and the nucleic acids of sample if the sequences are complimentary. The hybrids can be of high or low stringency depending on the degree of sequence matching; high stringency hybrid will form when the target agent is present. The probes can be modified with optical or electric labeling as signal indicator. Optical detection with fluorescent or chemiluminescence is a well-established and widely used technique. This method is of high sensitivity but requires bulky peripheral instruments such as laser and photodiode for light source and detector.

Along with this study, an enzyme-based electrochemical biosensor is under development. The sensor surface is modified to have high affinity to specific bio-molecules, for example, the biotin and avidin are conjugate and have high affinity when encountering each other. By anchoring the hybrids on the sensor surface and washing away low stringency hybrids and residual probes, the analyte can be examined for the existence of target agents. The enzyme catalyzed reaction forms a product species, whose increment can be electrochemically detected by amperometry.

To complete the task of biological agent detection, the processes of probe introduction and hybridization control are also included in the sample preparation module. Figure 7 shows the results of *E. coli* detection with the sensitivity down to 10^5 cells. The detection limit can be improved by the microfluidic system with electrochemical workstation for low-noise current measurements. Deionized water and the *Bordetella* cells were also tested with the electrochemical sensors; the output signals from both negative controls are lower than the signals from *E. coli* samples.

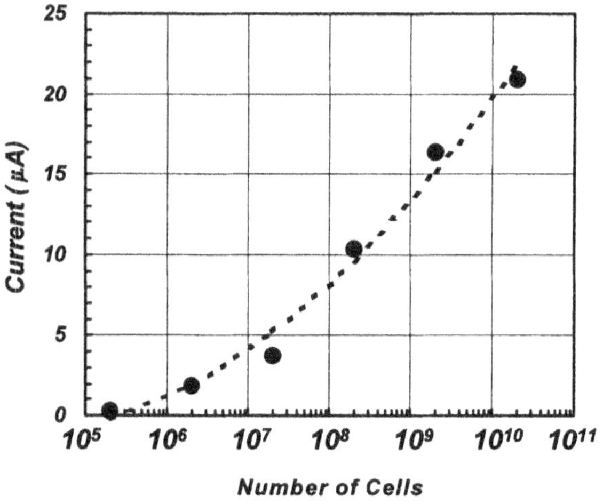

***Figure** 7; Preliminary tests of* E. coli *detection with electrochemical sensor.*

5. CONCLUSIONS

By incorporating micromachining technique and microfluidics, a compact microfluidic system has been designed for biological agent detection as an effort toward laboratory automation and miniaturization. Microfluidics provides the principle to accomplish sample preparation of fluid delivery, handling and process control in micro scale. A compact system has been built to exhibit the feasibility of system integration with airborne particle collection module and electrochemical DNA sensor. With its flexible configuration, the system can be programmed to complete tasks from single operation to complete protocol.

The micromachined thin, perforated membrane filters are designed and fabricated for airborne particle collection. The pressure dissipation can be estimated with the established design rule, low power consumption has been achieved by increasing the opening ratio. Airborne particle collection is a fundamental operation in air pollution monitoring to analyze its composition. The proposed design with μ-filter filtration can be integrated into the ventilation system in hospital or public transportation center for air quality control.

A series of molecular biology experiments from nucleic acid extraction to biological agent detection have been carried out with the modularized microfluidic system which the MEMS-based fluidic devices shall be incorporated into. The goals of automation and miniaturization have been achieved with the microfluidic system for biological agent detection; positive identification of *E. coli* is obtained with synthesized oligonucleotide probes and electrochemical DNA sensors. Substituting the fluid handling devices with the MEMS-based devices currently under development, a monolithic MEMS-based microfluidic system will be implemented based on the achievements of this study.

ACKNOWLEDGEMENT

This work is supported by the DARPA MICROFLUMES program under Naval Ocean System Center Contract N66001-96-C-83632.

REFERENCES

Chiu, C.W.; 1998; "Micro-channel flow studies and MEMS flow systems"; Master Thesis of the Mech. & Aerospace Engin. Dept. UCLA, USA.

Lee, S.W.; Yang, J.M.; Tai, Y.C.; Ho, C.M.; 1999; "Electrostatically active microfilters for automated airborne particle collection"; The 10th Int. Conf. on Solid-State Sensors and Actuators (Transducers '99), Sanday Japan.

Grosjean, C.; Tai, Y.C.; 1999; " A thermopneumatic peristaltic micropump"; The 10th Int. Conf. on Solid-State Sensors and Actuators (Transducers '99), Sanday Japan., 1776-1779.

Grosjean, C.; Yang, X.; Tai, Y.C.; 1999; " A practical thermopneumatic valve"; 12th IEEE Int. On MEMS (MEMS '99), Orlando, Florida., 147-152.

Wang, X.Q.; Lin, Q.; Tai, Y.C.; 1999; " A parylene micro check valve"; 12th IEEE Int. On MEMS (MEMS '99), Orlando, Florida., 177-182.

Yang, J.M.; Grosjean, C.; Tai, Y.C.; Ho, C.M.; 1997; "A MEMS thermopneumatic silicon membrane valve"; The 10th IEEE Int. Conf. on MEMS (MEMS '97), Nagoya, Japan.

Yang, X.; Yang, J.M..; Wang, X.Q.; Meng; E.; Tai, Y.C.; Ho, C.M.; 1998; "Micromachined membrane particle filters"; The 11th IEEE Int. Conf. on MEMS (MEMS '98), Heidelberg, Germany.

Yang, J.M.; Yang, X.; Ho, C.M.; Tai, Y.C.; 1999; "Prediction of the pressure drop through micromachined particle filters"; Tech. Proc. Int. Conf. on Modeling and simulation of Microsystem MSM (MSM '99), San Juan, Puerto Rico, 546-549.

Artificial shark skin on its way to technical application

W. Hage, D.W. Bechert, M. Bruse[1]

German Aerospace Center (DLR), Institute of Propulsion Technology, Dept. of Turbulence Research, Mueller-Breslau-Str. 8, D-10623 Berlin, Germany, e-mail: wolfram.hage@dlr.de
[1]*German Aerospace Center (DLR), Division of Wind Tunnels, Göttingen, Germany*

Abstract: The present paper gives an overview of the riblet research with the main emphasis on the european research in this field, which was motivated by the observation of the structure of the scales of fast swimming sharks. The drag reducing effect of this structure, which consists of fine streamwise aligned ribs, was shown experimentally and theoretically. Furthermore, an explanation how riblets do reduce the turbulent skin friction is given. The results of a parameter optimization of the riblet geometry is shown. Several aspects of the riblet film application on an long range aircraft are considered. The main focus lays here on the effects of an angle of yaw on the riblet performance.
Keywords: riblets, shark skin, drag reduction, yaw effects

1 WALL SHEAR STRESS REDUCTION

Drag reduction of a moving vehicle can be achieved by wall shear stress reduction and, in some cases, by separation control. This paper will focus on turbulent wall shear stress reduction. Obviously, laminar flow offers the lowest attainable wall shear stress. It has been introduced successfully on glider aircraft wings. On commercial aircraft, however, the problems associated with the implementation of laminar flow on swept transonic wings are at present not yet completely solved. Even with laminar flow being maintained on a major percentage of the wings, most of the aircraft surface will still be exposed to turbulent flow. Thus, turbulent wall shear stress reduction is obviously important in all scenarios of future aircraft design.

1.1 Artificial shark skin or "riblets": The mechanism of drag reduction

About a decade ago, it had become clear that a turbulent boundary layer on a surface exhibiting longitudinal ribs can develop a lower shear stress than that on a smooth surface (Walsh 1980 & 1990, Nitschke 1984). Whereas the American scientists invented ribbed surfaces on the basis of fluid-dynamical reasoning, parallel work in Germany was motivated by observations on shark skin (Reif & Dinkelacker 1982, Nitschke 1984, Reif

A. Gyr et al. (eds.), Science and Art Symposium 2000, 169-176.

1985, Dinkelacker et al., 1988). When we started working in this field, we first confirmed the effect in our wind tunnel using a direct shear force measurement (Bechert et al.; 1985). Subsequently, we developed, together with a group from the University of Naples, a theoretical model (Bechert & Bartenwerfer 1989, Bechert et al.; 1990, Luchini et al.; 1991). This model will be outlined in the following.

In order to obtain generally valid data, quantities are defined in dimensionless wall units. For instance, the distance y perpendicular to the wall takes the form $y^+ = y \cdot u_\tau / \nu$. This is a Reynolds number defined with the velocity $u_\tau = \sqrt{\tau_0 / \rho}$, where τ_0 is the average wall shear stress. ρ and ν are density and kinematic viscosity, defined as usual. In normal circumstances, u_τ assumes a value of a few percent of the free-stream velocity.

The strong exchange of momentum in a turbulent boundary layer is produced by high speed flow lumps approaching the surface ("sweeps") and by low speed flow moving away from the surface ("ejections") into the high speed regions of the flow. Regions of impinging flow ("sweeps") mostly occur in regions of elevated pressure, whereas "ejections", i.e., where the flow moves away from the surface, correspond mostly to regions of lower pressure. This exchange of fluid normal to the surface generates the enhanced shear stress of a turbulent flow because high speed flow is decelerated efficiently when it is swept towards the surface. By contrast, such an exchange normal to the surface does not occur in a laminar flow where the streamlines are essentially parallel and the flow lacks such violent local events.

It is also obvious that local events as "sweeps" and "ejections" require fluid motion in the lateral z-direction. Thus, hampering w-velocities in the z-direction will reduce momentum transfer and skin friction. It is plausible that that can be achieved with ribs on the surface which are aligned in the mean flow direction. On the other hand, it is known that surfaces exhibiting protrusions higher than about $y^+ \approx 5$ do increase the wall shear stress (Schlichting 1979). However, for protrusions smaller than $y^+ \approx 3$-5, ribs or other roughnesses are imbedded in the viscous sublayer. In this layer, very close to the wall, any fluid behaves like a highly viscous fluid, e.g., like honey. Therefore, it is admissible to calculate the flow around very small ribs with a viscous theory. Under viscous flow conditions, it turns out that the ribbed surface appears as a smooth surface located at a virtual origin, see Figure 1. However, the location of the origin, i.e., its elevation above the groove floor depends on the flow direction. For cross flow on the ribs, the virtual origin is closer to the rib tips than for longitudinal flow. The difference between these two heights, Δh, we call the "protrusion height difference". The existence of this difference has interesting consequences: If a flow motion is generated by a fluid lump in a plane at a height y^+ above the surface, the fluid lump experiences a higher resistance if it moves laterally than if it moves in the longitudinal direction. In this way, the cross flow is hampered by the ribs, as desired, and thus ribs do indeed reduce momentum transfer and shear stress. Therefore, for an optimization of the shear stress reduction, we have to select a ribbed surface which exhibits a maximum difference of the virtual origins for longitudinal and cross flow. As the theory shows (Luchini et al.; 1991), the maximum height difference is obtained for very thin blade-like ribs. The rib height should be $h \geq$

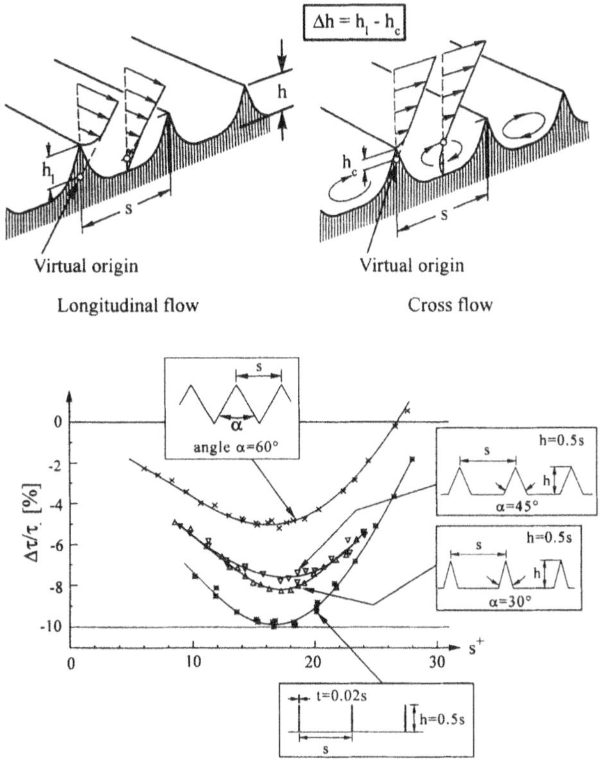

***Figure. 1**. Longitudinal and cross flow on a ribbed surface (above) and drag reduction performance of various rib geometries (below).*

0.6 s, where s is the lateral rib spacing. For this value of the rib height, the maximum difference of the elevation of the origins, or the protrusion height difference is Δh_{max} = 0.132 s. Careful experiments in our oil channel (Bechert et al.; 1992) with an adjustable blade rib height (Bechert et al.; 1997) have shown that the optimal rib height is actually slightly lower, at h = 0.5 s. With this surface, a turbulent shear stress reduction of 9.9 % below that of a smooth surface has been achieved (Bechert et. al 1997).

Figure 1 shows the data of optimal blade rib surfaces as compared to "riblets" with triangular cross-section, which had previously been considered as optimal. Clearly, it is difficult to manufacture surfaces with blade ribs. Therefore, we devised wedge-like ribs which still produce an impressive wall shear stress reduction, see Fig. 1. Ribbed surfaces such as the ones shown in Fig. 1 perform well over a certain s^+ range, which corresponds to a particular velocity range. By selecting an appropriate spacing s of the ribs, one can adjust the ribbed surface to the flow conditions at hand. In Fig. 1, $\Delta\tau$ is the difference between shear stress τ on the ribbed surface and τ_0 on a smooth reference surface, i.e., $\Delta\tau = \tau - \tau_0$. Negative values of $\Delta\tau/\tau_0$ refer to drag reduction and positive values to an increase of drag.

2 TECHNOLOGICAL APPLICATIONS ON AIRCRAFT

At present, artificial shark skin, i.e., riblet film, is on its way to implementation on long-range commercial aircraft. On long-range aircraft (i) the fuel costs contribute perceptively to the direct operating costs and (ii) the fuel weight exceeds by far the payload.

Consider an application on an Airbus A340-300. What are the implications of riblet film to the increase of its economical performance? The contribution of the skin friction to the total drag of this aircraft is about 48 %. This skin friction contribution can be reduced with our optimized riblets with trapezoidal grooves by 8.2 %. If the whole aircraft could be covered with riblet film, that would yield a total drag reduction of the aircraft of about 4 %. However, not the whole surface of the aircraft can be coated with riblet film for various reasons: Dust erosion at the leading edges of the wings and in the vicinity of the landing gear has, over greater time intervals, an effect like sand blasting. In addition, at the leading edges of the wings, the riblet film would interfere with the de-icing system and it would be incompatible with the laminar flow there. Locations where fuel and/or hydraulic fluid may come in contact with the plastic riblet film should also be avoided. Obviously, the windows cannot be covered either. Thus, only about 70 % of the aircraft can be coated with riblet film. In addition, in some places, riblet spacing and alignment will be suboptimal.

On the other hand, some of the roughnesses on the aircraft surface will be covered and thus be smoothened by the film. Moreover, a reduction of the wall friction ensues a slightly thinner boundary layer which, in turn, causes a reduction of the form drag in the rear part of the fuselage. This means that a 3 % reduction in total drag of the aircraft is probably achievable.

The weight of the riblet film is in the order of the weight of the paint which it replaces, i.e., 100-250 kg, depending on the percentage of the surface being covered. Additional technological issues, like durability, UV radiation tolerance, etc., have been solved in the meantime. One particular concern has fed the skepticism of innovation-dreading business people: At present, it takes one week to coat an aircraft with riblet film. During that time, the aircraft earns no money. Plausible as it sounds, it is indeed a spoof argument. Clearly, it is possible to coat the aircraft in steps parallel to other mechanical or maintenance work. In addition, if the riblet film replaces the paint, the time required for painting and the time for attaching the riblet film do not differ significantly. Moreover, the producer of riblet film (3M, St. Paul, Min., USA) has developed a riblet film which can be removed more conventiently than paint. The perspective of having an aircraft with a dirt-repellent film surface, which was observed since the riblet film was attached to a long-range aircraft, may be an additional attractive item in terms of maintenance costs.

The basic data of the A340-300 long-range aircraft are:

empty weight	126 t
fuel	80 t
payload, 295 passengers	48 t
max. take-off weight	254 t

Currently, the share of the fuel costs is about 30 % of the direct operating costs. Thus, one would save about 1 % of the direct operating costs by a 3 % reduction of the fuel consumption (which would be attributable to a 3 % total drag reduction). More important, however, is that about 2.4 tons of fuel can be replaced by additional payload. Therefore, the latter could be increased by 5 %. That would be equivalent to 15 more passengers. Consequently, including fuel savings the airline could make up to 6 % more money per flight. That would add up to something in the order of US$ 1 million more profit earned by one aircraft during one year. Incidentally, that would be roughly equivalent to what has been spent in terms of research money on that same issue altogether. Now one wonders whether or not this research has been too expensive?

Apart from the above application and considering the many requests that we have obtained in the meantime, we would like to stress the following point here: Drag reduction by rib-like surfaces makes sense only where turbulent wall friction represents an important contribution to the fluid-dynamical losses. However, in other cases like for instance car aerodynamics, which is governed by separated flow and form drag, the application of riblet film would be useless.

2.1 Yaw angle effects on optimized riblets

Because due to the technical constraints mentioned above, only 70% of the aircraft

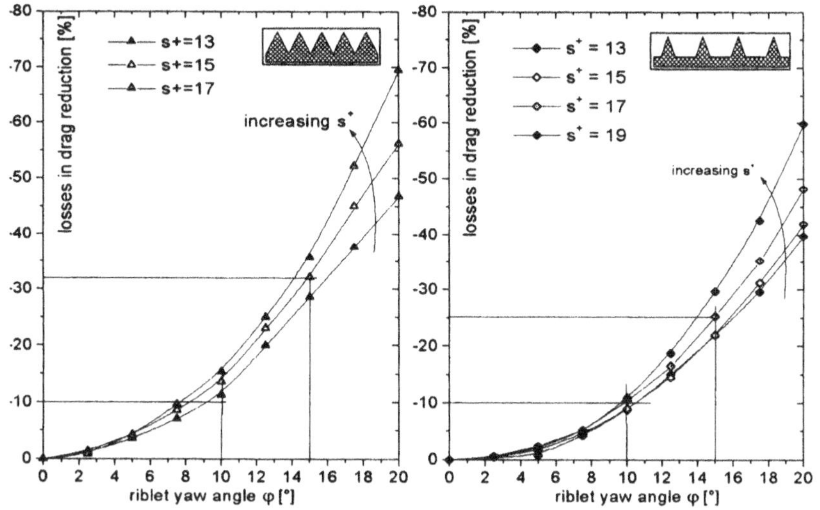

Figure 2 Drag data of V-grooved riblets (left) and trapezoidally grooved riblets (right)

surface can be coated with riblet film. Therefore, it is desirable to obtain the maximum performance of the riblet film coating. Previous work has shown that it is important to align the riblet film with the flow pattern on the aircraft in order to obtain an optimal result of the riblet application (Viswanath 1999, Gaudet 1989).

With the present experiments, the influence of the yaw angle on riblets with different riblet geometries is examined. V-grooved riblets with a maximum drag reduction of about 6 %, developed by M.J.Walsh, NASA Langley and the optimized trapezoidally grooved riblets with a maximum drag reduction of 8.2 % were tested. The experiments were carried out in the Berlin oil channel. The different riblet test plates were mounted on a turntable, which was inserted into a rectangular test plate. The riblet surface is moveable during the experiments. With a differential shear stress balance we compare directly the forces which occurs on the test plate and on a smooth reference plate. Thus we obtain a very high measurement accuracy of ±0.3%. The different riblet geometries were tested with a yaw angle up to ± 90°. With increasing yaw angle, one observes a degraded performance of the riblets (see Figure 2). The trapezoidally grooved riblets are less sensitive to misalignment as compared to V-grooved riblets. At the same angle of yaw, they are showing lower losses in drag reduction. For smaller yaw angles, the losses in drag reduction of the trapezoidally grooved riblets seem to be independent of the choosen nondimensional lateral riblet spacing s^+. This latter statement is valid for the s^+-range between 13 and 19, i.e., near the optimum of the riblet performance.

REFERENCES

(Bechert & Bartenwerfer 1989) Bechert, D.W. & Bartenwerfer, M.; "The viscous flow on surfaces with longitudinal ribs"; J. Fluid Mech., 206, pp. 105-129.

(Bechert et al., 1990) Bechert, D.W.; Bartenwerfer, M. & Hoppe, G.; "Turbulent drag reduction by nonplanar surfaces - A survey on the research at TU/DLR Berlin"; IUTAM-Symposium on "Structure of turbulence and drag reduction", Zürich 1989, Springer-Verlag, Berlin.

(Bechert et al., 1997) Bechert, D.W.; Bruse, M.; Hage, W.; van der Hoeven, J.G.Th. & Hoppe, G.; "Experiments on drag-reducing surfaces and their optimization with an adjustable geometry"; J. Fluid Mech., 338, pp. 59-87.

(Bechert et al., 1985) Bechert, D.W.; Hoppe, G. & Reif, W.-E.; "On the drag reduction of the shark skin"; AIAA-Paper 85-0546.

(Bechert et al., 1992) Bechert, D.W.; Hoppe, G.; van der Hoeven, J.G.Th. & Makris, R; "The Berlin oil channel for drag reduction research"; Exp. in Fluids, 12, pp. 251-260.

(Bruse et al., 1993) Bruse, M.; Bechert, D.W.; van der Hoeven, J.G.Th.; Hage, W. & Hoppe, G.; "Experiments with conventional and with novel adjustable drag-reducing surfaces" In: "Near-wall turbulent flows" (eds. R.M.C. So, C.G. Speziale & B.E. Launder), pp. 719-738, Elsevier, Amsterdam.

(Dinkelacker et al., 1988) Dinkelacker, A.; Nitschke-Kowsky, P. & Reif, W.-E.; "On the possibility of drag reduction with the help of longitudinal ridges in the walls"; IUTAM-Symp. on Turbulence Management and Relaminarization, Bangalore, India 1987 (eds. H.W. Liepmann & R. Narasimha), Springer-Verlag, Berlin.

(Gaudet 1989) Gaudet, L.; "Properties of Riblets at Supersonic Speeds", Applied Scientific Research, Vol. 46, No. 3

(Luchini & Pozzi 1997) Luchini, P. & Pozzi, A.; "Computation of three-dimensional Stokes flow over complicated surfaces (3D riblets) using a boundary-independent grid and local correction", 10th European Drag Reduction Working Meeting, Berlin 19-21 March 1997.

(Luchini, et al., 1991) Luchini, P.; Manzo, F. & Pozzi, A.;"Resistance of a grooved surface to parallel flow and cross-flow"; J. Fluid Mech., 228, pp. 87-109.

(Luchini, et al., 1992) Luchini, P.; Manzo, F. & Pozzi, A.;"Viscous eddies over a grooved surface computed by a Gaussian-integration Galerkin boundary-element method, AIAA Journal 30, pp. 2168-2170.

(Nitschke, 1984) Nitschke, P.; "Experimentelle Untersuchung der turbulenten Strömung in glatten und längsgerillten Rohren"; Max-Planck-Institut für Strömungsforschung, Göttingen, Bericht 3/1983.(1983). Trans.: "Experimental investigation of the turbulent flow in smooth and longitudinally grooved tubes. NASA TM 77480 (1984).

(Reif & Dinkelacker, 1982) Reif, W.-E. & Dinkelacker, A.; "Hydrodynamics of the squamation in fast swimming sharks"; Neues Jahrbuch für Geologie und Palaeontologie, Abhandlungen Band 164, pp. 184-187. E. Schweizerbart'sche Verlagsbuchhandlung.

(Reif, 1985) Reif, W.-E.; "Squamation and ecology of sharks"; Courier Forsch.-Inst. Senckenberg, Nr. 78, pp. 1-255, Frankfurt am Main.

(Schlichting, 1979) Schlichting, H.; "Boundary layer theory"; Transl.: J. Kestin, 7th Edition, McGraw-Hill, New York.

(Viswanath, 1999) Viswanath, P.R.; "Riblets on Airfoils and Wings: a Review"; AIAA paper 99-3402

(Walsh, 1980) Walsh, M.J,; "Drag reduction of V-groove and transverse curvature riblets". In: "Viscous flow drag reduction" (ed. G.R. Hough), Progress in Astronautics and Aeronautics, Vol. 72, pp. 168-184, AIAA, New York.

(Walsh, 1990) Walsh, M.J. Riblets. In: "Viscous drag reduction in boundary layers" (eds. D.M. Bushnell & J.N. Hefner), Progress in Astronautics and Aeronautics, Vol. 123, pp. 203-262, AIAA Washington.

Medusoid Vortex Rings

P. Koumoutsakos, J. Walther and J. Sagredo
Institut for Fluid Dynamics - ETH Zentrum ,
Zürich, CH-8092,Switzerland
petros,walther,julian@eniac.ethz.ch

Abstract: A high resolution three dimensional viscous vortex method is applied to the simulation of vortex rings. The ring is found to transition under specific perturbations generating a turbulent main "puff" and a medusa-like structure. The inner part of the ring remains laminar as it is revealed by these simulations.
Keywords: Vortex Rings, Vortex Methods, Medusas.

1. SIMULATIONS OF VORTEX RINGS

A high resolution vortex method is implemented for the simulations of vortex rings at Re = 4000. The calculations are performed in a periodic domain using a Vortex-In-Cell algorithm with a 192^3 grid and over 1 million particles. A key computational aspect is the use of high order moment conserving schemes for particle-grid projections and particle remeshing [Cottet and Koumoutsakos, 2000]. The initial configuration (see sketch in the following figure) consists of a vortex ring (green) perturbed by two smaller rings/bracelets (white). A second, thinner vortex ring (orange) of smaller strength is placed below the main configuration and acts to spread the disturbance azimuthally (I). The isovorticity surfaces (colored by velocity magnitude) show that the two rings interact so as to produce a Widnall type instability (II) and a repeated winding of the vortex cores (III, IV). At a later stage (V), various levels of isovorticity surfaces (VI) in the main core reveal a rather well defined ring-like inner structure, surrounded by a sea of small scale vortical filaments. The final structure has a strong resemblance to living medusas. This suggests an underlying physical mechanism in the growth and construction of these creatures based on similar principles as those of the formation of turbulent vortex rings.

REFERENCES

[Cottet and Koumoutsakos, 2000] Cottet, G.H.; Koumoutsakos, P. *Vortex Methods : Theory and Practice*. Cambridge University Press ; ISBN 0-521-62186-0

A. Gyr et al. (eds.), Science and Art Symposium 2000, 177.

Computational Anatomy of Human Lung Based on the Fluid Dynamics

[1] ***Ryuji Takaki and*** [2] ***Hiroko Kitaoka***
[1] *Tokyo University of Agriculture and Technology, Koganei, Tokyo 184-8588, Japan*
takaki@cc.tuat.ac.jp
[2] *Faculty of Medicine, Osaka University, Suita, Osaka 565-0871, Japan*
kitaokah@image.med.osaka-u.ac.jp

Abstract: A lung airway system is constructed numerically by setting up an algorithm based on geometrical requirements, the fluid dynamics and anatomical data. The algorithm contains several parameters to be adjusted, and an evaluating function is introduced to find an optimal result. The resulting airway has more than 20,000 branches and resembles the real one quite well. An application of the result to medicine is made to establish a quantitative method to detect diseases. A concept of computational anatomy is proposed.
Keywords: lung airway, branching system, optimal design, space filling, anatomy

1. INTRODUCTION

Anatomy is a field of medicine to observe human organs and to describe their structures. It is a basic and important field, but there is a general impression that it is going to be out-of-date. However, if structures of human organs are analyzed by the use of computers, new perspective will be opened.

Recent development in image analysis techniques from 3D-CT data enables to provide precise data of structures. Here, a new problem arises how to extract meaningful information from a large stuck of 3D data. An example of this problem is the detection of diseases of the lung. It is generally difficult to find cancers from a X-ray photograph, because it is a 2D projection of 3D structure. On the other hand, although realistic visualization of CT data are available, medical doctors must be trained for judging from 3D images. It is, therefore, necessary to develop a quantitative method to identify diseases and to assist medical doctors. For this kind of researches a certain model organ is necessary, which is constructed artificially.

The purpose of this paper is to introduce a recent attempt by the present authors to construct an airway system numerically by setting up an appropriate algorithm to assure optimal design. Similar research was made in the past [Weibel, 1963; Horsfield et al,.1971; Kitaoka et al, 1997]. But, no attempt to construct realistic 3D model has been made until recently by Kitaoka et al. [1999].

In the real lung each terminal branch of the airway supplies air to the smallest unit of the lung called "acinus". The inner structure of the acinus is different from that of

A. Gyr et al. (eds.), Science and Art Symposium 2000, 179-184.

the airway, and is thought as a tight packing of small spheres called "alveolus". The present authors have tried to construct this structure, but it will not be presented in this paper.

2. METHOD OF SIMULATION

The algorithm generating a 3D airway structure is set up according to both geometrical and fluid dynamical principles which should be satisfied at any branching generation.

The former is composed of rules for shapes of branches, *i.e.* locations, lengths, thickness and directions of branches after bifurcation, and rules for the regions governed by branches, *i.e.* its volumes and the contour surfaces. It is required that the air should be supplied equally to any region, hence the air flow rate through a branch should be proportional to the volume of the region. Another requirement is a homogeneous distribution of terminal branches within the entire region of the lung.

The fluid dynamical principles are, first, the flow rate conservation and the momentum conservation during branching (according to the latter the mother and daughter branches lie on the same plane). Secondly, the viscous dissipation of energy at the branching should be minimized, which leads to a relation

$$d_0^{\,n} = d_1^{\,n} + d_2^{\,n}, \tag{1}$$

between the diameters of the mother and daughters, where n=3 [Murray, 1926; Suwa et al, 1963; Kamiya et al, 1974], and n=2.8 from experiment.

We introduce the flow dividing ratio r ($0 \leqq r \leqq 0.5$), so that the flow rates in daughters are r and $1-r$ (see Figure 1). Kamiya et al. [1974] also derived from the minimum energy dissipation the following formula for the branching angles:

$$\cos\theta_1 = (1 + r^{4/n} - (1-r)^{4/n}) / 2r^{2/n}, \quad \cos\theta_2 = (1 + (1-r)^{4/n} - r^{4/n}) / 2(1-r)^{2/n}. \tag{2}$$

In addition to the above principles we need to assign some anatomical conditions. The mode of branching is dichotomous. The ratio of branch length to its diameter should be given according to anatomical data. The outer shape of the lung is given, where a realistic lung shape is used including boundaries of the thorax, the thoracic vertebrae, the diaphragm, the heart, the descending aorta, etc. The location of the main trunk (the initial condition for branching) should be specified. The algorithm based on these requirements are summarized below.

Rule 1: Branching is dichotomous.
Rule 2: The parent and its two daughters lie in the same plane (branching plane).
Rule 3: The flow rate is conserved after branching.
Rule 4: The region governed by a parent is divided into two daughter regions by a plane (space dividing plane) perpendicular to the branching plane.
Rule 5: The flow dividing ratio r is equal to the volume dividing ratio of daughter

regions.
Rule 6: Diameters and branching angles of the two daughter branches are determined by Eqs. (1) ands (2).
Rule 7: The length of each daughter branch is three times its diameter.
Rule 8: The branching plane is perpendicular to the preceding branching plane.
Rule 9: The branching process terminates whenever the flow rate becomes less than a specified threshold value or the branch extends out of its own region.

Adjustable parameters are introduced, such as a correction of branching angles, the length-diameter ratio, corrections of the angle between successive branching planes and that between the branching and dividing planes.

A resulting airway tree is evaluated by calculating the following five quantities. (1) Sum of flow rates through terminal branches which terminated because of extending out of their own region, (2) variation of the density of terminals, (3) variation of acinar volumes, (4) ratio of the total airway volume to the total lung volume, (5) energy loss per terminal. An evaluation parameter, i.e. a performance index *P.I.*, is given by the simple sum of these five quantities. We seek the optimal airway by minimizing *P.I.*

3. RESULTS OF SIMULATION AND APPLICATION

The optimal 3D model of the human airway tree is shown in Figure 2. This model resembles the real airway remarkably. It has 27,306 terminal branches, and the mean generation numbers down to terminals is 17.6 ± 3.4 varying from 8 to 32. The lung volume in the present model is 6,388 m*l* and the corresponding total airway volume is 174 m*l*. The mean volume of acini is 0.211 ± 0.076 m*l*. These values are consistent with anatomical data.

As an interesting application of the present result, we can make up pathological airway systems. Cancer or inflammation in the lung attracts ambient tissues to itself with different strengths, thus parts of airway branches near to these diseases are deformed differently. Medical doctors usually observe these deformations and make diagnosis. In the present analysis the deformations can be calculated based on the elasticity theory, hence deviations of branch orientations are estimated theoretically.

Figure 3 shows parts of simulated airways in three cases, normal, cancerous and inflammatory. Although distinctions among them are hard to recognize, the averaged angles between the branches and the radii pointing from the centers of disease areas show a clear distinction between these diseases (Figure 4).

4. CONCLUDING REMARKS

The success of the present lung model suggests that the relatively simple algorithm proposed here is sufficient for creating complicated branching structures. In other words, apparently complicated structures can be results of simple algorithms. Since the present algorithm is associated with space division, it will be applied to any technical problem to design a hierarchical space division associated with material transport. As examples, branching systems within a sphere and a cube can be constructed (shown as a poster).

In many scientific papers the phrase "bridge between form and function" appears and plays a role to provoke challenges for basic studies. However, there arise difficulties, such as how the functions should be expressed scientifically, and which characteristics of the forms should be observed. At present an effective way to overcome these difficulties is to construct organs artificially and to estimate "functions" mathematically. We would like to propose to call this type of research "computational anatomy". The present work is expected to be an example of it.

The present authors would like to express their cordial thanks to Prof. B. Suki of Boston University and to Prof. H. Itoh of Kyoto University for their helps during this work.

REFERENCES

Horsfield, K., Dart, G., Olson, D.E., Filley, G.F., and Cumming, G., 1971; "Models of the human bronchial tree"; *J. Appl. Physiol.*, 31, pp.202-217.

Kamiya, A., Togawa, T. and Yamamoto, A., 1974; "Theoretical relationship between the optimal models of the vascular tree"; *Bull. Math. Biol.*, 36, pp.311-323.

Kitaoka, H. and Zuki, B., 1997; "Branching design of the bronchial tree based on a diameter-flow relationship"; *J. Appl. Physiol.*, 82, pp.968-976.

Kitaoka, H., Takaki, R. and Suki, B., 1999; "A three-dimensional model of the human airway tree"; *J. Appl. Physiol.*, (to appear)

Murray, C.D., 1926; "The physiological principle of minimum work"; *I. Proc. Natl. Acad. Sci. USA*, 12, pp.207-214.

Suwa, N., Niwa, T., Fukasawa, H. and Sasaki, Y., 1963; "Estimation of intravascular blood pressure gradient by mathematical analysis of arterial casts"; *Tohoku J. Exp. Med.*, 79, pp.168-198.

Weibel, E.R., 1963; "Morphometry of the Human Lung"; Academic Press.

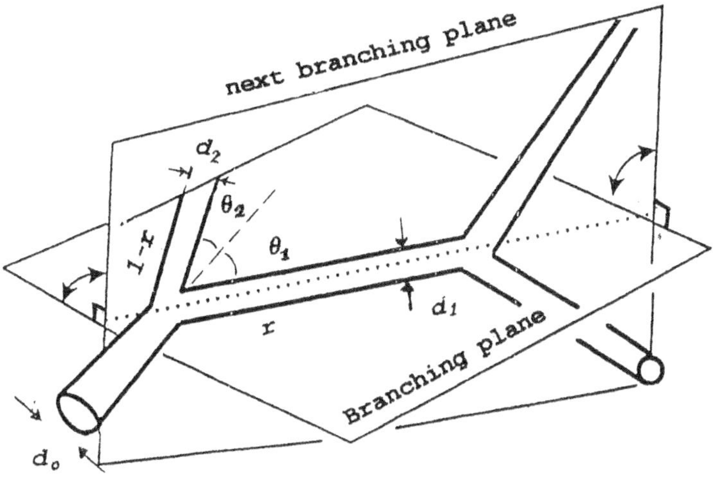

Figure 1; *Sketch of geometrical definitions.*

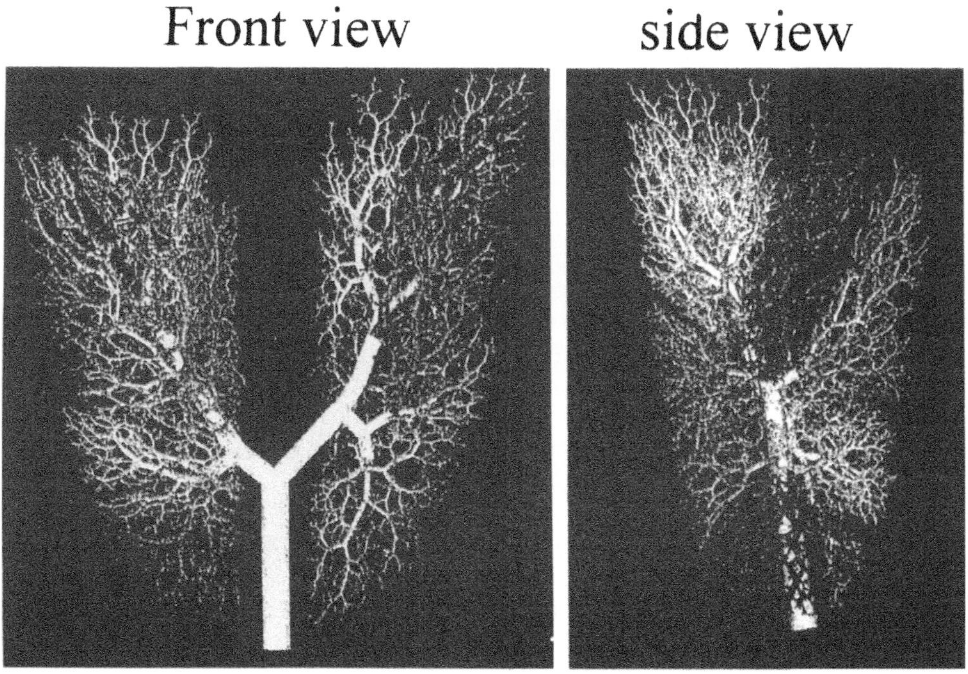

Figure 2; *Resulting 3D airway tree with the best choices of parameters. Left: anterior view, Right: right lateral view. The parameters to produce this model are slightly different from those used in the previous study [Kitaoka, et al, 1999].*

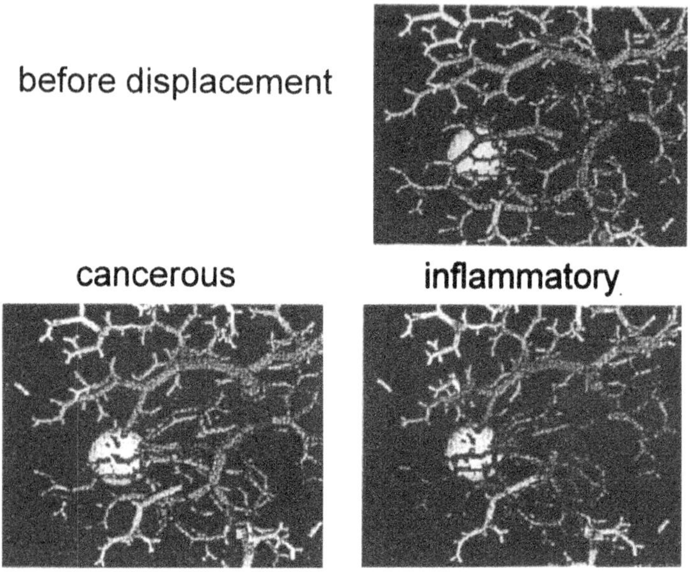

Figure 3;. *Parts of airways in three cases.*

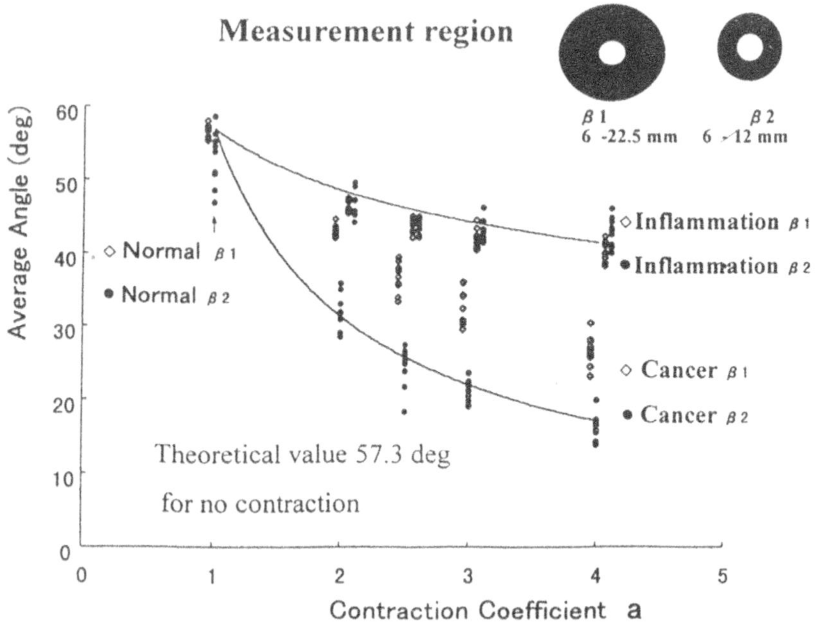

Figure 4; *Averaged angles between the branches and the radius pointing from the center of disease areas vs. strength of deformation. Isotropic distribution yields 57.3 deg. for the average, and smaller values indicate alignment towards the center. Data show separation between cancerous and inflammatory cases.*

PARADOXES OF FISH-LIKE PROPULSION

Lyudmyla Koryenna
Institute of Hydromechanics NASU, 8/4 Zheliabov St., Kiev, 03057, Ukraine
lkor@rdfm.freenet.kiev.ua

Abstract: A fish-like propulsion interpreted by mechanisms based on form and friction drag may produce some unlikely conclusions about the phenomenon. The body is moved by a traveling wave: the thrust, is a useful force, it arises from the form drag; the friction drag due to this "wave" is similar to the friction on a body with a moving boundary. These was concluded from the kinematics of the wave-type motion of the body.
Keywords: fish, wave, propulsion, thrust, friction.

1. INTRODUCTION

This work is based on two sets of experimental data: (1) measured thrust of a wave-like deformable plate, by the author; (2) kinematic parameters of fish and dolphin movements obtained by high-speed filming by Kozlov. Both works were carried out at the Institute of Hydromechanics National Academy of Sciences of Ukraine.

If the body of a fish participates in oscillatory motion, or if oscillatory movements are concentrated only in a part of the tail with a powerful fin, the motion in both cases can be characterized identically by two effects: 1) a traveling wave propagating along the axis of the body from head to tail, 2) the wave moves backward along the body faster than the body moves forward (i.e. the wave velocity is greater than that of the translatorial motion of the body). The observed identity of the two cases is the basis of the present discussion of hydrodynamic effects of a 'traveling wave' .

2. RESULTS AND DISCUSSION

First some theory. The coordinates xyz are fixed on a body which moves uniformly with velocity V relative to stationary fluid are denoted by the coordinate $\xi\eta\zeta$, see Fig.1 If x is a horizontal axis and y is a normal axis directed upwards, then the equation of a traveling wave can be written as

$$y = A(x)\sin(\omega t + \beta x + \varphi_0) \qquad (1)$$

A. Gyr et al. (eds.), Science and Art Symposium 2000, 185-190.

where A(x) is the amplitude, $\omega=2\pi f$ the circular frequency; $\beta=2\pi/\lambda$ is the wave number; λ is the wavelength, f the frequency; and φ_0 the phase angle. The constant relative wave velocity is $C=\lambda f$.

The motion of the traveling wave type propulsion can be demonstrated as follows. A "rigid sinusoid" is interacting with a flexible leaf. The 'sinusoid' plays the role of the wave, and the leaf plays the role of a fish or a wave-like deformable body. When the 'sinusoid' is traveling inside the leaf, all elements of the leaf are subjected to transverse oscillations as a fish. The velocity of the 'sinusoid' inside the leaf is the wave velocity C. A propulsed body moves with the velocity V, V< C in opposite direction to the traveling wave.

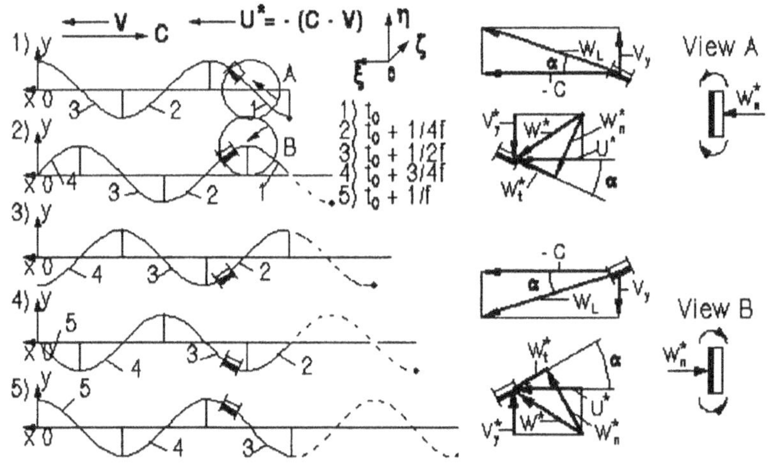

Figure 1; A kinematic scheme of a deformable body under a wave-like propulsion

This mechanistic model is shown in Figure 1. An infinite 'sinusoid' travels along the deformable body (five states of the 'sinusoid' for one oscillation along the body are shown by the solid line) and the leaf position together with its wake is shown by a dashed line. By observing the inverse motion, each element of the deformable body moves along the 'sinusoid' with the velocity W_L. An asterisk denotes the inverted motion. For example, a velocity of the water in a flume $-V= V^*$.

Thus in the flow, there is a body having a form of a 'sinusoid', so that at every moment one "marked" element of the 'sinusoid' deforms. In such a way, the boundary of the 'sinusoid' moves along this body.

The thrust of the wave-like deformed plate was measured in a three-dimensional setup especially designed for this purpose. As a body a 2 mm thick horizontally aligned rubber plate was fabricated, with a length L=400 mm and H=100 mm wide. Seven П-shaped supporting and controlling rods were fixed to the plate on each sides. Their phase-shifted harmonic oscillations produced a wave traveling along the plate. The thrust was measured depending on V, f, A(x) and L/λ [Korennaya, L.I., 1997]. The Reynolds numbers based on L were within a range of $R=VL/\nu=(1.25-4.6)\times10^5$. The Strouhal numbers based on l were $S_\lambda = f\lambda/V$, i.e. the ratio of the velocities C/V, ranged from 0.64 to 21. The thrust force was measured only in cases $C>V$.

The studies should be focused on the regimes of the plate when the thrust on it drops to zero since these regimes are analogous to those of a fish in motion. In Figure 2, the range of Strouhal numbers for the plate is within the range as found for aquatic animals. The results for the plate are compatible with the experimental data obtained for fishes and dolphins. Hydrodynamic forces are fully modelled using a kinematic approach and the effect of a moving surface, which will be discussed below. Fish mucus and dolphin compliant skin properties can be expected to expand the range of the Strouhal numbers for aquatic animals according to Figure 2.

The segments 1, 2, 3, 4, 5 of the wave in Figure 1 located between extreme points of the 'sinusoid' (between crests and valleys of the wave), operate similarly to an oar blade. They are the elements of propulsion they push the fluid in a direction opposite to the motion of the body. A part of the energy is spent for the creation of the traveling wave and is transformed into work of the thrust force.

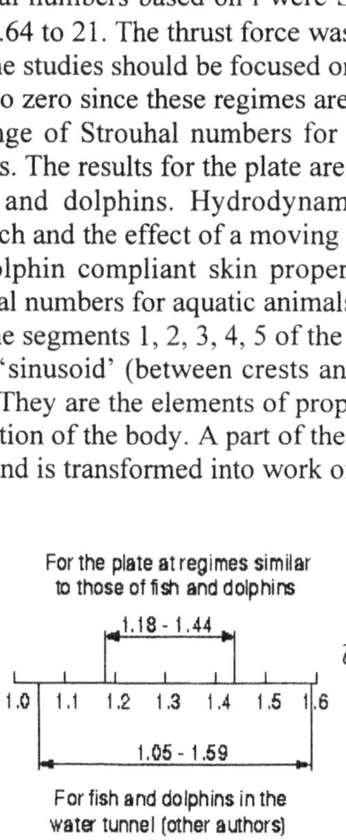

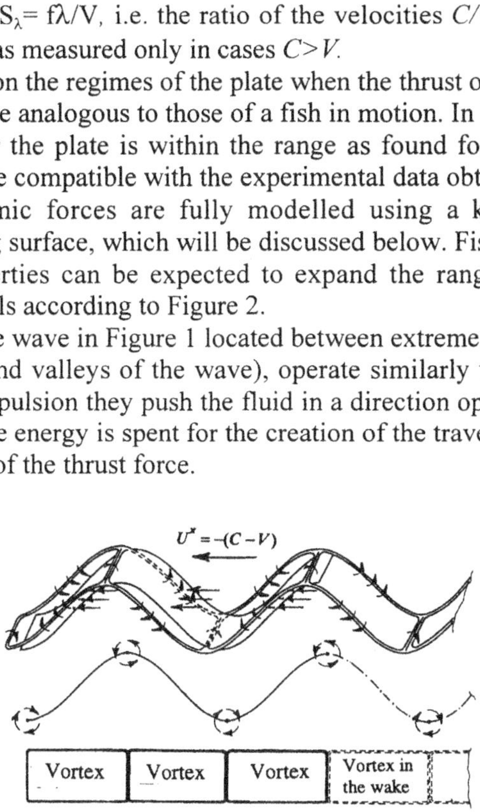

***Figure 2**:The scales for the Strouhal numbers $S_\lambda = C/V$*
Figure 3: Sketch of a vortex model on a wave-like plate

Using fundamental hydromechanics, it is possible to construct a vortex model of the wave-like propulsion (Fig. 3). Vortices emerge from the working elements near extreme locations of the 'sinusoid' and near edges of the working elements. This vortex model of the fish-like propulsion is an important issue to develop an adequate theory of the wave-like propulsion (similar to the theory for wings).

When a body moves in a fluid, the direction of the drag force coincides with the direction of the inverse motion. In inertial coordinates $\xi\eta\zeta$, Figure 1, the flow velocity in inverted motion U*= -(C-V) has the same direction as V. Thus, the thrust necessary for the locomotion with velocity V results from the form drag on the 'sinusoid', i.e. from the pattern of the vortices on the working elements. When V= C, the velocity U* and the thrust force are equal to zero.

The form drag of the 'sinusoid' causes a thrust resulting in a fish-like propulsion. However, friction forces produce internal losses. It was assumed above that the surface of the 'sinusoid' is moving. One can assume that the boundary layer is vanishing if the surface of the body has a velocity equal to the free-stream velocity, and the flow around the body is a potential flow.

Let us introduce the ratio of the velocities W_L of the moving boundary and the tangential component of the free-stream velocity W_t,: $p=\frac{W_L}{W_t^*}=\frac{C+V_y dy/dx}{C-V-V_y dy/dx}$, Figure 1, as the parameter of the moving boundary [Koryenna L.I., 1998]. In the propulsion regime of the 'sinusoid' in the extreme points $p_{y(\max,\min)}=\frac{C}{C-V}>0$, i.e. the surface is moving in flow direction because W_L and W_t have the same direction. The case of $p>0$ occurs at any point of the 'sinusoid', except for $y=y_{(\max,\min)}$, if

$$C-V>V_y dy/dx \tag{2}$$

and opposite to it for p<0 or

$$C-V<V_y dy/dx \tag{3}$$

Water animal	L [m]	V [m/s]	$R=$ $=VL/\nu$	C [m/s]	$C-V$ [m/s]	λ [m]	A [m]	f [Hz]	p *at y=0*
Tursiops Truneadus	2.60	2.34	$6\cdot10^6$	3.12	0.78	1.90	0.15	1.64	389
Belone	0.48	1.05	$4.5\cdot10^5$	1.21	0.16	0.22	0.03	5.58	-2.55
Pomatomus Saltatrix	0.42	1.72	$6.5\cdot10^5$	1.95	0.23	0.37	0.05	5.27	-2.58
Sarda Sarda	0.16	1.12	$1.6\cdot10^5$	1.25	0.13	0.13	0.02	9.58	-2.68
Cristivomer Namaycush	0.21	1.33	$2.5\cdot10^5$	1.48	0.15	0.15	0.02	9.74	-2.54

Table 1: *Parameter **p** of a moving boundary for the caudal fin of a dolphin (Tursiops Truneadus) and four fish species.*

Available kinematic data of water animals were used, some of them given in Table 1. It should be mentioned that for the caudal fin at y= 0, four fish species are characterised by values of p= (−2.54) – (−2.68) which differs essentially from p = 389 for a dolphin. Thus the scheme of a boundary moving in a flow direction, p>0, is realized for a dolphin ($R>R_{cr}$ in Table 1); and the scheme of a surface moving opposite direction, p<0, is realised for a significant part of a fish body ($R<R_{cr}$ in Table 1).

Boundary-layer velocity profiles are superimposed over a wave-like body for a physical interpretation as illustrated in Figure 4. For reliability, the two regimes were chosen. All velocity profiles are given in the same scale. The laminar boundary-layer Blasius profile was used to construct velocity profiles (*a*) for W^*_t and (*b*) for W_L. The (*c*) velocity profile is the superposition of (*a*) and (*b*). This resultant (*c*) profile is given in coordinates fixed at the 'sinusoid' element. To turn back to coordinates fixed at the moving surface of the 'sinusoid', each element of the body and the adjacent fluid had a velocities W_L and directed opposite to W_L, the (*d*) profile. The dot-and-dash curve in the resulting profiles is a laminar boundary-layer Blasius profile.

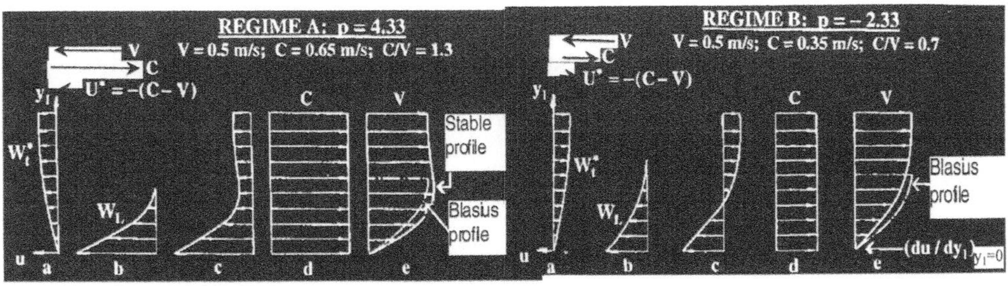

Figure 4: Velocity profiles in a boundary layer of a wave-like moving body, $y = y_{max;min}$.

Let us compare the (*e*) profiles at p>0 and p<0.

Regime A: The resulting profile (*e*) at p>0 is wider compared with the Blasius profile and is similar to a profiles of a stable type [Schlihting, H., 1974]. Laminar-turbulent transition is delayed in this case. In Table 1 dolphins show up with p=389>0 and $R>R_{cr}$. Hence, the velocity profiles over a dolphin should be stable, and the transition in the boundary layer is delayed. Indeed, according to [Romanenko, Ye.V. 1986], the amplitudes of the pressure fluctuations in a boundary layer of the dolphin are of $R>R_{cr}$ type in case of a passive motion (inertial), the level of the pressure fluctuations correspond to a developed turbulent flow; however in case of an active motion, this level is considerably lower and corresponds to an "underdeveloped" turbulent flow.

Regime B: In case $p<0$, the velocity gradient $(du/dy_1)_{y1=0}$ is less than compared with the velocity gradient $(du/dy_1)_{y1=0}$ on a flat plate. According to Newton's rheology, the local shear stress is proportional to $(du/dy_1)_{y1=0}$. Therefore the friction drag should be slightly less than for a flat plate. In Table 1 it is shown that for the four fish species, $p<0$ and $R<R_{cr}$. Hence, the local shear stress and the friction drag for fishes should be smaller than those known for a flat plate.

3. CONCLUSIONS AND COMMENTS

The kinematically "traveling wave" is a concept in understanding the hydrodynamics of the fish-like propulsion, see also Lighthill, 1975. It is argued that by introducing a wave ratio parameter the efficiency of this mechanism can be characterized Thus it was estimated that in nature, the transition from laminar to turbulent flow in a boundary layer is shifted towards higher Reynolds numbers, and the reduction in frictional drag occurs at at lower Reynolds numbers as estimated by a non-oscillating boundary.

The wave-like deformation of flexible body can be used not only to explain the produced thrust, but also to understand the drag reduction of the boundary layer flow. Therefore, the nature proposes an elegant way to solve the problem of producing a thrust force at minimum friction force by a controlled "wave". The correlation between the parameters of a wave as given by (2) and (3) can be applicable in engineering problems.

4. REFERENCES

Korennaya, L.I.; 1997, "Distinctive Features of the Wave Plat (or Fish) Motion"; In: *AGARD Rep.827, High Speed Body Motion in Water*, pp. 6/1-9.

Koryenna, L.I.; 1998, "Boundary Layer Control at Wave-Like Swimming; In: *Proceedings of the International Symposium on Seawater Drag Reduction, 22-23 July 1998*, pp. 257-259; Newport, Rhode Island.

Lighthill, J.; 1975, "Mathematical Biofluiddynamics". SIAM Reg. Conf. Ser. In Appl. Math. **17**.

Romanenko, Ye.V.; "1986, The theory of fish and dolphin swimming", Moscow, Nauka.

Schlichting, H.; 1974, "Theory of Boundary Layer"; Moscow, Nauka.

Two-phase flow processes in porous media producing geometric patterns

Insa Neuweiler and Wolfgang Kinzelbach
IHW, ETH Hoenggerberg, 8093 Zürich
neuweiler@ihw.baum.ethz.ch, kinzelbach@ihw.baum.ethz.ch

Abstract: In two-phase flow processes in a porous medium where one fluid displaces another immiscible one various patterns of the displacing front and the fluid configuration after breakthrough can be observed. Under certain conditions the displacement process is stable, leading to a regular and well predictable fluid distribution. However, very often the flow conditions lead to unstable displacement and irregular, complex shapes evolve. In this article, the nature and the development of the different shapes will be discussed.
Keywords: Two-phase flow, porous media, viscous fingering, DLA, invasion percolation

1. FLOW PROCESSES IN POROUS MEDIA

Flow processes which take place in a porous medium are usually very slow. We can therefore exclude turbulent flow behaviour caused by inertia forces. If we neglect buoyancy forces as well, the flux of a one-phase flow process is described by Darcy's law, which says that the flow velocity q is proportional to the driving pressure gradient ∇P. The proportionality constant includes the impact of the fluid properties as well as the medium properties,

$$\vec{q} = -\frac{K}{\mu}\vec{\nabla}P. \tag{1}$$

μ is the fluid viscosity and K is the permeability of the porous medium. K contains the impact of the pore space geometry and the soil properties and is on the large scale an empirical quantity, although approaches exist to derive the permeability of a medium from its pore scale properties (see e.g. [Bear, 1972]).

We consider a flow process involving two immiscible fluids. In case of independent flow of both fluids at steady state, where each fluid occupies its own part of the porous medium, we can consider the flow as two one-phase flow processes, each taking place in a medium with reduced pore space. If one fluid occupies more pore space, it has better accessibility to the medium and therefore a higher permeability.

A. Gyr et al. (eds.), Science and Art Symposium 2000, 191-200.

In this case the permeability is a function of the fraction of the accessible pore space which is occupied by the fluid, a quantity called saturation and denoted with S. We assume Darcy's law for each fluid separately

$$\vec{q_i} = -\frac{K(S_i)}{\mu_i}\vec{\nabla}P_i, \tag{2}$$

where i is an index for the fluid. The permeability is here no longer a constant but depends on the fluid saturation. The shape of this function has to be obtained empirically (e.g. [Bear, 1972]). In equilibrium, the pressure difference of the two fluids is given by the capillary pressure. On the pore scale this is the pressure difference of the fluids across the interface between them. It is given by Young's equation

$$P_c = P_{\text{nonwetting}} - P_{\text{wetting}} = \frac{\sigma \cos\phi}{r_1 + r_2}. \tag{3}$$

σ is the surface tension between the fluids, ϕ is the wetting angle and r_1 and r_2 are the radii of curvature. On the larger scale a similar kind of phenomenon is obtained. In this case, the capillary pressure is a monotonically decreasing function of the wetting phase fluid saturation

$$P_c = P_{\text{nonwetting}} - P_{\text{wetting}} = P_c(S). \tag{4}$$

For a displacement process, these considerations are no longer strictly applicable, however, the equations (2) and (4) are in general also applied to describe a displacement process. This is not justified theoretically, however the concept has proven to be adequate in many laboratory experiments (see e.g. [Marle, 1981]).

The flow process is determined by two forces, the viscous force which shows in the pressure gradient generated by the constant inflow. The second one is the capillary force, which acts in principle on the pore scale but yields an effect on the larger scale as well. Depending on the fluid properties and on the relation of these two forces, the displacement process can be stable, leading to a compact and regular pattern of the displacement front, or it can become unstable, leading to complex geometrical patterns (for reviews see e.g. [Homsy, 1987], [Feder and Jossang, 1995] or [Meakin, 1993]). The ratio of the two forces, pressure gradient and capillary forces, is quantified by the socalled capillary number N_{Ca}

$$N_{Ca} = \frac{q\mu_2}{\sigma \cos\phi}. \tag{5}$$

2 is the index for the displacing fluid. In this form the capillary number is defined on the pore scale from Young's equation, however equivalent expressions can also be found on the larger scale. The second dimensionless quantity which determines the characteristics of the flow process is given by the fluid properties. Despite their wetting properties, influencing the capillary pressure forces, the viscosities of the

fluids are important for the flow process. One can see from Darcy's law (2), that the more viscous fluid exerts a larger resistance to a pressure gradient than the less viscous one. In a single pore, where a sharp interface between the fluids exists and the permeability is constant, the number characterizing the situation is the viscosity ratio

$$M = \frac{\mu_2}{\mu_1}. \tag{6}$$

On a larger scale, where we have averaged over the pore space and the permeability depends on the saturation, the number is given by the mobility ratio

$$M = \frac{\mu_1 \, k_{r,2}(S)}{\mu_2 \, k_{r,1}(S)}. \tag{7}$$

In the following we will consider a displacement process, where one fluid displaces another immiscible one, in a rectangular two-dimensional porous medium. One fluid is injected with constant flow rate at one boundary of the medium. The opposite boundary is held at constant pressure of the displaced fluid phase.

2. VISCOUS DOMINATED FLOW

In the case where the capillary forces can be neglected in comparison to the viscous forces ($N_{Ca} \to \infty$), the displacement process is determined by viscous forces. The pressure gradient in the more viscous fluid is larger than that in the less viscous fluid. This effect is quantified by the mobility ratio.

The important mechanisms of the displacement can be classified by examining a displacement process in a Hele-Shaw cell, i.e. a cell consisting of two narrowly spaced parallel plates, which at the beginning is filled with one fluid. The fluid is displaced by another fluid, injected at one boundary. The flow can be described by Darcy's law, where the permeability is constant and depends on the width of the cell. A Hele-Shaw cell is therefore often considered as a simplified model for a porous medium. In our case, this is not exactly justified, as the flow in the Hele-Shaw cell is described by the movement of the interface between the two fluids. In a porous medium the interface will be very complex. On a larger scale, we will not see an interface, but rather a transistion zone, where the saturation of the displacing fluid changes from 1 to 0. However, the principal mechanisms can be explained with this example in a very simple way.

The flow process in a Hele-Shaw cell can be described by the movement of the interface line between the two fluids. Depending on the viscosity ratio the pressure gradient in one phase is larger than in the other. In the extreme case, we can neglect one pressure gradient compared to the other one ($M \to 0$ or $M \to \infty$) and consider the pressure in this phase as spatially constant. Initially the interface between the two fluids in the flow configuration considered here is a straight line which can be assumed to be aligned with the y-direction (compare Figure 1). We now assume a small disturbance of this interface at location $y = y_d$, where the interface moves a little faster compared to the rest of the line. In the case where the more viscous fluid displaces the less viscous one, this leads for a short time to a smaller pressure gradient on the line $y = y_d$. The velocity of the interface at $y = y_d$ is therefore a little slower than at the rest of the interface which leads to a reduction of the disturbance.

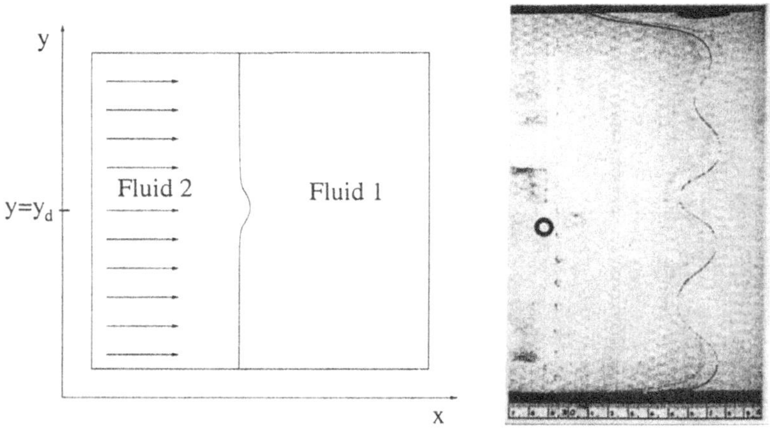

Figure 1; Displacement process in a Hele Shaw cell,
left: sketch, right: picture from an experiment by Park et al. [1984]

However, if the less viscous fluid displaces the more viscous one, the same disturbance leads to a higher velocity of the interface at $y = y_d$, as the pressures at the interface and at the outlet are assumed to be constant. The disturbance is therefore enhanced, and a finger of the interface develops. If the less viscous fluid displaces the more viscous one, the displacement process is inherently unstable. In the Hele-Shaw cell there are two mechanisms counteracting each other. On one hand there is the viscous force, which in the case $M < 1$ drives the fingering. On the other hand there is the capillary force, which tries to keep the interface between the two fluids as small as possible and damps the fingering. Whether the displacement process becomes unstable therefore depends on the capillary number as well as on the mobility ratio of the displacement process and can be analyzed in a stability analysis ([Saffman, 1986] and [Saffman and Taylor, 1958]).

In a porous medium the mechanisms are more complicated, however in principle the same mechanisms are important. The viscous forces in the case $M < 1$ lead to viscous fingering and the capillary forces counteract the fingering mechanism. In a porous medium there is an additional phenomenon, counteracting the fingering process: the transition zone of the saturation which has a stabilizing effect ([Yortsos and Huang, 1986]).

To model viscous unstable flow on the pore scale by solving the flow equation in each pore, the scale on which the instabilities evolve, is very difficult if not unfeasible. Network models, which are not based on the flow equations, have been established to model unstable flow phenomena. It has been stated by many authors, that viscous dominated flow can be simulated very well by diffusion limited aggregation (DLA), developed by [Witten and Sander, 1983] (e.g. [Trantham and Durnford, 1999], [Paterson, 1984] or [Maloy et al., 1987]). DLA describes a growth process of a cluster, which would in our example be the body of the displacing fluid. At the beginning of the growth process, there has to be an initial body of the fluid defined on a grid. The growth of the cluster is caused by particles, which start very far from the cluster and perform a random walk on the grid. As soon as the particle gets in touch with the displacing fluid cluster, the cluster grows by occupying the site of the particle. After this step the next particle starts its random walk. It has been shown that diffusion limited aggregation models the solution of a Laplace equation on a grid. It has therefore been successfully applied to processes described by the Laplace equation such as dielectric breakdown. The displacement of one fluid by another one in a Hele-Shaw cell can also be described by a Laplace equation if we assume that both fluids are incompressible ($\vec{\nabla} \cdot \vec{q}_i = 0$). However, in the case of flow in a porous medium, this is no longer true. It is not clear from a theoretical point of view, why DLA should be applicable to model viscous unstable flow in a porous medium ([Feder and Jossang, 1995]). However, there is some evidence that it is a suitable model (compare Figure 2)

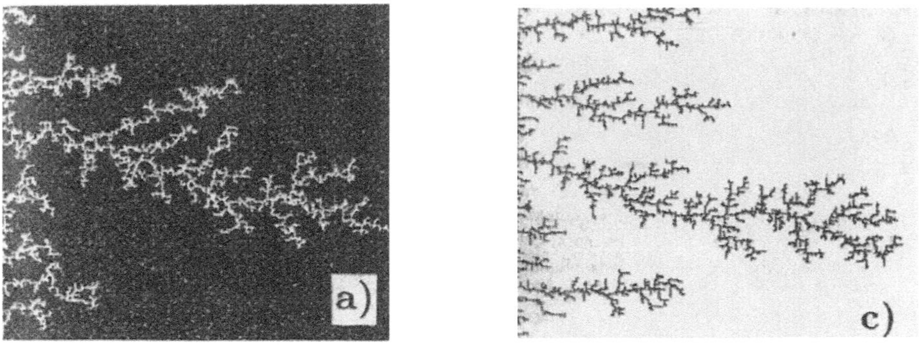

Figure 2; Displacement process dominated by viscous forces, left: laboratory experiment, right: DLA simulation, taken from Lenromand [1989]

3. CAPILLARY DOMINATED FLOW

In the case, where the flow process is so slow, that viscous forces can be neglected against capillary forces, the flow process in the porous medium can become unstable in a very different fashion. If the flow process is very slow, we can assume that the pressure gradients in each phase are equilibrated instantaneously and the pressure in each fluid is constant. The pressure in the displaced fluid is given by the pressure at the outlet. If the nonwetting fluid displaces the wetting one, the pressure in the displacing fluid is increasing due to compression caused by the inflowing fluid. No pore is invaded, as long as the pressure difference between the fluids is smaller than the smallest capillary pressure at the interface, resisting the invading fluid. As soon as the pressure difference exceeds the smallest capillary pressure, the corresponding pore is invaded, while the pressure in the invading fluid is decreased due to relaxation of the pressure. If the wetting phase displaces the nonwetting one, there is no resistance due to capillary pressure. However, if the flow process is determined by the capillary forces, the displacement will also take place pore by pore and the pore with the highest capillary pressure is invaded at each time step.

We can consider two extreme cases. On one hand, we assume that the pore size distribution is spatially homogeneous, but we have patterns on the large scale which consist of pores which are preferable for the invading fluid. Then the displacement will take place on these patterns and the cluster of the displacing fluid will reproduce the patterns. This can be seen in Figure 3 where the results of a laboratory experiment of displacement of water by air in an artificial porous medium, modeled as a Hele-Shaw cell with one rough surface, is shown ([Neuweiler, 1999]). On the left hand side, the aperture distribution is shown, and on the right hand side the distribution of water and air is given.

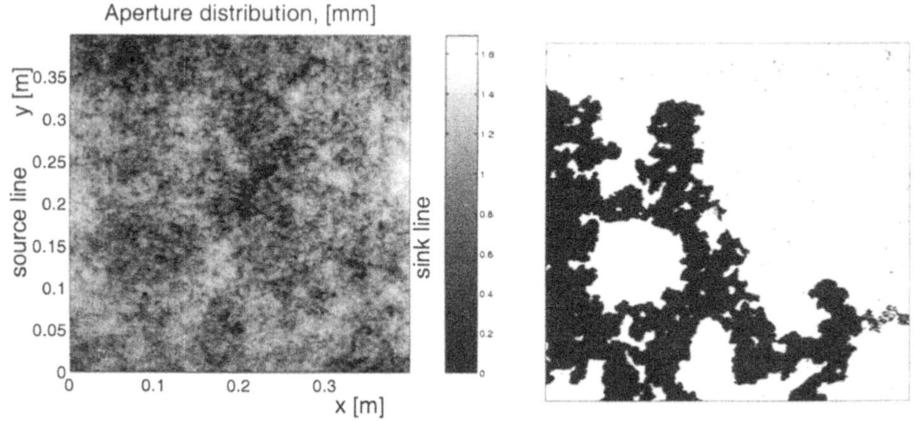

Figure 3; left: aperture field,
right: air cluster produced on this field by a displacement process, black is air

One can see clearly, that the air distribution is reproducing the structures which are already inherent in the porous medium. On the other hand we can consider the case where the pore size distribution does not show any patterns, which means the distribution is uncorrelated or has a correlation length which is small compared to the size of the medium. We would expect that the flow is stable and does not show any structures or patterns. That this is not the case can be seen in Figure 4, where the results of an experiment similar to the one above are shown for a porous medium with a correlation length which is 1/100 of the medium size ([Neuweiler, 1999]).

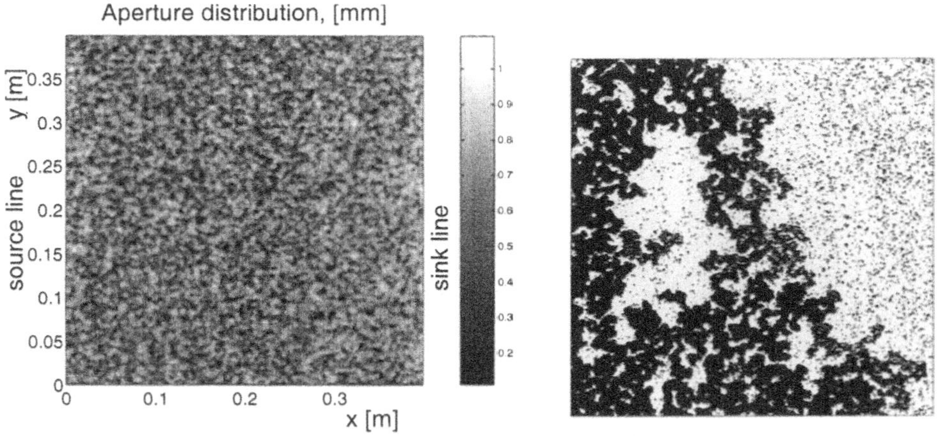

Figure 4; left: aperture field,
right: air cluster produced on this field by a displacement process, black is air

Although the pore size distribution does not show any patterns, the air body consists of clusters of very different sizes. This behaviour can be understood if we look again at a network model. It has been confirmed in many experiments ([Shaw, 1986], [Furuberg et al., 1988], [Glass and Yarrington, 1996], [Lenormand and Zarcone, 1989]) that invasion percolation (developed by [Wilkinson and Willemsen, 1983] and [Dias and Wilkinson, 1986]) is an appropriate model to describe a two-phase flow displacement process in the capillary dominated regime on the pore scale. Invasion percolation is a process defined on a grid consisting of sites and bonds connecting the sites. Random numbers are assigned either to the bonds or to the sites so that we speak of bond and site percolation respectively. The IP model which corresponds to the experiment described above is a site percolation model. In a site oriented invasion percolation process each site of a lattice has a probability that it can be invaded. This probability is interpreted as inverse resistance of the pore and for two-phase immiscible displacement is proportional to the capillary entry pressure. In general the probability is randomly distributed. At the beginning of the percolation process all sites of the lattice are occupied by the displaced fluid. Some sites are connected to a 'source'. Usually these are the sites at one of the boundaries. As the percolation begins, the site with the highest probability out of the set of sites that are connected to the source is occupied by the invading fluid. Its neighbouring sites now become connected to the source. This procedure is repeated until one site gets connected to the 'sink', usually the boundary opposite to the 'source'. The cluster of the invaded sites is called percolation cluster. Site clusters which are occupied by the displaced fluid and are surrounded by invaded sites or invaded sites and the boundary (excluding the 'sink'-boundary), are 'trapped' and can no longer be invaded. This mechanism reflects the incompressibility of the displaced fluid as it cannot be moved if there is no connection to the 'sink' where it can leave the grid. An inherent assumption is, that the pressure gradients never get high enough to move a whole cluster of trapped fluid. It is evident that invasion percolation with trapping describes the displacement process explained above (compare Figure 5).

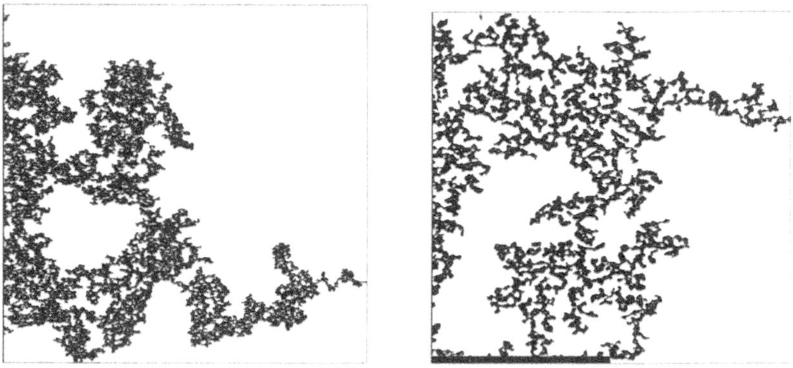

Figure 5; Cluster produced by invasion percolation ([Neuweiler, 1999]), left: field from Figure 3, right: field from Figure 4

One important property of the clusters produced by invasion percolation on an uncorrelated field is that these clusters are fractals (e.g. [Stauffer, 1992]). The fractal properties of percolation clusters in general and for invasion percolation clusters is discussed in a number of books and articles. For invasion percolation on a two-dimensional grid the fractal dimension has been determined by [Furuberg et al., 1988] or [Lenormand and Zarcone, 1985] as 1.82. Consequently, the concept of a representative elementary volume (REV), which is explained e.g. in [Bear, 1972] is not applicable for such flow processes, as has already been stated by [Lenormand and Zarcone, 1989]. The REV concept is based on the requirement that a variable, such as the fluid saturation, which has been averaged over a spatial volume, turns out to be constant and independent of the averaging volume if the averaging volume is larger than the REV. However, fractal structures are inherently nonstationary. Fractal clusters also have the property to include inner clusters of all sizes smaller than the total cluster size. This explains the existence of patterns on all sizes in the air cluster of the experiment described above. It is remarkable, that these patterns evolve although they are not caused by the pore structure of the porous medium but just by the displacement process itself. In a real porous medium it depends on the correlation structure whether fractal patterns can develop or whether the flow is taking place on structures provided by the porous medium itself.

The characteristics of the flow process are determined by the capillary number and the mobility ratio. In [Lenormand et al., 1988] a phase diagram of this dependency is proposed, which is based on numerical as well as laboratory experiments.

REFERENCES

Bear, J., 1972; *Dynamics of Fluids in Porous Media*, Dover Publications Inc., New York

Berkowitz, B., Ewing, R.P., 1998; "Percolation theory and network modeling applications in soil physics"; *Surveys in Geophysics (19)*, pp. 23-72

Dias, M.M., Wilkinson, D., 1986; "Percolation with trapping"; *Journal of Physics A 19*, pp. 3131-3146

Feder, J., Jossang, T., 1995; "Fractal patterns in porous media flow"; In: *Fractals in Petroleum Geology and Earth Processes*, ed.: C.C. Barton and P.R. LaPointe, pp. 179-226; Plenum Press, New York

Furuberg, L., Maloy, K.J., Feder, J., 1996; "Intermittend behaviour in slow drainage"; *Physical Review E (53)*, pp. 966-977

Furuberg, L., Feder, J., Aharony, A., Jossang, T., 1988; "Dynamics of invasion percolation"; *Physical Review Letters, 61(18)*, pp. 2117- 2120

Glass, R.J., Yarrington, L., 1996; "Simulation of gravity fingering in porous media using a modified invasion percolation model"; *Geoderma 70*, pp. 231-252

Homsy, G.M., 1987; "Viscous fingering in porous media"; *Ann. Rev. Fluid Mech. (19)*, pp. 271-311

Lenormand, R., 1989; "Applications of fractal concepts in petroleum engineering"; *Physica D 38*, pp. 230-234

Lenormand, R., Zarcone, C., 1989; "Capillary fingering: Percolation and fractal dimension"; *Transport in Porous Media 4*, pp. 599-612

Lenormand, R., Touboul, E., Zarone, C., 1988; "Numerical models and experiments on immiscible displacement in porous media"; *Journal of Fluid Mechanics, 189*, pp. 165-187

Marle, C.M., 1981; *Multiphase flow in porous media*; Editions Technip, Paris

Maloy, K.J., Boger, F., Feder, J., Jossang, T., Meakin, P., 1987; "Dynamics and structure of viscous fingers in porous media"; In: *Time-Dependent Effects in Disordered Materials*, ed.: R. Pynn and T. Riste, pp. 111-138; Plenum Press, NY

Meakin, P., 1993; "The growth of rough surfaces and interfaces"; *Physics Reports 235(4/5)*, pp. 189-289

Park, C.W., Gorell, S., Homsy, G.M., 1984; "Two-phase displacement in Hele-Shaw cells: experiments on viscously driven instabilities"; *Journal of Fluid Mechanics, 141*, pp. 257-287

Paterson, L., 1984; "Diffusion-limited aggregation and two-fluid displacements in porous media"; *Phys. Rev. Lett. 52*, pp. 1621-1624

Saffman, P.G., 1986; "Viscous fingering in Hele-Shaw cells"; *Journal of Fluid Mechanics 173*, pp. 73-94

Saffman, P.G., Taylor, F.R.S., 1958; "The penetration of a fluid into a porous medium or Hele-Shaw cell containing a more viscous fluid", In: *Proceedings of The Royal Society London A 245*, pp. 312-329

Shaw, T.M., 1986; "Movement of a drying front in a porous material"; In: *Mat. Res. Soc. Symposium Proc. (73)*, ed.: C. J. Brinker, D. E. Clark and D. R. Ulrich, pp. 215-223

Stauffer, D., 1992; *Introduction to percolation theory*, Taylor & Francis, London

Trantham, H., Durnford, D., 1999; "Stochastic aggregation model (SAM) for DNAPL-water displacement in porous media"; *Journal of Contaminant Hydrology, 36*, pp. 2076-2080

Wilkinson, D., Willemsen, J.F., 1983; "Invasion percolation: a new form of percolation theory"; *Journal of Physics A (16)*, pp. 3365-3376

Witten, T.A., Sander, L.M., 1983; "Diffusion limited aggregation", *Physical Review B, 27*, pp. 5686-5697

Yortsos, Y.C., Huang, A.B., 1986; "Linear stability analysis of immiscible displacement including continously changing mobility and capillary effects: Part I - simple basic flow profiles"; *SPE Reservoir Engineering*, pp. 378-390

Flow patterns in ballet and fluid dynamics

Tatyana S. Krasnopolskaya [1,2], ***Valeria Melechko*** [3]
[1] *Ukrainian Institute of the Environment and Resources*
13, Chokolivsky Blvd,
Kiev 01180, Ukraine
[2] *Eindhoven University of Technology,*
Gebouw Cascade, P.O. Box 513,
5600 MB Eindhoven, The Netherlands
tanya@kiev.phys.tue.nl
[3] *Microsoft Corporation, Multimedia Division,*
One Microsoft Way,
Redmond, WA 98052, USA
valmel@microsoft.com

Abstract: In this paper, we will discuss the similarities between patterns in fluid dynamics and ballet. Both fluid dynamics and ballet are concerned with flowing, continuous motion -- of fluid and of the human body, respectively. Furthermore, ballet scenes are often highly symmetric, as are cellular patterns in fluid dynamics. We present a series of images of flow patterns in a partially filled horizontally rotating cylinder (including two new cellular patterns), and of scenes from famous ballets such as "Swan Lake" and "The Nutcracker". We will also talk about multimedia and the use of high - tech tools in fluid dynamics and ballet, in creating and capturing the images.
Keywords: ballet, multimedia, flow patterns, partially filled horizontally rotating cylinder.

1. INTRODUCTION

In this paper, we intend to show patterns occurring in fluid dynamics and ballet and point out the similarities. The topic of our paper may seem unusual to the audience gathered at SCART 2000, and may thus require some explanation, as similarities between visual patterns in fluid mechanics and in ballet, as far as we know, have not been previously discussed. We will talk about how those similarities in patterns stem from the parallels between fluid dynamics and ballet.

We present experimental patterns in fluid inside a partially filled horizontally rotating cylinder. The experiment discussed here was inspired by the one described in [Thoroddsen and Mahadevan, 1997], with several important differences in experimental

A. Gyr et al. (eds.), Science and Art Symposium 2000, 201-210.

set-up and additional patterns and results, which will be noted below. One can read that excellent paper for a further review of literature on this subject.

The ballet scenes that are included here are taken from great classical ballets such as "Swan Lake", "The Nutcracker" and "Scheherazade" because those are some of our favourite ballets, and because we think the audience will be familiar with them.

2. BALLET AND FLUID DYNAMICS

We will start by discussing the parallels that exist between the science of fluid dynamics and the art of ballet. These will help explain the similarities between the patterns in fluid dynamics and ballet that will be presented later.

Both fluid mechanics and ballet are concerned with flowing, continuous motion - of fluids in one case and of human bodies in the other. This fluidity of motion often is what makes ballet pleasant to watch, even for audiences that are not necessarily experts in fine details of the dancers'technique.

One of the reasons for the fluidity of dancers' movements is the fact that the motions of dancers' bodies are guided by music -- and music is a wave phenomenon. Since there are innumerable similarities between music and fluid dynamics (that no doubt will be explored in some of the talks presented at SCART 2000), music thus gives us a direct link between ballet and fluid dynamics. As a rule, ballet aims to have perfect harmony between the music and dancers'motions, striving to express the music through dancers' bodies and thus producing patterns on stage similar to those produced by waves. According to Encyclopaedia Britannica, "the highest level of balletic achievement now requires the music and choreography to become extensions of each other, to be heard and seen on equal levels of perception"([Goodwin, 1999]).

Ballet seeks perfect harmony in yet another way - through symmetry. "The design of classical ballet is traditionally symmetrical in the shapes made by the dancers' bodies, in the grouping of the dancers on stage, and even in the structure of the whole dance", says Encyclopaedia Britannica ([Mackrell, 1999]). Symmetry is one of the cornerstones of classical ballet, and the structures of *pas-de-trois* (dance of three people together), *pas-de-quatre* (dance of four people), *pas-de-cinq* (dance of five), and so on, and the movements of the *corps de ballet* (large group of dancers, all performing the same motions, often in the background) follow it strictly. Encyclopaedia Britannica mentions that, "for example, if two principal dancers perform a *pas-de-deux* (a dance for two), other dancers on stage remain still, are arranged in symmetrical framing patterns, or move in harmony with the main dancing, not distracting from it. Even when large groups of dancers move, they are usually arranged in formal lines or circles."([Mackrell, 1999]). Symmetry and uniformity of execution of movements and poses is what makes a perfect *corps de ballet*, since, traditionally, the role of the *corps de ballet* was to serve as a "resonator" of the feelings of the main characters, and for that all the members of *corps de ballet* had to be in sync. Figure 1 presents a scene from the ballet "Scheherazade", by

M. Fokine, music by Rimsky-Korsakov, which demonstrates the symmetry in movements and poses of *corps de ballet*, here consisting of three dancers.

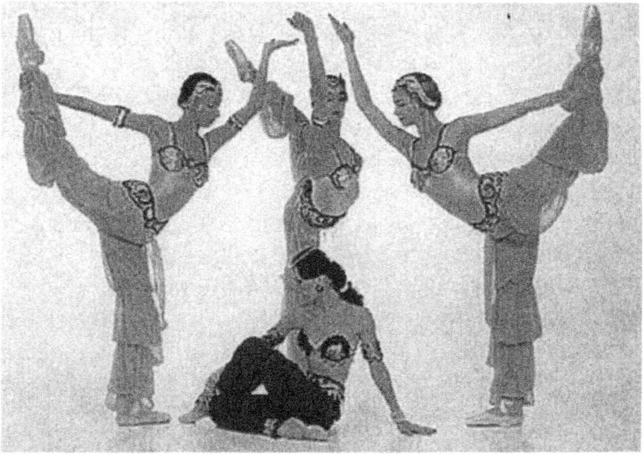

Figure 1; *Scene from "Scheherazade", illustrating symmetry in ballets.*

Cellular patterns formed in fluid dynamics experiments are also often symmetric. It happens in particular because the experimental set-up possesses a high degree of symmetry (in our case, it is a circular cylinder rotating around its axis). The symmetry is evident in experimental images (see, for example, Figure 4 on the color plate "Flow patterns in ballet and fluid dynamics"), just as it is in the scenes from ballets presented later in this paper (Figure 7). Different symmetries seen in ballets and our experiments will be discussed further in Section 6.1.

3. TECHNOLOGY AND MULTIMEDIA

Another parallel between ballet and fluid dynamics is the use of technology and multimedia: both fluid dynamics and ballet use technology tools to create and capture images. We believe this topic warrants special discussion, as it is the technological advances of today that enable us to do this analysis of patterns found in fluid dynamics and ballet.

In modern science, multimedia and technology, and especially computers, play a big role, from making computer simulations of mixing (for instance, [Krasnopolskaya et al., 1999]) to word processing and plotting graphs. As a further example, the experimental photographs of a partially filled horizontally rotating cylinder presented in this paper were taken with a digital camera. We believe the audience is familiar with the use of technology tools in the field of fluid dynamics, and will therefore move our discussion to technology in ballet.

Ballet may be the ultimate "multimedia" art form, us ing the co-operation of motions of human body, music, costumes, lighting, the stage, and stage decorations to convey its message and tell a story. It is now possible, through films, photographs and computer images, to store the wealth of choreography of past and present performances. This is evident in the scenes presented in this paper (Figures 1, 7, 8, 9), as the ballets from which they are taken were all created at least eighty years ago ("Swan Lake" and "The Nutcracker", for instance, are more than a hundred years old). Before, such information was only transmitted from person to person, from dancer to dancer, without any kind of permanent record. So it would be impossible for us to meaningfully discuss the topic of patterns in ballet and fluid dynamics, as we would lack all the visual aids! Ballet as an art form has embraced multimedia even further, as to the records of music, the drawings of costumes and stage decorations are now added the records of stage lighting and of motion through space and time. Moreover, ballet now uses technology not only in performance itself, but also in the creation process, as choreographers, composers and set designers use computer tools to try multiple variants and select the best one.

4. THE EXPERIMENT

4.1 Experimental Set-up.

The experimental apparatus is a modification of a device used by [Johnson, 1988], and later by [Zhang, 1995] and [Thoroddsen and Mahadevan, 1997]. The set-up consists of a circular Plexiglas cylinder (L = 43 cm, R = 5 cm), connected through its axis with a belt driver, which is in turn connected to a manually controlled variable speed motor. The motor can gradually change its angular velocity w from 0 rpm to 800 rpm (revolutions per minute). The angular velocity was determined directly by reading it off the display on the motor. Through a hole in the end side of the cylinder, the liquid was injected with a syringe.

Three different kinds of liquid were used in the experiment: ordinary tap water; demi water (distilled, without salts water, non-viscous, μ = 1 cP); and viscous fluid, made of mixtures of demi water and glycerin in various proportions. The viscosity thus varied from μ = 1 cP to μ = 35 cP. The viscosity and type of fluid is noted when a particular pattern is discussed. As mentioned in [Thoroddsen and Mahadevan, 1997], the surface tension was fairly independent of the mixture proportions.

The observed flow phenomena were independent of the direction of rotation. The volume filling fraction η was varied between 1.0% and 50%. The filling fraction is also noted for each pattern presented here.

4.2 Dimensionless Parameters

Several forces act on a small parcel of fluid in the middle horizontal cross-section of the cylinder: gravity, viscous forces, centrifugal force and surface tension. They are given by:

$$F_{gravity} = \rho g h L R \partial \theta \tag{4.1}$$

$$F_{viscous} = \mu \frac{wR}{h} L R \partial \theta \tag{4.2}$$

$$F_{centrifugal} = \rho w^2 R L R \partial \theta \tag{4.3}$$

$$K_{surfacetension} = \sigma_k \tag{4.4}$$

where R is the cylinder radius, L is cylinder length, ρ is the density of the fluid, μ is the viscosity, σ_k is the surface tension, w is the angular velocity, θ is the angle, g is the acceleration due to gravity and h is the mean film thickness. From this, one gets the following dimensionless numbers: the Reynolds number R_e, based on the mean film thickness and the cylinder velocity, obtained by dividing (4.3) by (4.2),

$$R_e = \frac{\rho w R h}{\mu} \tag{4.5}$$

the Froude number F_r, obtained by dividing (4.3) by (4.1),

$$F_r = \frac{w^2 R}{g} \tag{4.6}$$

and the capillary number C_a, obtained by dividing (4.2) by (4.4)

$$C_a = \frac{\mu w R}{\sigma} \tag{4.7}$$

As noted in [Thoroddsen and Mahadevan, 1997], the Reynolds number as defined in (4.5) "represents the ratio of the inertial to the viscous forces only when there is a pool at the bottom of the cylinder so that there is a velocity gradient inside the fluid. At large enough angular velocities, the fluid moves rigidly along with the boundary and there is no relative motion and the effective Reynolds number is zero."

5. EXPERIMENTAL IMAGES

In an ideal case, along the axis of the rotating cylinder there is no excitation, so there shouldnt be any symmetry breaking. But in reality, small imperfections in the experimental setup lead to the appearance of cellular structures (patterns) along the axis. The patterns presented in this paper were observed in all three types of fluids. It was noted that the angular velocity at which a particular pattern first appeared depended not only on μ, but also on η.

The following patterns were observed: six vortices on the fluid free surface, in demi water - glycerin mixture, μ = 35 cP, at w = 24 rpm, η = 50 % (Figure 2); "head of a snake", in demi water – glycerin mixture, μ = 35 cP, at w= 134 rpm, $\eta < 10\%$ (Figure 3); "shark teeth" along the liquid pool, in demi water - glycerin mixture, μ = 35 cP, at w = 124 rpm, $\eta < 10\%$ (Figure 4 on the color plate "Flow patterns in ballet and fluid dynamics"); "fish -like" patterns in demi water, at w = 511 rpm, $\eta > 6\%$ (Figure 5 on the color plate 'Flow patterns in ballet and fluid dynamics'); "curtains" in tap water, at w = 711 rpm, η > 6% (Figure 6 on the color plate "Flow patterns in ba llet and fluid dynamics").

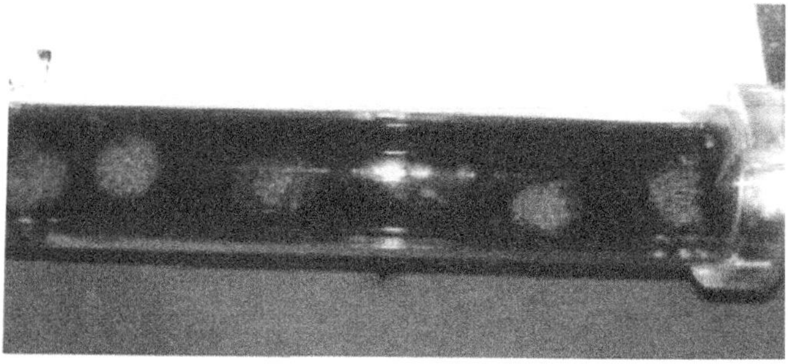

***Figure 2**; Six vortices.*

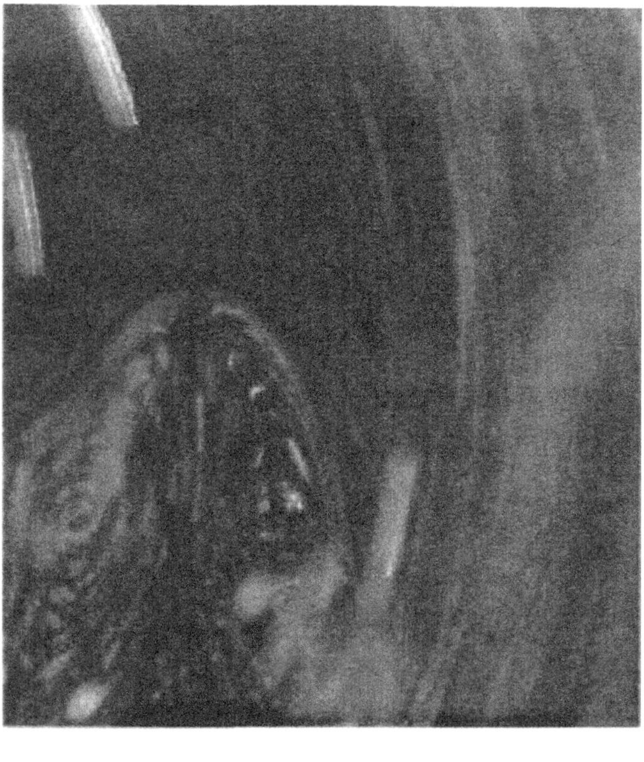

***Figure 3**; "Head of a snake" pattern.*

5.1 New results

Of these, two patterns, to the best of the authors' knowledge, have never been observed in similar experiments: six vortices (Figure 2) and "head of a snake" (Figure 3). They were obtained by adding two different types of particles, of bigger and smaller radius, to the demi water – glycerin mixture. The particles of bigger radius are rotating at the center of each vortex in Figure 2. The particles of smaller radius make up the "head of a snake" in Figure 3, while thick bright lines on that picture are produced by trajectories of the bigger particles.

6. COMPARISON WITH SCENES FROM BALLETS

If one then looks at the scenes from classical ballets, presented in Figure 7 - 9, one sees a striking visual similarity between them and the experimental photographs presented above. This similarity is particularly apparent between the following pairs of images, which will be discussed below in more detail: Figures 2 and 7; Figures 5 (on the color plate) and 8; and Figures 6 (on the color plate) and 9.

One of the reasons these pairs of pictures look similar is symmetry, as was mentioned in Section 2 -- even though they may not have the same symmetries, as defined below. Another reason is a prevalence of horizontal or vertical lines, as well as round shapes, in the particular pairs of pictures.

6.1 Symmetries

The most common symmetries present in a ballet are left-right translational, meaning that dancers on the left side of the stage do the same things as on the right side of the stage; front-back translational, meaning the dancers at the back of the stage follow the motions of the people at the front of the stage exactly; left-right reflectional, meaning the movements of the people on the left side of the stage are mirror images of those on the right side of the stage; diagonal reflectional, when the motions of the dancers are symmetric with respect to a plane whose line of intersection with the surface of the stage coincides with a stage diagonal; what may be called "even-odd reflectional" symmetry, where the movements of people in even-numbered lines of the corps de ballet (2, 4, 6, 8, etc.) are mirror images of those of people in odd-numbered lines (1, 3, 5, 7, etc.); and radial symmetry, when the dancers are moving in a circular or radial fashion.

Below, in pointing out the symmetries present in the patterns obtained experimentally, we will always assume we are looking at the cylinder from the side, seeing its entire length L in front of us. This is the best viewing position for making comparisons with ballet patterns, allowing us to use the definitions introduced above, since like in a cylinder with $L > R$ (as we have in our experiment), the side of the theatre stage that is facing the viewer is almost always longer than the perpendicular side.

6.2 Individual comparisons

The patterns presented in Figure 2 and Figure 7 have the same symmetries -- left-right translational and front-back translational. Another visual similarity between these pictures is that the vortices in Figure 2 are round in shape and vertical in orientation, and the dancers' round tutus are also seen here almost vertically. This roundness is underscored by how the dancers' outstretched arms on top visually "balance out" their straight legs on the bottom. Moreover, the center of each round fluid vortex in Figure 2 is a solid particle, and the center of each airy tutu, a costume that calls to mind the lightness of a snowflake, is a dancer's body – in both cases, the element serving as the center of the round shape is different from its immediate surroundings.

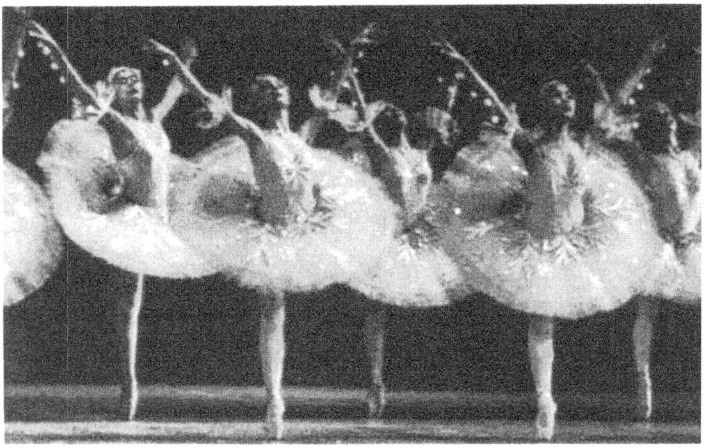

***Figure** 7; Dance of Snowflakes from 'The Nutcracker".*

What makes Figure 5 on the color plate "Flow patterns in ballet and fluid dynamics" and Figure 8 similar is the overall pattern of many small "wavelets" – "fish" in Figure 5, dancers' bodies and bowed heads in Figure 8. Also noteworthy is the horizontality in both cases, made particularly apparent in Figure 8 by the near-horizontal line of overlapping dancers' tutus. Moreover, "swans" arms and legs, extended behind them, make seemingly continuous, flowing lines that are very similar to the "fish" in Figure 5. Figure 5 has left-right translational and front-back translation symmetry, and Figure 8 has left-right reflectional symmetry.

Figure 6 on the color plate and Figure 9 look similar because of the prevalent verticality in both pictures – "curtains" versus dancers' raised and stretched out arms and legs. According to Encyclopaedia Britannica, "the long vertical line is [a] basic feature of ballet: verticality in the upright stance of the body, in the high leg extensions, and above all in the aerial quality of the movement" ([Mackrell, 1999]). Figure 6 has left-right translational symmetry, and Figure 9 has diagonal symmetry with respect to the diagonal of the "swans" (top right to bottom left corner of the stage) and with respect to the other stage diagonal (top left to bottom right corner).

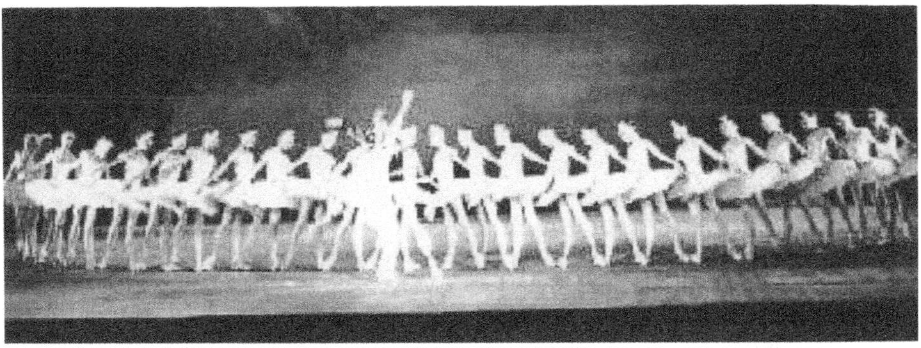

Figure 8; *Scene from "Swan Lake", Act 2.*

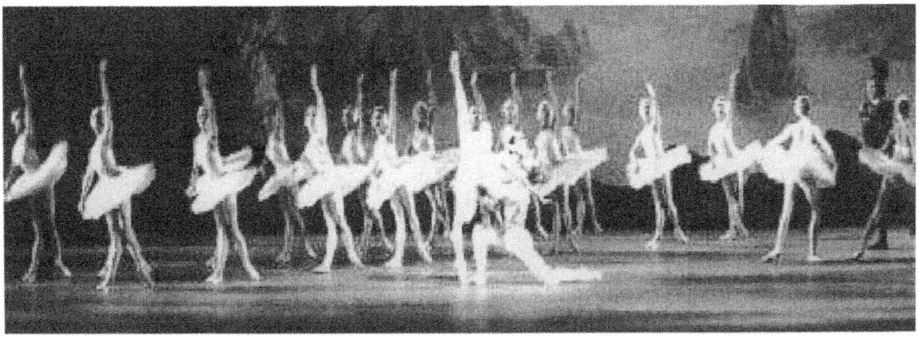

Figure 9; *Scene from "Swan Lake", Act 3.*

And, of course, all the images presented in this paper are also united by something else – their beauty!

7. CONCLUSION

In this paper, we have discussed the parallels between fluid dynamics and ballet. We have shown experimental photographs of a partially filled horizontally rotating cylinder, including two new cellular patterns obtained by adding solid particles to the fluid, and have compared them to scenes from classical ballets We hope the similarities between these give the audience at the Science and Art 2000 conference new insights into the world of science and art.

8. ACKNOWLEDGEMENTS

The experiment described in this paper was conducted during the stay of the first author at the Fluid Dynamics Laboratory, Eindhoven University of Technology (TUE), The Netherlands; this visit was supported by Eindhoven University of Technology. We are greatly indebted to Professor GertJan van Heijst for his hospitality and unfailing kindness in coming to our aid with suggestions and advice.

REFERENCES

Gordon, N.; 'Theatre music – Music for ballet'; In: *Encyclopaedia Britannica* (1999), online at www.britannica.com

Johnson, R.E; 'Steady -state coating flows inside a rotating horizontal cylinder'; In: *Journal of Fluid Mechanics* 190 (1988), pp. 321-332

Krasnopolskaya, T.S; Meleshko, V.V; Peters, G.M.W; Meijer, H.E.H.; "Mixing in Stokes flow in a wedge cavity"; In: *Eur. J. Mech. B/Fluids*, 1999, vol. 48, N 5, pp. 793-822

Mackrell, J.R.; 'Dance – Theatre dance – Ballet'; In: *Encyclopaedia Britannica* (1999), online at www.britannica.com

Thoroddsen, S.T.; Mahadevan, L.; 'Experimental study of coating flows in a partially-filled horizontally rotating cylinder'; In: *Experiments in Fluids* 23 (1997), pp.1-13

Zhang, H.B.; 'Selected topics in viscous flow: Coating flow with rotation'; PhD Dissertation, University of Illinois at Urbana-Champaign

Postcomputational Visualisation of Baroclinic Wave Drift

Valeri Goriatchev, Nikolay Ivanov, Evgueni Smirnov
Tver Technical University, St. Petersburg Technical University
A. Nikitina, 22, Tver, Russia, 170026
valery@tversu.ru

Abstract. Vortex related phenomena are difficult to interpret, and artist techniques are needed to express the artificial "frozen" life of simulated fluids and flames. Applications with swirled or twisted 3D spatiotemporal objects require various levels of detail rendering and visibility culling. The present contribution is aimed at a practice of up-to-date visualization techniques. Methods of data interpretation are discussed on the basis of results obtained by numerical simulation of low-Pr convection in a vertical rotating annulus.
Keywords: Computational Fluid Dynamics, swirl, stream visualization, baroclinic waves.

1. INTRODUCTION

Numerical simulation of complicated fluid flows is a rapidly developing branch of computer applications. Up-to-date simulations in 3D computational fluid dynamics (CFD) require sophisticated visualization tools, and other methods to facilitate the study of dynamic behavior of liquids and gases. In general these methods go back to the technique of Leonardo Da Vinci, who made sketches of paths of small bubbles or fine particles of sand dropped into water stream. This technique was developed in CFD visualization to analyze the velocity vector field using streamlines that can be extended to planar or spatial objects, such as ribbons and tubes.

The present paper covers artistic interpretation of data obtained via numerical simulation of unsteady buoyancy-driven convection of low-Prandtl-number fluid in a vertical rotating annulus. Visualization of circulating and/or twisting liquid structures in this flow requires wide spectrum of different approaches.

One advanced method is the technique that is developed in impressive artist manner. It can be used to intensify the researcher's attention to variations of such field characteristics as vorticity or variable scalar magnitude. The success of image perception of 3D swirled flow depends on the details of rendering: light spot gradient on the stream-tubes surface, colour of the pixel brush strokes, "shower-glass" effect, etc.

A. Gyr et al. (eds.), Science and Art Symposium 2000, 211-216.

2. LOW-PRANDTL-NUMBER CONVECTION IN ROTATING ANNULUS

Flow in a rotating cylindrical annulus has been a subject of considerable interest and was investigated in a great number of laboratory experiments as a basic model for a variety of buoyancy-driven motions in planetary and stellar atmospheres and liquid cores [Fowlis et al., 1965]. In most of the experiments liquids with moderate and high Prandtl numbers were used, the remarkable exception is a contribution [Fein et al., 1976], where a hydrodynamics of rotating annulus filled with mercury ($Pr = 0.0246$) was investigated. Convection in annuli filled with low-Pr-number liquid has become of great interest due to its importance for various applications, such as semiconductor crystal growth processes and planetary cores.

The present work deals with an annular ring of liquid confined between two concentric vertical cylinders and two flat rigid lids. The annulus rotates counter-clockwise at a constant rate about a vertical axis. Basic ratios are $R_o : R_i = 3:1$, $H : R_i = 2:1$, where R_o is the outer radius, R_i is the inner radius and H is the height of the annulus.

Convection in an annulus of given geometry is controlled by three dimensionless parameters: the Rayleigh number $Ra = g\,\beta\,\Delta T R_i^3 / \nu\, a$, the Rossby number $Ro = V_h / \omega\, R_i$, and the Prandtl number $Pr = \nu / a$. Here, g is the gravity acceleration, β is the thermal expansion coefficient, $\Delta T = T_{max} - T_i$, T_i is the inner cylinder temperature, ν is the kinematics viscosity, a is the thermal diffusivity, and $V_h = (g\beta\, \Delta T R_i)^{1/2}$ is the reference velocity.

Governing equations are the three-dimensional unsteady Navier-Stokes equations, written with the Boussinesq approximation in the rotating reference system, the continuity equation, and the energy equation.

No-slip conditions on the annulus walls are imposed. Axisymmetric wall temperature distributions with $\Theta = 0$ on the inner cylinder, linear variation over the bottom from $\Theta = 0$ to $\Theta = 0.5$ and over the outer cylinder from $\Theta = 0.5$ to $\Theta = 1$ are adopted. Here $\Theta = (T - T_i)/(T_{max} - T_i)$ is non-dimensional temperature. The upper rigid wall is treated as adiabatic one. For case presented in this study the determining parameters are $Ra = 3.38x10^4$, $Ro = 6.67$, $Pr = 0.015$.

The governing equations were solved by applying a numerical method based on the second-order finite-volume discretization with a cell-centered arrangement of all variables. An upwind second-order scheme was applied to compute convective parts of fluxes. The pseudo-compressibility approach was employed to introduce a pressure-velocity linkage through the continuity equation. To obtain a time-dependent solution, the pseudo-compressibility technique was used at every physical time step. Physical time discretization was carried out with a second order scheme.

The computations were performed with a recent version of well-validated Navier-Stokes code described elsewhere [Smirnov, 1993]. This code is a core part of the Hypermedia Information and Calculation System SELIGER [Goriatchev et al., 1997].

Time-dependent numerical solutions were obtained using two grids with nodes clustered to the wall to resolve higher gradients of the flow field in the near-wall layers. The first (coarse) grid consists of 61 ×21×21 nodes, and the second (fine) one has 121x41x41 nodes. For both runs, the initial fluid temperature was taken as the average of the minimum and maximum temperature of the annulus walls, whereas the initial velocity was set to zero.

Having used the coarse grid, a regular baroclinic wave structure develops after a transient stage. The structure consisting of six waves drifts slowly in the direction of the global rotation. The time of one complete revolution of the structure in the rotating reference system is approximately equal to $480\ R_i / V_h$. Analyzing the behavior of the fluid at a few monitoring points, one can detect also a weak temporal modulation of the major waves.

Computations with the fine grid produce irregular fluctuations of the fluid, and actual patterns of convection look extremely complicated. This is in qualitative agreement with the experimental observations [Fein et al., 1976] for a lot of the determining parameters. However, application of a filtering procedure to the data results again in a six-wave structure similar to that obtained with the coarse grid.

3. VISUALIZATION OF BAROCLINIC WAVES

The results of simulation were processed using the LEONARDO system, the graphics post-processor of the SELIGER. Some LEONARDO functions have been recently filtered in the manner of impressionist art [Goriatchev, 1997]. The last version of this software has been created on the OpenGL basis.

Details of convection were investigated using different mapping and some transparent matter visualization functions with imitation of dye streamlines and energy-diffused droplet trajectories.

Figure 1 shows distributions of representative variables over the computational domain. Levels of the Z - component of velocity are given for two vertical sections together with the mapping of shadow contours of the same variable onto the temperature iso-surface of $\Theta = 0.5$. Using similar approaches the CFD-user can analyze links between correlated variables effectively.

In Figure 2, pictures (1) and (2) display three levels of temperature and velocity (featured by arrow glyphs) in orthogonal sections of annulus and a scene of the precession of velocity Z-component iso-surfaces. The semi-transparent surface-rendering method was used to display iso-surfaces. It can be combined with glass filtration of velocity field intensity interpolated by shading on iso-surface. This method can add more fine local transparency and attractive by reinforce visualization that is illustrated by picture (3) in Figure 2.

Moving frozen-into-stream particles were used to visualize the fluid circulation. A path for each point was computed, maintaining a continuous tangent to the flow field. Individual points were identified by color-coding to distinguish vortex zones where different points could be separated or mixed.

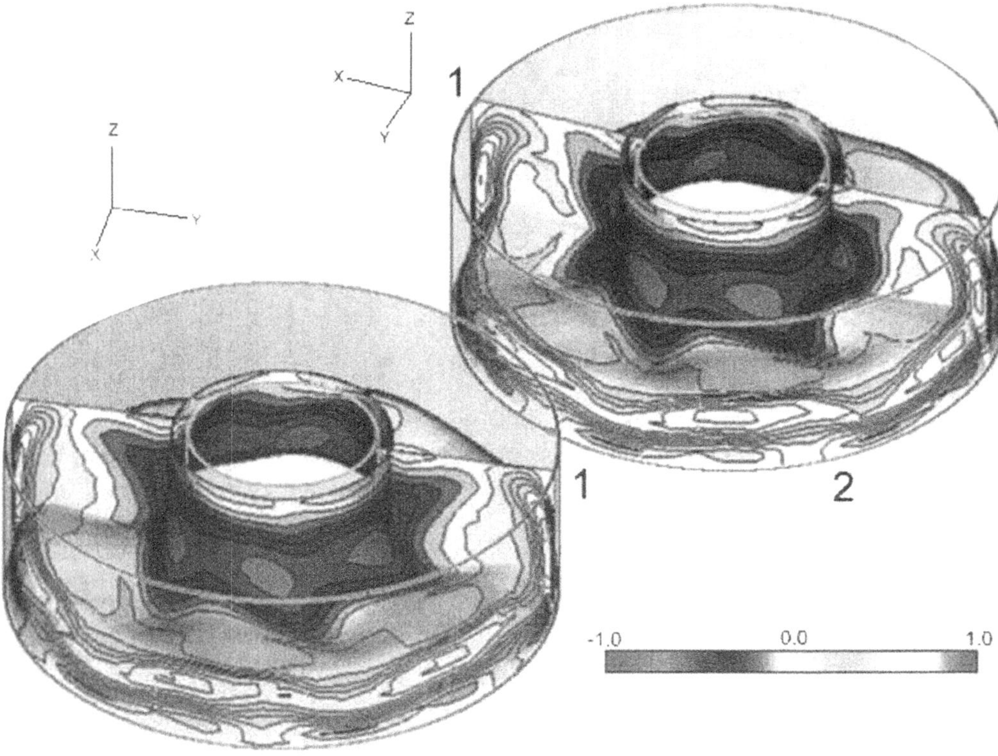

Figure 1. *Visualization of vertical velocity and temperature fields*

Levels of Z *- component of velocity are given:*
(1) for two meridional sections;
(2) mapped also onto the temperature iso-surface of $\Theta = 0.5$

Our experience with result data processing has shown that tangled flocs and balls of trajectories of weightless particles does not give a clear understanding of topology of pattern and should be completed by additional cutting of iso-surface volumes of processed values in the scene. Additionally, the streak lines technique was used to

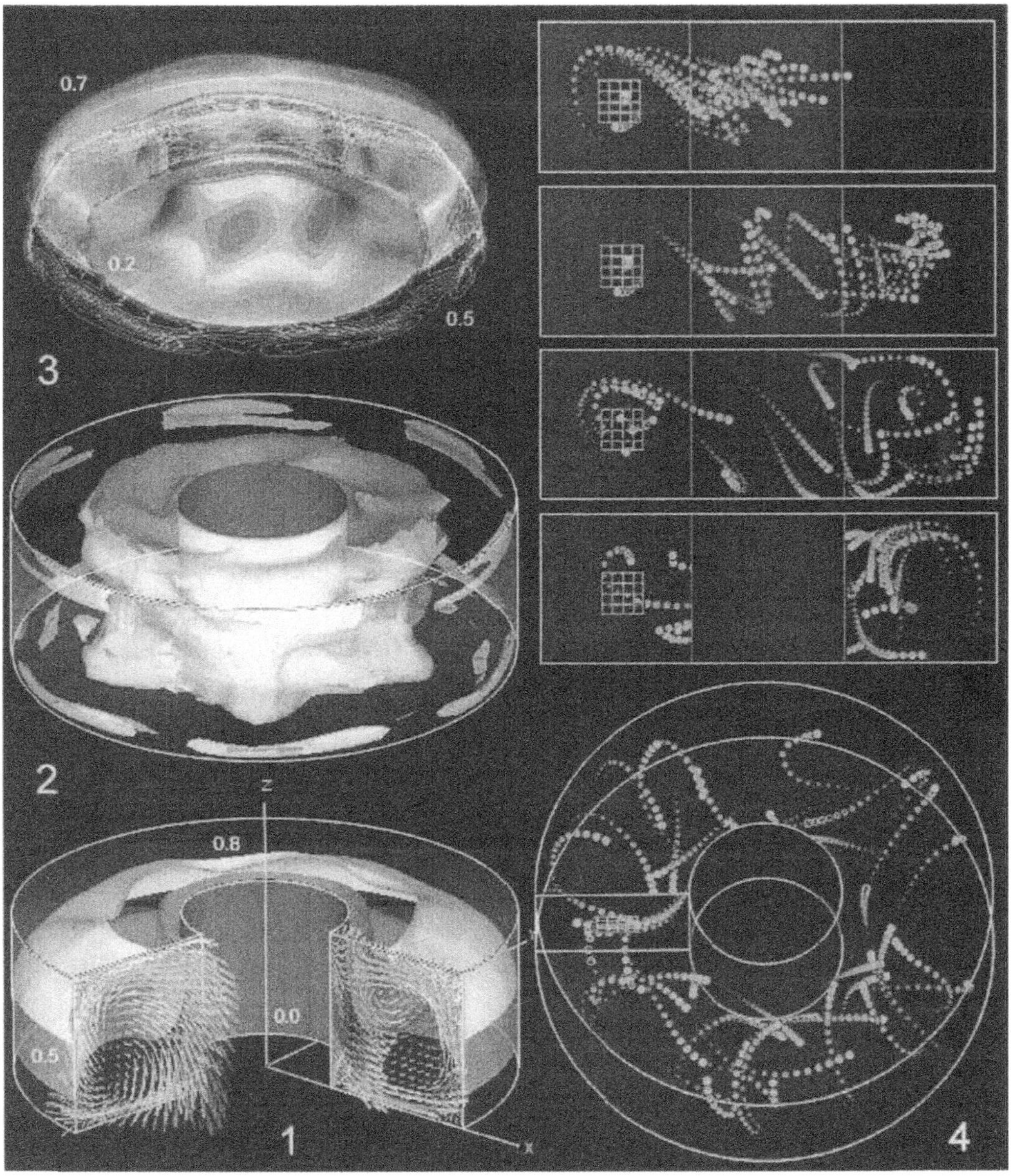

***Figure 2.** Frames of convection with baroclinic waves*

(1) temperature levels of $\Theta = 0.5,\ 0.8$ *and velocity distribution in meridional sections; (2) precession of the iso-surface correspondent to* -0.05 *level of* Z *- component of velocity; (3) filtered mapping of velocity shadow contours onto temperature iso-surface; (4) path traces of diffused drops inside the annulus*

represent the stream. Paths of diffused drops emanated from a discrete set of start points were used to animate the image of stream. Some frames of such movie are given in Figure 2 (4).

CONCLUSIONS

Over a wide range of determining parameters, buoyancy-driven convection in rotating annuli results in a rather complicated time-dependent behavior of velocity and temperature fields. A drift circulation with regular wave-structures may attract the researcher's attention not only as a mechanical process, but also as an art object.

Successful image perception of objects connected with fluid stream depends on as quality rendering of such objects, as well as on a reinforcement of various important fine details of stream. Additionally to getting esthetic feelings, use of various artist visualization tools with a painted look at the object image can help the researcher to better understand the process under simulation.

ACKNOWLEDGEMENTS

The work has been performed under grants # 15-GR-98 and # 96-17-7 of the Ministry of Education of the Russian Federation.

REFERENCES

Fowlis W.W., Hide R. "Thermal convection in a rotating annulus of liquid: effect of viscosity on the transition between axisymmetric and non axisymmetric flow regimes", In: *J. Atmos. Sci., Vol. 22*, pp.541-559, 1965

Fein J.S., Pfeffer R.L. "An experimental study of the effects of Prandtl number on thermal convection in a rotating, differentially heated cylindrical annulus of fluid", In: *J. Fluid. Mech., Vol. 75*, pp. 81-112, 1976

Smirnov E.M. "Numerical simulation of turbulent flow and energy loss in passages with strong curvature and rotation using a three-dimensional Navier-Stokes solver", *Report on the "Research in Brussels'92" grant*, Dept. Fluid Mech., VUB, p.101, 1993

Goriatchev V.D., Smirnov E.M. "A Computer and Information System for Computer Fluid Dynamics: SELIGER", In: *Proc. Of the 15th IMACS World Congress on Scientific Computation, Modelling and Applied Mathematics*, 09.24-29.1997, Berlin, Vol.3, pp. 47-53, 1997

Goriatchev V.D. "CFD art visualisation technique" In: *Book of abstract of the SCART 97 "Science & Art"*, 07.21-25.1997, Berlin, pp. 51-53, 1997

Art, History, Self-Organization and Coherent Structures in Fluid Mechanics

S.P.Bardakhanov, S.A.Kozlov*
Institute of Theoretical and Applied Mechanics, Novosibirsk 630090 RUSSIA
E-mail: bard@itam.nsc.ru

Abstract: The analogy of events of art and science are discussed from the point of view of self-organization. The examples of painting and photography art, nanotechnology, history of mankind and coherent structures' origination in fluids are used to show that the perception both in science and art is probably based on the aspiration to educe the organized structures from the chaotic behavior of the environment.

Keywords: synergetics, coherent structures, nano-science, history of mankind.

1. INTRODUCTION

The idea about unity of all processes in a nature is not new, and only the inability of the person to operate by complete system of the laws, determining processes in the world environmental us, forces the people to dismember all occurring on a separate science and areas of activity. At the same time the aspiration grows to comprehend the most general laws. In scientific activity one of stages of realization of this aspiration was onset of a science, named "Synergetics", about development of the order from chaos. In arts, having business with sense perception, from the prehistoric times each individual tried to realize a process of order origination from chaotic conditions, or on a background of the chaotic phenomena. It is possible to speak, that the efforts of the artists, musicians and other creative persons on an establishment of displays of harmony around us is other side of the same process of study of the order onset phenomenon, of which the scientists are involved.

The displays of harmony in a nature in different historic times and in various areas of human activity were titled differently. The authors of present paper think that at the given stage of science development the carrier of concept of the order is a concept of structures. It is obvious, that this concept is fundamental in a nature. And more, the concept of structure can be used for the processes' description of the whole human society. At the same time, it is hard to give the definition of structure, and in basic it is intuitive or associative. Most full the importance of structures, apparently, is realized in fluid mechanics, where the ordered formations, found out in various turbulent flows have received the name "coherent structures".

* - Co-author of Section 3

A. Gyr et al. (eds.), Science and Art Symposium 2000, 217-226.

In this paper the attempt is done, according to tasks of a conference, to present a some general sight on a problem of occurrence of organized structures, using the associative approach, on which the creation and recognition in art is based. For an illustration of a generality of the phenomena examples from art, history, results of experimental researches in area of nanostructures and on onset and development of coherent structures in some turbulent free shear flows are presented.

2. EXAMPLES OF SELF-ORGANIZATION OCCURRENCE

2.1 Arts

Famous painting "Black square" [Kazimir Malevich, 1915, 1990], which it seemed very easy to place here, is simultaneously both simple and complex object. On the one hand, the form of a masterpiece, created by the artist, is extreme simple, and on the first sight on it there is nothing except a square field of a rich black paint, inserted into a framework. Therefore it is possible to speak, that on it a symbol of initial chaos is represented, in which it seemed is completely absent any order. However by long its observation the sensation of order presence clearly arises, it is possible to tell more precisely, that the "anticipation" of order appears. So the Malevich's painting represents a vivid example of the simultaneous imagination of chaos and order, and we can examine "Black square" as a certain philosophical idea that the structures are everywhere.

As it was already marked at the previous conference, the liquid motion represents one of the most beneficial objects, allowing getting much creative achievements in art. By confirmation of this were a number of presentations at the Conference in Berlin.

A new set of pictures of well-known Siberian photoartist [Marina Svinina, 1999] also concerns to the sphere of art. One of picture, representing structures, arising on surfaces of water, one can see on a Figure 1. As it is a photo, and special methods of processing were not applied to it, the image represents a real picture of fluid flow. At the same time aesthetic effect in this picture presents, as well as in other photos of this author, illustrating the infinite variety of the forms of ordered structures in a liquid.

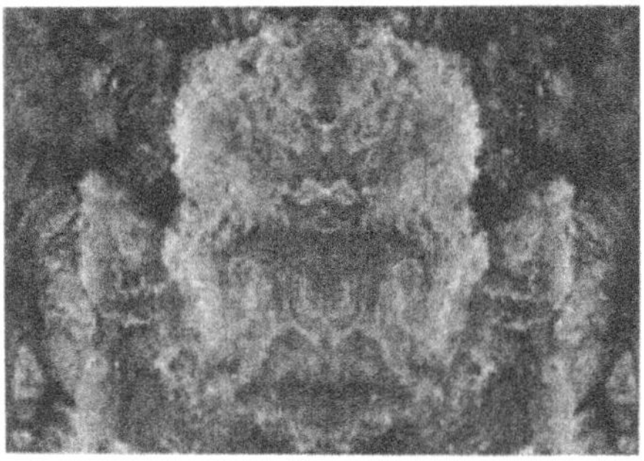

Figure 1. Photograph of the structures on the water surface [Marina Svinina, 1999].

The level of feeling of structure depends on scale. Let's assume that we have the image of a Figure 1 in computer representation, for example, in a format TIFF. If we magnify it, with each step of magnification the size of a dot will grow, and eventually the sensation of structure will be lost. If we did not know, that each magnified dot is a part of a structure on a Figure 1, we would treat a set of these dots as chaos, and the aesthetic aspect would disappear.

Such a thing happens with art subjects, in which with change of scale the aesthetic effect, and consequently, the "structure" of creation, disappears. However from the point of view of a science the order in it still exists, but already at the other level, having its new "aesthetic" structure.

2.2 Nanostructures

The last fact can be well illustrated by an example from nanostructures area, which is also close to one of the authors. On a Figure 2 photos made through on electronic microscope are presented. The purpose of a study was to investigate the structure of a fine powder of amorphous silica.

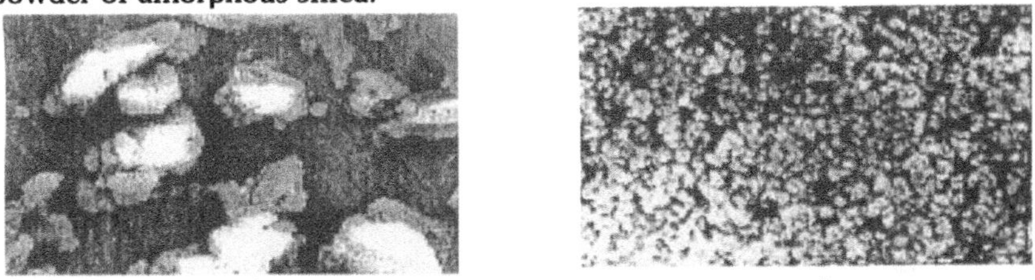

a) b)

Figure 2. Photos of a structure of amorphous silica powder. a) - Magnification 440, b) - magnification 10000.

If to speak about the scientific and technological aspect, the powder had been obtained as a result of evaporation process of solid natural quartz. After the subsequent condensation the particles fell out from high-temperature aerosol. For the evaporation the electron beam accelerator of continuous action, with system of focusing of a concentrated beam in an atmosphere with density of power outside of vacuum up to 5000 kW/sq.cm, was used. The installation allows to get the temperatures, sufficient for evaporation any known nowadays materials with rate of heating higher then 1000 degree per second. The powders, consisting of ultrafine primary particles with the sizes about 100 nanometer, have properties, essentially distinguished from properties of an initial material and allowing of unusual application [Bardakhanov, 1996a, b, 1997].

These fine powders are beautiful in themselves, and with use of pure raw material have white color. It is important, that process of particle formation out of high-temperature vapor consists of several stages, on which the structures of various scales are formed. In a photo with relatively low magnification (the Figure 2a) it is possible to see final conglomerates, which formed as a result of self-organizing with transition from a molecular level to agglomerates. In turn these structures consist of structures of the smaller size (Figure 2b). If first have the different shape, among second particles prevail mostly uniform ones, and they have the approximately identical sizes. The similar processes take place with evaporation of initial substance by others means, for example, with use the powerful CO_2 - laser [Bardakhanov et al., 1992]. From the point of view of aesthetic perception it is possible to speak, that by consideration of photos similar of given on Figure 2, as well as in a case with "Black square" the sensation of order also arises, though it is difficult to define in what it consists. If to proceed to fluid mechanics, the environment being initially a high-temperature vapor, then undergoes a number of stages of self-organization.

One more aesthetic effect in the given example is beauty of possible technological process [Lucashov, Bardakhanov et al., 1996], in which chemical transformations, emission of unpleasant substances and the other ecologically unpleasant phenomena are absent.

2.3. History

With use of concept of structures can be constructed the hypothetical theory of development of civilizations on a planet the Earth. In this case we will consider structures as the organized phenomena, which are displayed as waves. The basic moments here are following.

1. The whole history of mankind consists of history of wars.
2. Any war is displayed as more or less large movement of the peoples, in it participating.
3. The movements of the peoples are considered as waves, or fluctuations, which have a period.
4. The waves are originated by some source having earth or space origin, which set the periodicity in a movement of the peoples.
5. The sources are located in one or several primary places on a planet.

We will look now, how the known historical data correspond with the statements listed above.

By our opinion one of major places on the Earth from the point of view of this theory is the area of Central Asia, to which we assume the territory of modern Mongolia and its vicinities. Let's take in consideration an interval of time in 5 thousand years, which includes more or less known civilizations on our planet from the point of view of western history. We can see that the majority of the historical facts testify to powerful influences on them of the peoples, which came from their east borders. And all these were the nomadic peoples. So were with Ancient Egypt, State Shumer, Ancient Europe, under last one we assuming the area up to Ural Mountains.

On the other hand, data known us from a history of Asia show that the territories, not concerning to Central Asia, in particular of modern China, Korea, were also exposed to periodic intrusions from north (for China), from Northwest (for Korea), or from continent, that is from west for Japan. It is known that the territory of Siberia was always subjected to strong pressure from the south. The territories located to the south or to the east from above mentioned were also exposed to invasions, sources of which also were events in Central Asia.

In western history the most known such a wave was Great Resettlement of the Peoples, period, when the masses of nomads moved trough Europe, and these peoples served as the subsequent basis of modern European nations. From the most known peoples it is possible to mention Skiths, Huns, Goths, Turks, Mongolian peoples and others. However it is obvious, that number of other nomadic peoples, native land of which was Central Asia, has played a not smaller role in formation of a modern civilization. The simply historical science at the present has not discovered yet full enough this role, first of all due to insufficient studies in archaeology and written sources of these areas [Danilov (ed.), 1996]. In particular, only archaeology discoveries, which started one hundred years ago by Russian scientist Talko-Grintsevitch, have shown, that the civilization of nomads, were not wild horde of the supernumeraries on arena of a world history.

In our opinion in a documentary history, since the middle of first millennium BC the four "Great waves" happened of the most significant resettlements of the peoples. The first had been started by Hunnu people, about which it is known very little. However for Europe their role is known through others, for example, through Skiths' appearance in steppe of East Europe. It is doubtless, that in Asia they were widely known in time. The second wave has caused the breakdown of Roman Empire in IV-V centuries of our era [Golubtsova (ed.), 1988]. The third wave, also started from Asia, consisted of the movements of the Turk peoples in VII-IX centuries [Gumilev, 1993]. And at last, fourth - Mongolian conquests, the detailed scheme of which is given on a map in [Edwards, 1996]. A period of occurrence of these waves (or structures?!) it is possible to estimate by 500-800 years. The rather conditional scheme of propagation of these waves one can see on a Figure 3.

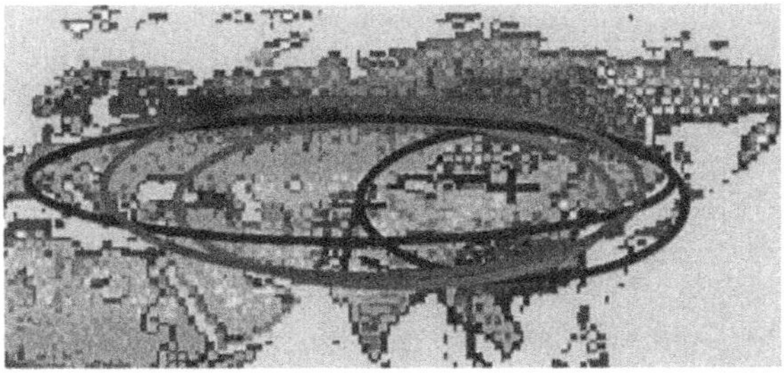

Figure 3. The scheme of Great Resettlements of the Peoples waves propagation. Black - Hunnu (V century BC), dark blue - Huns (IV century A.D, pink - Turks (IX century A.D), red - Mongols (14 century A.D).

Certainly, the real picture of events hardly will be constructed in a near future. Coming back to physics, it is possible to assume that approximately with such periodicity something occurred in Central Asia, what was a source of excitation of "structures". The discussion of a nature of this source falls outside the ranges of this paper. However, it is necessary to note that the specified area has a number of special features, displaying, for example, in the geological data. What concerns the dissemination of this theory to farther times, as far as we know, there are the proofs that "waves" also happened in so-called pre-historic times? For example, it was shown that the people, who come from Central Asia, occupied America from North. And the largest were the resettlements of 50 and 20 thousand years ago.

According to this speculative theory it is possible to apply the physical approach to a history and to simulate Great Resettlements of the Peoples by waves. Probably, during propagation of these "waves" the new structures - or new peoples were born? It is possible to name as a joke, but from this theory, for example, "geopolitical" consequences follow: If the whole modern mankind originated from one place on the Earth (and it is in accord with Bible!), whether it is not the time to stop the international conflicts?!

3. COHERENT STRUCTURES IN TURBULENT SHEAR FLOWS

We will proceed now from lofty reasoning to more real subjects, namely to onset and development of structures in turbulent shear flows.

For the fluid mechanics the attractive aspect of existence of ordered formations or coherent structures in turbulent flows is an opportunity of turbulent flow control, as the structures can serve as the objects, by means of a manipulation of which it is possible to control the whole flow. At the same time there are no many researches, in which the aspiration to realize knowledge of fine structure of turbulent flow in flow control is precisely traced. Apparently, it is connected, first, to difficulties of recognition of coherent structures, second, with necessity of a finding of effective ways of influence on them, and, third, with necessity to separate the contribution of changes in coherent structures in change of the mean flow.

We carried out researches of processes of onset and development of coherent structures and influence on them of the various excitation in different flows, for example, in the wakes of well flown bodies [Yanenko et al., 1984] and in swirling flows [Bardakhanov et al., 1995]. In the given work we will be limited by consideration of these processes in resonant conditions, meaning, that our basic task is representation of the data, having the relation to self-organization processes in the flow field.

The aeroacoustic resonance, for the first time described by [Parker, 1966], is an example of flow, in which the above-stated difficulties take place. However they can be overcome, as the coherent structures are well educed, it is possible effectively to influence them, and these influences essentially affect on the whole flow, hence, the coherent structures have essential meaning for the flow in whole. Two configurations of bluff bodies are examined. First - thick plate with rounded leading and trailing edge (the Figure 4a), the edges are parallel each other, with turbulent boundary layers on both sides of a plate.

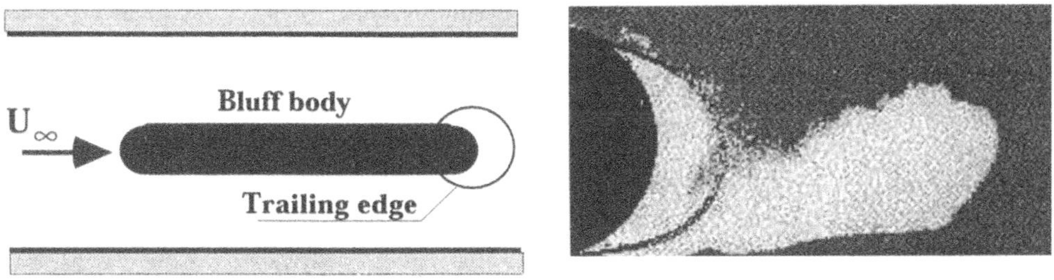

a) b)

Figure 4. Coherent structures in turbulent wake behind a thick plate. a) - Scheme of experiment, b) - image of structure in the region of trailing edge.

As up to relatively recent time the turbulent regime of flow was considered chaotic, it was supposed the absence of any order in the beginning of the wake, where the pretty large separation occurs. However the mechanism of self-organization is forcing the flow to generate the coherent structures, which are displayed, for example, with hot-wire measurements. At the same time it is possible to see their image with application of methods of visualization (Figure 4b) [Bardakhanov et al., 1987]. In this photo we see the section of a wake, but such a picture exists in all sections along trailing edge of a plate, and structures represent cylindrical tubes with concentrated vorticity. From the point of view of art we see some objects, which reflect fine structure of environment. It is hard to get any quantitative information from the image of these objects, however as well as in a case with "Black square" the sensation of the order arises.

At the same time the shape of structures can be obtained quantitatively. On a Figure 5a the results of computer visualization of flow in aeroacoustic resonance regime are presented as isolines of velocity pulsation.

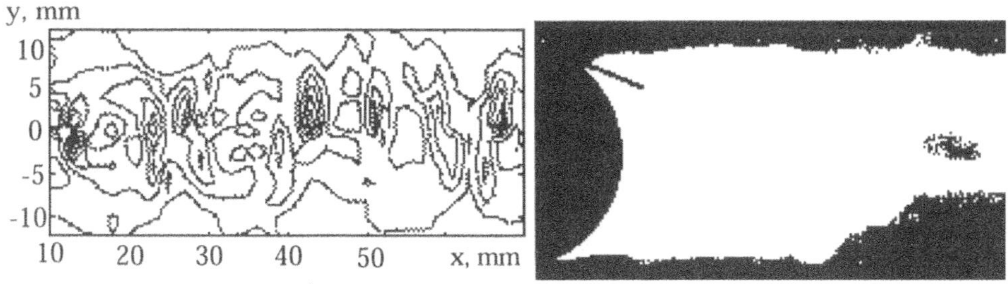

a) b)

Figure 5. a) Computer visualization of coherent structures in a wake behind the plate. b) Suppression of structures with simultaneous excitation from two acoustic sources with determined difference of phases between them.

The realizations of hot-wire signal were summarized with use of a signal from an acoustic microphone as the trigger [Bardakhanov et al., 1998]. Using in such a way we can examine structures in more details, and in principal we can continue the analysis of a degree of order at a level of smaller scales, by analogy with nanostructures. At the same time the aesthetic effect is kept and in it the mutual crossing of a science and art again can be stated.

However a kind of structures, and their existence, depends on external excitation. With the appropriate phase relation between sources influencing on flow it is possible completely to suppress the generation of structures (Figure 5b). It the certainly very simplified model, however in it the idea is presented that the external influences can cause the transition from the order to chaos.

The chaos can be intermittent with the order. Let's assume, that we study a wake behind a body, which is made as a cross of two plates [Bardakhanov et al., 1999], such as on a Figure 4a, but their thickness slightly differs. In this case wake behind each plate will generate structures with close frequencies. If the flow velocity will reach the range, in which the resonance occurs for the both plates, the structures in the wake will be modulated. The illustration of this regime as a trace of hot-wire signal is given on a Figure 6. It is possible to say, that the beatings are observed and at the certain moments of time and in certain points of space the structure are absent. The images of any other intermediate flow states, if they could be obtained, would probably have artistic meaning.

The study of the wake behind a circular disk [Bardakhanov et al., 1999], the section of which was identical to one represented at Figure 4a, have also shown the generation of coherent structures in this essentially three dimensional flow.

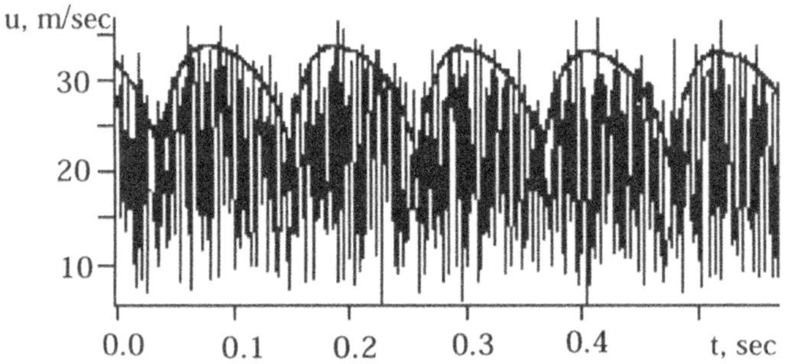

Figure 6. Modulation of velocity pulsations in coherent structures.

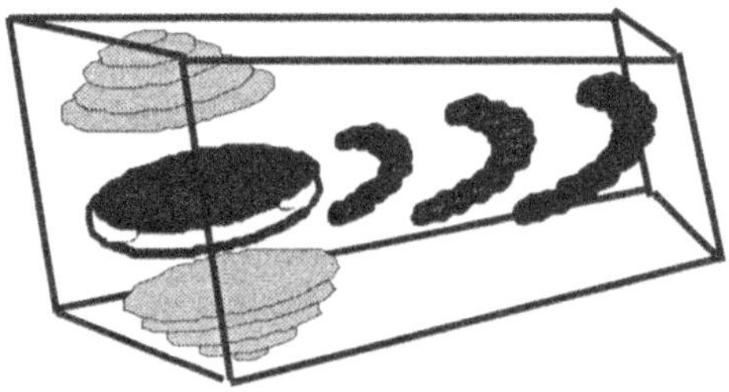

Figure 7. Coherent structures in the wake behind a thick circular disk at aeroacoustic resonance regime

But their frequencies were different in various regions along trailing edge. It was displayed that the flow was almost chaotic. However when the flow velocity achieved the value corresponding to one at which the capture of structure by eigen acoustic frequency of the channel was possible, the flow self-organization started. Structures began to shed with the same frequency (the acoustic oscillations differ on 180 ° at the upper and bottom volumes), and they looked like tubes and their axis were at the equal distances from the circular trailing edge of the disk (Figure 7). Whether sense of this scheme does not coincide with idea about Great Waves of Resettlements of the Peoples given on a Figure 3?!

4. CONCLUSIONS

In the present paper we tried to connect the various phenomena from the general point of view that the self-organization of environment is observed there and as a result the organized structures originate. In fluid mechanics such structures are named as "coherent". In turn the existence of structures and also the skill to distinguish them and to educe the order from chaos makes, in opinion of the authors, the basis of creative activity including the art.

5. REFERENCES

Bardakhanov, S.P.; 1996a "The formation of amorphous silica powder in quartz vaporization"; In: *Abstracts of Conference on "Chemical and Mechanical Processes in Composite Materials and Structures"*, pp. 178-180, Moscow *(in Russian)*

Bardakhanov, S.P.; 1996b "The formation of ultrafine amorphous silica at quartz vaporization process", In: *Program of Conference on "Ultrafine Powders, Materials and nanostructures"*, p. 8, Krasnoyarsk *(in Russian)*

Bardakhanov, S.P.; 1997 "The formation of fine silica powder after vaporization of quartz"; In: *Abstracts of V International Conference on "Computer-Aided Design of Advanced Materials and Technologies (CADAMT-97)"*, pp. 88-89, Baikal Lake

Bardakhanov, S.P.; Belay, O.V.; 1999 "Control of coherent structures shedding and generation of sound in bluff-body wake flow", In: *Proceedings of IUTAM Symposium on "Mechanics of Passive and Active Flow Control"*, pp. 223-228, Kluver Academic Publishers; ISBN 0-7923-5928-3

Bardakhanov, S.P.; Dandaron, G.-N.B.; Ivanchenko, A.I.; et al. (12 authors); 1992 "Amorphous silica"; *Report of Sc.-Tech. Com. Center "North of Russia"*, Novosibirsk *(in Russian)*

Bardakhanov, S.P.; Kozlov, S.A.; 1999 "Coherent structures and aeroacoustic resonance in flow behind coin-like disk"; In: *Abstracts of International Symposium "Actual Problems of Physical Hydroaerodynamics"*, p. II-16; Novosibirsk

Bardakhanov, S.P.; Kozlov, V.V.; 1987 "Onset and development of coherent structures in turbulent shear flows", In: *Perspectives in Turbulence Studies*, pp. 154-187, Springer-Verlag

Bardakhanov, S.P.; Kozlov, V.V.; 1995 "Control of instability waves development in separated swirling flows", In: *Proceedings of Royal Aeronautical Society Conference "High Lift and Separation Control"*, p.30.1-30.14; London

Danilov, S.V. (Ed.); 1996 "*Abstracts of International Conference "100 years of Huns Archaeology. Nomadism: Last, Future in Global Contest and in Historical Perspective. Huns Phenomenon"*, Part I, II, Ulan-Ude

Edwards, M.; 1996 "Genghis Khan", In: *National Geographic*, pp. 2-37, vol. 190, No. 6

Golubtsova, E.C. (Ed.); 1988 "*History of Europe*", volume 1, Moscow *(in Russian)*

Gumilev, L.N.; 1993 "*Ancient Turks*"; Moscow *(in Russian)*

Lucashov, V.P.; Bardakhanov, S.P.; Salimov, R.A.; Korchagin, A.I.; Fadeev, S.N.; Lavrukhin, A.V.; 1996 "A way to obtain the ultrafine silica powder, device for its realization and ultrafine silica powder"; Russia Patent No. 2067077

Malevich, Kazimir; 1915; *Tretiakov Gallery*, Moscow

Malevich, Kazimir; 1990; "Kazimir Malevich 1878-1935. The Exhibition at National Gallery of Art, September 16 - November 4, 1990"; *Publication of National Gallery of Art*, Washington

Parker, R.; 1966 "Resonance effects in wake shedding from parallel plates: some experimental observations", *Journal of Sound and Vibration*, vol. 4, No. 1, pp. 62-72

Svinina, M.; 1999 "Baikal Myths"; *Internet - http://ms.irk.ru/miff.htm*

Yanenko, N.N.; Bardakhanov, S.P.; Kozlov, V.V.; 1984 "Transformation of acoustic oscillations into the eddy ones in turbulent flows", In: *Proceedings of IUTAM Symposium on "Turbulence and Chaotic Phenomena in Fluids"*, pp. 427-432; North-Holland Publishers

An experimental and numerical investigation of the free vibration of an elastic cylinder in a cross-flow

R. M. C. So, Y. Liu and Y. Zhou
Department of Mechanical Engineering, The Hong Kong Polytechnic University
Hung Hom, Kowloon, Hong Kong
mmmcso@polyu.edu.hk

Abstract: This paper examines the flow-induced vibrations of an elastic cylinder in a cross-flow at two different reduced velocities, one near and one away from synchronization, and attempts a comparison between measurements and numerical calculations. It is the interaction between the flow and the motion of the cylinder that is the subject of this investigation. The present approach solves the flow field using a finite element method and models the cylinder motions using a two-degree-of-freedom structural dynamics model. An experiment to measure the vibration amplitude of the freely vibrating cylinder has been carried out at the same reduced velocity, damping factor and mass ratio as the numerical calculations. The measured drag, transverse displacements and their spectral characteristics are compared with the numerical results. Good agreement between measurements and calculations is obtained.
Keywords: flow-induced vibration, comparison, measurement, calculation.

1. INTRODUCTION

Elastic structures exposed to a cross-flow are found in a variety of engineering applications, including offshore marine platforms, tall buildings in an urban area, bridges, smokestacks and a wide variety of industrial engineering problems. These structures are subject to flow-induced vibrations due to vortex shedding, and possibly resonance at some critical speed where the vortex shedding frequency is approximately equal to the structural natural frequency. When that happens, the structure will undergo violent vibrations and possible failure as a result. An example is the collapse of the Tacoma Bridge during a violent windstorm. The cost associated with a typical failure, such as the Tacoma Bridge, could easily run into millions of dollars, not to mention the social disruption that the accident created. Thus, extensive efforts have been made to understand the physics of flow-induced vibrations and to use the knowledge thus gained to help design better structures. However, verifications of numerically calculated time series against experimental measurements have not been attempted before, partly because of numerical resolution and partly due to experimental difficulties (see e.g. Bearman 1995; Paidoussis 1998).

With the advent of computers and laser measurement techniques, it is now possible to resolve the flow around freely vibrating cylinders in a cross flow with accuracy

A. Gyr et al. (eds.), Science and Art Symposium 2000, 227-232.

(Zhou *et al.* 1999; So *et al.* 2000). Therefore, the present study proposes to compare the numerically and experimentally determined structural dynamics of a cylinder in a cross-flow. The objective is to verify the numerical approach, thus lending credence to the technique and help establish its credibility when applied to perform structural design analysis. This first attempt will focus on the free vibration problem only, leaving the forced vibration problem to a later investigation. Comparisons between the calculated and measured time series, their spectral characteristics, the natural frequency of the combined fluid-structure system and the shedding frequency are attempted.

2. NUMERICAL SIMULATION AND DATA ANALYSIS

The Navier-Stokes equations are solved using a finite element method and the non-linear coupling terms in the equations are treated separately, at different fractional time steps, by an operator splitting time stepping method. The method is suitable for both steady and transient problems and can readily be extended to include extra equations describing additional physical effects, such as the effects of the cylinder motions on the flow field and vice versa. The cylinder motions are accounted for by solving a two-degree-of-freedom dynamic equation

$$\ddot{\boldsymbol{Z}} + \frac{4\pi\zeta_s}{U_r}\dot{\boldsymbol{Z}} + \left(\frac{2\pi}{U_r}\right)^2 \boldsymbol{Z} = \frac{\boldsymbol{C}_f}{2M^*} \quad , \qquad (1)$$

where $\boldsymbol{Z}=(X/D)\boldsymbol{i} + (Y/D)\boldsymbol{j}$, X and Y denote the instantaneous displacements of the cylinder in the x- and y- directions, respectively, ζ_s is the dimensionless structural damping coefficient, $M^* = m/(\rho D^2)$ is the mass ratio, ρ is the air density, m is the mass per unit length, D is the diameter of the cylinder, $\boldsymbol{C_f}$ is the force coefficient, and U_r is the reduced velocity. The reduced velocity is an important parameter relating the structural vibration frequency to the characteristic length, D, and the free-stream velocity, U_∞. It is defined as $U_r = U_\infty / f_n^* D$, where f_n^* is the natural frequency of the stationary cylinder. The solution yields the vibration displacement and the velocity of the cylinder as they respond to the surrounding flow field. The incompressible flow solution procedure is coupled with the cylinder response, thus allowing the fluid-structure interactions to be adequately resolved at each time step.

Solving fluid-structure interaction problems generally involves moving computational domains and dynamic re-meshing. With this arrangement, the elastic cylinder is free to vibrate within the flow domain. Therefore, a deforming computational mesh is required to accommodate the arbitrary motion of the elastic cylinder. The cylinder surface is adjusted according to the motion of the cylinder by means of nodal displacement. In order to distribute the mesh deformation as uniformly as possible, i.e. to minimize local mesh deformation, a Laplacian equation of displacement is solved throughout the computational domain with the cylinder displacements as the boundary condition. The Laplacian equation and boundary condition can be written as $\nabla^2\boldsymbol{\delta} = 0$, where $\boldsymbol{\delta} = 0$ at the outside boundary and $\boldsymbol{\delta}|_w = \boldsymbol{Z}$ at the cylinder wall. The entire computational mesh is adjusted by a Laplacian interpolation, which is designed to map a mesh smoothly onto a reasonably similar shape, specified by the displacement of the elastic cylinder.

An auto-regressive moving average (ARMA) technique is used to analyze the calculated and measured time series. Details of the technique can be found in Mignolet and Red-Horse (1994). This allows the frequency contents of the signal to be examined in detail and the statistics to be determined with accuracy. The use of this technique to analyze time series derived from flow-induced vibration problems has been attempted by Jadic *et al.* (1998) and So *et al.* (1999) and their results were compared with those obtained from the Fast Fourier Transform method. The method was found to be most suitable for flow-induced vibration problems (Jadic *et al.* 1998). Therefore, the present study also used this technique to analyze the time series.

3. THE MEASUREMENTS

The measurements were taken from the experiments of So *et al.* (2000). All experiments were carried out in a suction-type wind tunnel. The length of the square test section (*0.35m* by *0.35m*) was *0.5m*. A circular cylinder with D = *6*mm was mounted in the mid-span of the test section at a location that was *20cm* downstream of the test section entrance. The cylinder was mounted to simulate a fixed-fixed end support. Thus, the cylinder deflection at the support is essentially zeroed. Various materials, such as acrylic, bronze, glass and steel, were used to fabricate the cylinder. In the actual experiments, the choice was made to match U_r and M^* chosen for the calculations or vice versa. The U_∞ could be varied from a low of about *2m/s* to a high of ~ *26m/s*. The corresponding Reynolds number $Re = U_\infty D/\nu$ varied from *800* to *9000*, where ν is the kinematic viscosity of air and M^* is estimated to be *445*. The free stream turbulence intensity was measured at 0.5%. A range of U_r was investigated, covering situations close to resonance to far away from resonance. In the present comparison, two cases were chosen, one was near resonance and another far away from resonance.

4. COMPARISON OF CALCULATIONS WITH MEASUREMENTS

Two cases are compared; they are the cases where U_r = *4.5* and *15.0* with the corresponding Re = *994* and *3340*, respectively. An acrylic cylinder with M^* = *445* is specified, and the reduced damping parameter $S_g = 8\pi^2 St^{*2} M^* \zeta_s$ is about *41*. Here, $St^* = f_s^* D/U_\infty$ is the Strouhal number of the stationary cylinder. Through out this paper, the asterisk in f and St is used to denote the value for the stationary cylinder.

A comparison of the calculated transverse displacement Y time series of the cylinder with the experimentally measured result in the stationary region of the signal for the case U_r = *4.5* is shown in Figure 1. The Y is made dimensionless by D, while the non-dimensional time τ is defined as tU_∞/D, where t is time. It can be seen that the two signals track each other rather well. The root mean square of the measured Y is Y_{rms}/D = *0.00036,* while the corresponding calculated value is ~ *0.00043*. This gives a discrepancy of more than 16%. The calculated $(E_Y)_c$ of Y/D is compared with the measured $(E_Y)_m$ in Figure 2. Here, E is the ARMA power spectral density, which is defined as the square norm of the system function scaled by the sampling interval and white noise variance (Marple 1987). Since all calculated variables are dimensionless, the E thus deduced is also dimensionless. The sampling interval is identical for both the calculation and the measurement. The agreement between the major frequencies is excellent, thus indicating a complete agreement in the vortex shedding frequency and the natural frequency of the combined fluid-cylinder system. At this U_r, it is apparent that synchronization occurs, because the natural frequency of the fluid-cylinder system f_n and the shedding frequency f_s are identical and there are no other peaks in the spectrum. Although the calculated $(Y_{rms})_c$ is higher than that of the measured one, the $(E_Y)_c$ is lower than $(E_Y)_m$. This is because the measured signal includes higher frequency components and these higher frequency components also contribute to the power spectral density. The difference between the measured and calculated E_Y at lower frequencies could be attributed to wind tunnel noise contamination. When a high past filter (60Hz) is applied to the measured time series, the power spectral density at low frequencies decreases significantly and moves closer to the calculated result.

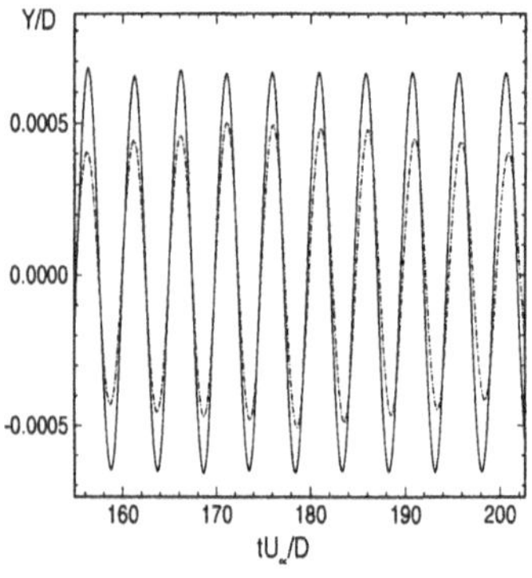

Figure 1. Comparison of the calculated and measured Y time series at $U_r = 4.5$; —— calculated; -·-·- measured.

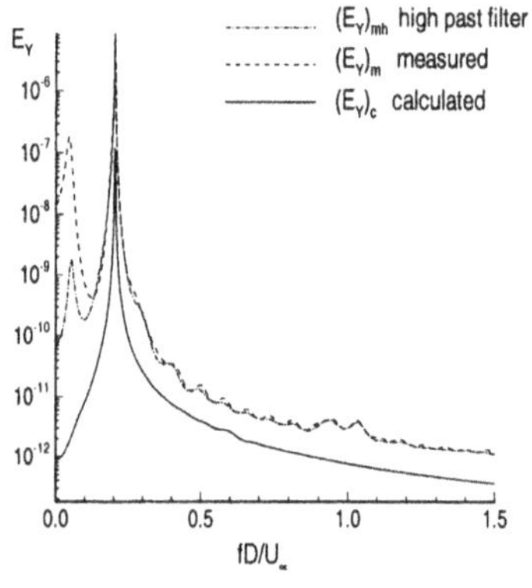

Figure 2. Comparison of the calculated and measured power spectral density E of Y/D at U_r = 4.5.

Comparisons of the calculated and measured Y obtained for the $U_r = 15$ case are shown in Figure 3. As expected, the two signals track each other quite well. The measured Y_{rms}/D is *0.00048*, while the corresponding calculated value is about *0.00052*. In this case, the discrepancy is only 7%, much smaller than that noted for the $U_r = 4.5$ case. The comparison of $(E_Y)_c$ with $(E_Y)_m$ is plotted in Figure 4. At this U_r, the cylinder vibrates at a frequency far away from f_s. The f_n and f_s determined from the calculated spectrum are coincident with those obtained from the measured spectrum (Table 1), thus, indicating that the numerical method is capable of resolving the fluid-structure interactions correctly, at least in the prediction of the major frequencies in the transverse displacement. Again, the discrepancies at lower frequencies could be attributed to contamination by wind tunnel noise. The effective damping ratio ζ_e, which is the sum of the structural damping ratio ζ_s and the fluid damping ratio ζ_f, is determined from the ARMA analysis. This calculated effective damping ratio ζ_e is in qualitative agreement with measurements (Table 1).

The time histories of the drag and lift coefficients, C_D and C_L, for the $U_r = 4.5$ case are given in Figure 5, while Figure 6 shows their corresponding power spectral densities. The dominant drag frequency is *0.4054*, which is exactly twice the f_s determined from the lift time series (Table 1). As shown in Table 1, the agreement between the calculated and measured mean drag coefficient is excellent. In the $U_r = 4.5$ case, the discrepancy is less than 5%, while in the $U_r = 15$ case, the discrepancy is only 1.3%. This is indicative of the accuracy of the calculation method.

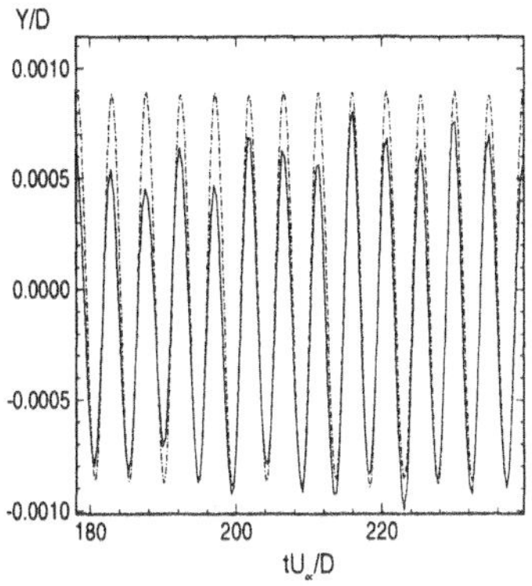

Figure 3. Comparison of the calculated and measured Y time series *at* $U_r = 15$; —— calculated, ---- measured.

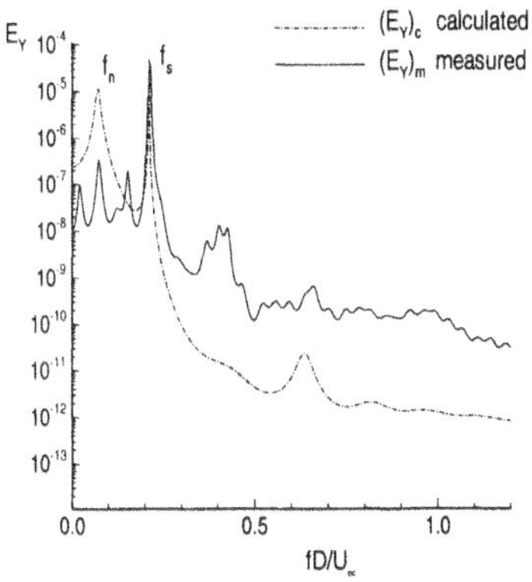

Figure 4. Comparison of the calculated and measured power spectral density E of Y/D at $U_r=15$.

Table 1. Comparison of the calculated and measured statistics.

	$U_r = 4.5$ ($Re = 994$)		$U_r = 15$ ($Re = 3440$)	
	Calculation	Experiment	Calculation	Experiment
Y_{rms}/D	0.00043	0.00036	0.00052	0.00048
X_{rms}/D	0.00002		0.000025	
$(C_D)_{mean}$	1.07	1.02	0.968	0.955
$(C_D)_{rms}$	0.064		0.092	
$(C_L)_{rms}$	0.67		0.71	
f_n	0.2027	0.2027	0.075	0.075
f_s	0.2027	0.2027	0.2117	0.2102
$3f_s$			0.6366	0.6533
ζ_e	0.00114	0.008	0.0369	0.0314

5. CONCLUSIONS

A comparison of the calculated and measured flow-induced vibration of a circular cylinder in a cross-flow has been attempted. The results show that good agreement has been obtained between the *Y* time series, its root mean square amplitude, the mean drag, the effective damping ratio, the shedding frequency and the natural frequency of the combined fluid-cylinder system. The good agreement applies to the resonance case as well as to the off-resonance case, thus, lending credence to the numerical analysis.

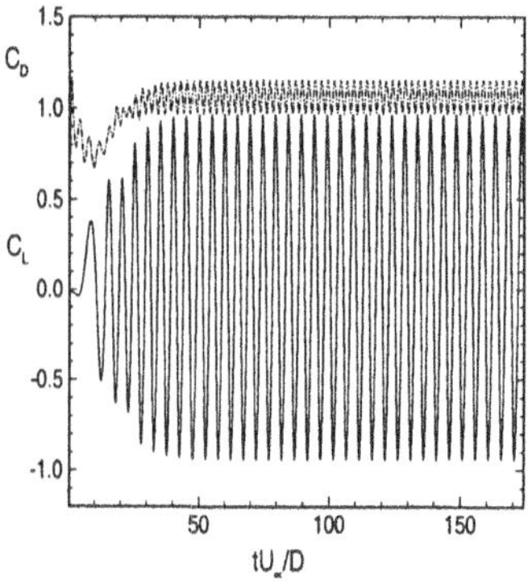

Figure 5. Calculated lift and drag coefficient time series at $U_r = 4.5$; - - - C_D ; —— C_L.

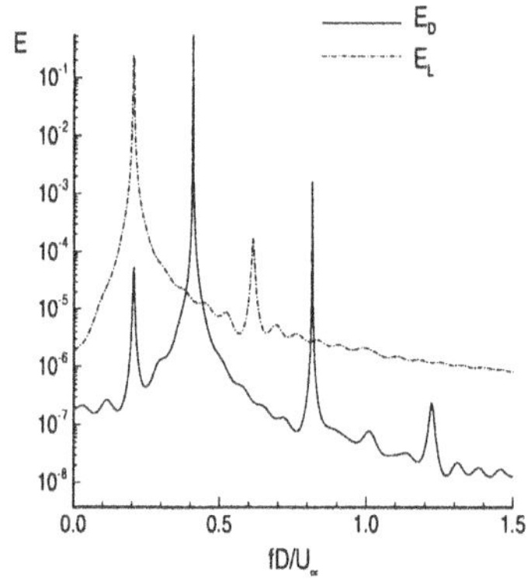

Figure 6. Spectral analysis of the calculated lift and drag coefficient at $U_r = 4.5$.

6. ACKNOWLEDGEMENTS

Support given by The Hong Kong Polytechnic University under Central Research Grant No. G-V396 and by the Research Grants Council of the Government of the HKSAR under Grant No. PolyU 5159/97E is gratefully acknowledged.

7. REFERENCES

Bearman, P. W. 1995 *Flow-Induced Vibration*. Proceedings of the Sixth International Conference on Flow-Induced Vibration, London, UK, 10-12 April 1995, Published by A. A. Balkema, Rotterdam.

Mignolet, M. P. and Red-Horse, J. R. 1994 ARMAX Identification of Vibrating Structures: Model and Model Order Determination. *Proceedings of the 35th Structures, Structural Dynamics, and Materials* Conference, Hilton Head, South Carolina, April 18-20, 1628-1637.

Paidoussis, M. P. 1998 *Fluid-Structure Interactions – Slender Structures and Axial Flow*. Vol. 1, Academic Press, London.

Jadic, I., So, R. M. C. and Mignolet, M. P. 1998 Analysis of Fluid-Structure Interactions Using a Time Marching Technique. *Journal of Fluids and Structures* **12**, 631-654.

Marple, S. L. Jr. 1987 *Digital Spectral Analysis with Applications*. Prentice Hall, NJ.

So, R.M.C., Jadic, I. And Mignolet, M.P. 1999 Fluid-Structure Resonance Produced by Oncoming Alternating Vortices. *Journal of Fluids and Structures* **13**, 519-548.

So, R.M.C., Zhou, Y. and Liu, M.H. 2000 Free vibrations of an Elastic Cylinder in a Cross Flow and their Effects on the Near Wake. *Experiments in Fluids*, to appear.

Zhou, C. Y., So, R. M. C. and Lam, K. 1999 Vortex-Induced Vibrations of Elastic Circular Cylinders. *Journal of Fluids and Structures* **13**, 165-189.

Elongational Behaviour of Dilute Polymer Solutions

Bernhard Gampert, Thomas Eich, Christoph Wilkes
University of Essen, Applied Mechanics
Schuetzenbahn 70, D-45127 Essen, Germany
bernhard.gampert@uni-essen.de

Abstract: Experimental results concerned with the elongational behaviour of dilute polymer solutions were presented. The elongational viscosity was measured as a function of the strain rate for an aqueous solution of the neutral polymer PAA-T19 and for an aqueous 0.2 mol/l NaCl solution of the ionic polymer Sedipur AF 701. Flow induced birefringence was monitored simultaneously in order to assess the orientation of the polymer molecules. A critical strain rate $\dot{\varepsilon}^*$ depending on the type of polymer solution separates two flow regions: a Newtonian region of strain independent elongational viscosity and a region with markedly and continuously increasing elongational viscosity. Calculations applying the FENE dumbbell model were shown to be in good accordance with the experimental results of the non-ionic solution. Experimental data for the pipe friction factor λ as a function of the *Re* number are presented for both solutions and are discussed in connection with the elongational viscosity results.
Keywords: elongational viscosity, opposed jets, drag reduction, dilute polymer solutions

1. INTRODUCTION

Dilute polymer solutions are particularly interesting for fundamental research because in a systematic study of the rheology of polymers they are the system adjacent to the Newtonian flow. Moreover such studies are of practical interest because the behaviour of highly dilute polymer solutions already varies greatly from that of Newtonian fluids for flows that possess a predominantly elongational character. Examples of this are laminar porous media flow and turbulent pipe flow.

In laminar porous media flow, Durst and Haas, 1981, the flow resistance increases markedly compared to the pure solvent beyond a critical flow rate. In the literature, Hinch, 1977, it is suggested that the extending nature of such flows could produce considerable stretching of the random-coil molecules resulting in a strong enhancement of the elongational viscosity from which follows an anomalous large increase in fluid drag.

In turbulent pipe flow of dilute polymer solutions the drag is strongly reduced in comparison to the drag of the Newtonian solvent, Gyr, 1990, Virk, 1985. In the literature it is argued that in turbulent flow fields strong extension rates appear

A. Gyr et al. (eds.), Science and Art Symposium 2000, 233-238.

and that by the stretching of molecules and the significant increase in elongational viscosity which results the production of turbulence is diminished. Accordingly the fluid drag is decreased. The altered turbulence structure linked with the reduced drag varies significantly in highly dilute and in transient network flow, Gampert and Rensch, 1996.

Explanations of the flow behaviour of polymer solutions remain hypothetical as long as the elongational flow of these solutions cannot be better understood. Up to now only a few measurements of the elongational viscosity for polymer solutions of low concentration have been published in the literature. Therefore an experimental apparatus was developed for measuring the elongational viscosity. Some of the results obtained for dilute polymer solutions are presented in this paper.

2. EXPERIMENTAL

In this laboratory a measurement system has been developed recently by which the elongational viscosity can be determined as a function of the elongation rate using the well known opposed jets principle, Cathey and Fuller, 1988, Keller et al., 1987. An optical system was applied with which flow induced birefringence was monitored simultaneously in order to assess the stretching and the orientation of the polymer molecules, Eich, 1999.

Controlled elongation rates can be achieved when the polymer solution is ejected simultaneously from two jets that face each other with a small separation. Such a system approximates pure hyperbolic flow with a stagnation point at its centre of symmetry.

The apparent elongational viscosity $\eta_{E,A}$ was calculated from the flow rate and the pressure drop measured across the jets. The pressure drop had to be corrected for the Bernoulli effect.

Aqueous solutions of ionic and non-ionic polyacrylamides were investigated. As the ionic polymer, Sedipur AF 701 supplied by the company BASF AG, was chosen. Sedipur AF 701 is 70% hydrolised, anionic and has a very high average molar mass of $\bar{M}_w = 14 \cdot 10^6$ g/mol. Sedipur AF 701 was dissolved in highly pure water. Sodium chloride with a concentration of $c_s = 0.2$ mol/l was added in order to partially neutralize the charges of the strongly ionic Sedipur AF 701.

The neutral polymer PAA-T19 was polymerised in this laboratory. It is highly pure, non-ionic, threadlike and has an average molar mass of $\bar{M}_w = 5.6 \cdot 10^6$ g/mol and a relatively low dispersivity, $\delta = 1.9$, see Eich, 1999.

The conventionally calculated critical value c_p^* for the upper concentration limit in the particle range or range of dilute polymer solutions based upon the close packing of coils depends upon the polymer solvent system. For the aqueous solution of PAA-T19 the critical concentration was $c_p^* = 800$ ppm and for the aqueous 0.2 mol/l NaCl solution of Sedipur AF 701 $c_p^* = 170$ ppm was obtained.

The stretching of polymer molecules in strong elongational flow fields might result in interactions between the polymer molecules appearing for concentrations

c_p below the conventionally defined critical concentration c_p^*.

The concentrations for the elongational viscosity measurements in this paper were $c_p = 250$ ppm for PAA-T19 and $c_p = 4$ ppm for Sedipur AF 701 and thus significantly below the critical concentrations. The shear viscosity measurements were carried out using the rotational viscometer Low Shear 30 from the company Contraves.

3. RESULTS AND DISCUSSION

Figure 1 shows the shear viscosity for shear rates of $\dot{\gamma} \leq 50\ s^{-1}$. For the 4 ppm Sedipur AF 701 solution the viscosity is almost indistinguishable from water and for the 250 ppm PAA-T19 solution only slightly increased in comparison to water. This means that also for higher shear rates of $\dot{\gamma} > 50\ s^{-1}$ the Sedipur AF 701 solution behaves Newtonian and the PAA-T19 solution nearly Newtonian.

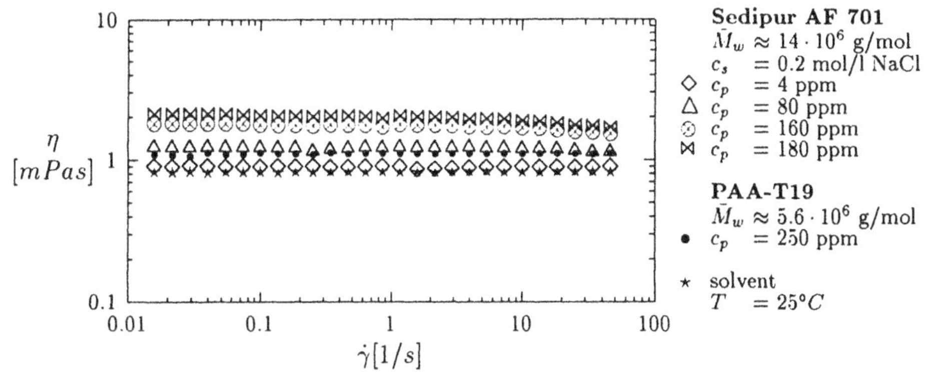

***Figure* 1**; *The shear viscosity as a function of the shear rate for different polymer solvent systems*

The experimental results shown in Figure 2 and Figure 3 illustrate that a critical strain rate $\dot{\varepsilon}^*$ depending on the kind of polymer solution separates two flow regions: a Newtonian region of strain independent elongational viscosity and a region with markedly and continuously increasing elongational viscosity.

For the PAA-T19 solution the critical strain rate at which the strong enhancement in elongational viscosity occurs is $\dot{\varepsilon}^* = 1600\ s^{-1}$. In the case of Sedipur AF 701 $\dot{\varepsilon}^*$ is shifted to a lower strain rate of $\dot{\varepsilon}^* = 500\ s^{-1}$ which can be attributed to the extremely high molar mass of this polyacrylamide.

The Trouton number Tr of the Sedipur AF 701 solution increases between $\dot{\varepsilon} = 500\ s^{-1}$ and $\dot{\varepsilon} = 3500\ s^{-1}$ from $Tr = 6$ to $Tr = 24$. The Trouton number of the PAA-T19 lies between $Tr = 4$ at $\dot{\varepsilon} = 1600\ s^{-1}$ and $Tr = 32$ at $\dot{\varepsilon} = 4000\ s^{-1}$. This demonstrates that both solutions have distinct elastic properties.

The results for the Sedipur AF 701 solution illustrate that a strong enhancement of the elongational viscosity can appear for remarkably low polymer concentrations.

Calculations applying the FENE dumbbell model of Bird et al., 1987 to the PAA-T19 solution were in good accordance with the experimental data, Eich, 1999. The measurements confirmed the onset behaviour of the elongational viscosity characteristic of the FENE dumbbell model. The continual increase of $\eta_{E,A}$ especially in the case of the Sedipur AF 701 solution is due to the large dispersivity of this polymer.

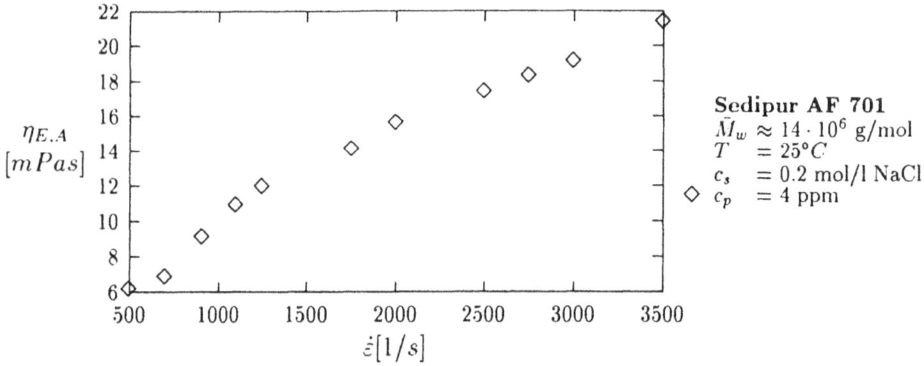

***Figure* 2**; *The apparent elongational viscosity of the Sedipur AF 701 solution*

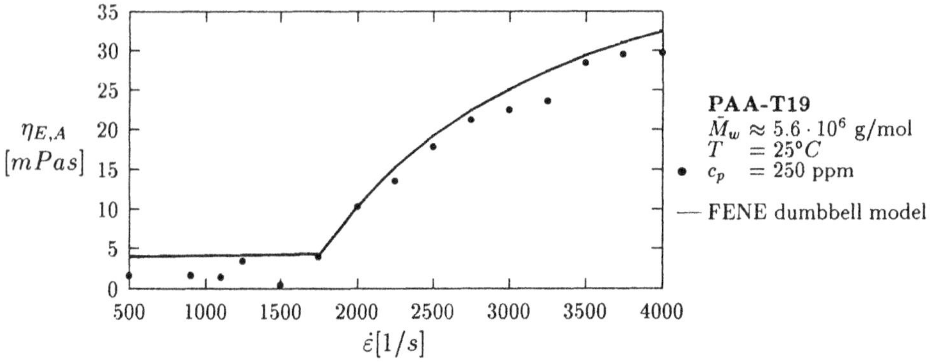

***Figure* 3**; *The apparent elongational viscosity of the PAA-T19 solution*

Figure 4 shows the light intensity due to flow induced birefringence as a function of the extension rate $\dot{\varepsilon}$ for the PAA-T19 solution. The birefringence signals illustrate the extension and orientation of the polymer molecules. Therefore the critical value of the strain rate at which the coil-stretch transition occurs should be about $\dot{\varepsilon} = 1500\ s^{-1}$ for the polymer solution investigated. As the threshold of the birefringence signals lies very near to the critical elongation rate of $\eta_{E,A}$ it can be concluded that the orientation of the macromolecules results in the increase of elongational viscosity shown in Figure 3.

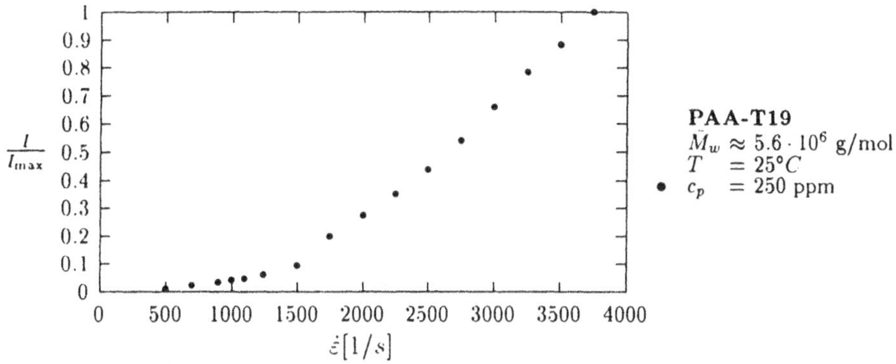

***Figure* 4**; *The light intensity as a function of the strain rate for PAA-T19*

Finally in Figure 5 results for the pipe flow friction factor λ as a function of the *Re* number obtained in a 6 mm diameter pipe are given for the PAA-T19 and the Sedipur AF 701 solution.

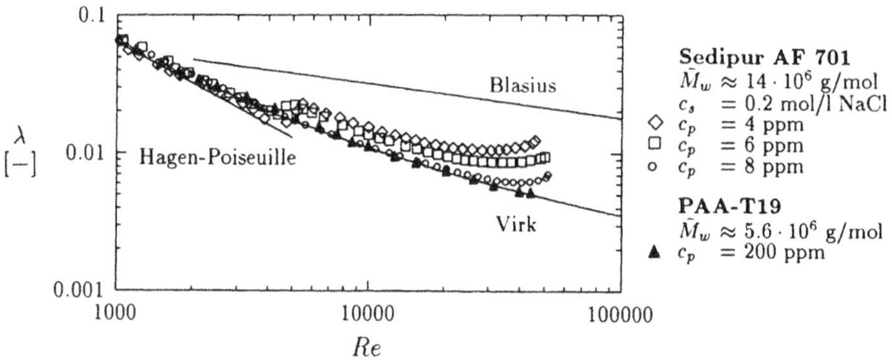

***Figure* 5**; *The friction factor as a function of the Reynolds number for different polymer solvent systems*

Immediately after the laminar-turbulent transition the PAA-T19 and the Sedipur AF 701 solution behave non-Newtonian. This means that at this *Re* number elongation rates above the critical values of Figures 2,3 should already be present in the flow fields. But because of the shear stress which appears in the pipe it could be that these critical values are reduced by preshearing.

The results presented in this paper demonstrate that solutions of low polymer concentration can show a pronounced elastic behaviour and can be strongly non-Newtonian as far as the elongational viscosity is concerned while the shear viscosity behaves Newtonian. The polymer concentrations were far below the conventionally calculated critical concentration c_p^* based upon the close packing of coils, especially in the case of Sedipur AF 701. However as the magnitude of coil expansion could not be determined the question could not be decided definitely whether the strong increases in elongational viscosity are due to the stretching out of isolated molecules or of overlapping chains.

4. ACKNOWLEDGEMENT

The authors wish to thank the University of Essen for its financial support throughout the cause of this project.

REFERENCES

Bird, R.B.; Curties, C.F.; Armstrong, R.C.; Hassager, O.; 1987; "Dynamics of Polymeric Liquids: Kinetic Theory"; Wiley & Son, New York

Cathey, C.A.; Fuller, G.G.; 1988; "Uniaxial and Biaxial Extensional Viscosity Measurements of Dilute and Semi-Dilute Solutions of Rigid Rod Polymers"; J. of Non-Newtonian Fluid Mech., 30, pp. 303-316

Durst, F.; Haas, R.; 1981; "Dehnströmungen mit verdünnten Polymerlösungen: Ein theoretisches Modell und seine experimentelle Verifikation"; Rheol. Acta, 20, pp. 179-192

Eich, T.; 1999; "Elastische Eigenschaften hochverdünnter Polymerlösungen"; PhD Thesis; University of Essen

Gampert, B.; Rensch, A.; 1996; "Polymer Concentration and Near Wall Turbulence Structure in Channel Flow of Polymer Solutions"; Proc. of the ASME Fluids Engineering Division Summer Meeting; San Diego (USA); pp. 129-136

Gyr, A.; 1990; "Structure of Turbulence and Drag Reduction "; IUTAM Symposium, 25.-28. July 1989, Zürich (Switzerland); Springer, Berlin/Heidelberg/New York

Hinch, E.J.; 1977; "Mechanical Models of Dilute Polymer Solutions for Strong Flows"; Phys. Fluids, 20, pp. s22-s30

Keller, A.; Müller, A.J.; Odell, J.A.; 1987; "Entanglements in Semi-Dilute Solutions as Revealed by Elongational Flow Studies"; Progr. Call. Polym. Sci., 75, pp. 179-200

Virk, P.S.; 1985; "The Drag Reduction Fundamentals: A Brief Recapitulation"; Biotechnology of Marine Polysaccarides, Third Annual Lecture and Seminar Papers, 1984, Cambridge (USA), pp. 149-193

Visions of Numerical Relativity 1999

Werner Benger
Konrad-Zuse-Zentrum für Informationstechnik Berlin
Takustrasse 7, D-14195 Berlin-Dahlem, and
Max-Planck-Institut für Gravitationsphysik, Potsdam
benger@zib.de
Hans-Christian Hege
Konrad-Zuse-Zentrum für Informationstechnik Berlin
hege@zib.de
Stefan Heusler
Arts of Science, www.artsofscience.de, *Cologne*
stefan@khm.de

Abstract: We present visualizations of recent supercomputer simulations from numerical relativity, exploiting the progress in visualization techniques and numerical methods also from an artistic point of view. The sequences have been compiled into a video tape, showing colliding black holes, orbiting and merging neutron stars as well as collapsing gravitational waves. In this paper we give some background information and provide a glance at the presented sequences.

Keywords: General Relativity, Data Visualization, Numerical Simulations

1 INTRODUCTION

More than 80 years ago Einstein's theory of general relativity revolutionized physics and replaced New-ton's theory of space, time and gravitation. Although Newtonian theory provided a very successful framework for centuries, astrophysical observations revealed small deviations, which could not be explained before general relativity was born. (Even an additional planet, called "Vulcan", had been predicted in order to correct for the unexplainable.) General relativity solved these mysterious phenomena and up to now all experimental evidence is in outstanding numerical coincidence with the theory.

On the other side, general relativity is much more complex than Newtonian theory, and suffers from the fact that no general way is known to solve Einstein's field equations analytically. While a couple of explicit solutions have been found, even relatively simple cases like the two-body problem still remain unsolved. For a long time analytic solutions were sought, without much success. Alternatives are:

- find approximate solutions with exact mathematical methods
- find exact solutions with approximate mathematical methods.

A. Gyr et al. (eds.), Science and Art Symposium 2000, 239-246.

The first approach is realized in perturbation theories, which cover linear and post-Newtonian field theories. However, these theories are applicable only for weak gravitational fields and break down in the strong field regime, to which e.g. collisions of black holes belong. The second approach employs numerical methods like finite differences for solving systems of partial differential equations. It is concentrated in the efforts of *Numerical Relativity*, a subfield that emerged during the 1960's, as computers became powerful enough to treat Einstein's field equations numerically [Hahn, Smarr, Seidel&Suen, Seidel].

However, due to limited computational power and not yet matured numerical methods the first attempts were restricted to simple cases, like axisymmetric collisions of black holes. In the 1990's big efforts were undertaken in the US, in particular at the National Center for Supercomputer Applications (NCSA) in Champaign, IL, among other sites, to achieve large steps forward. Some of these initiatives, like the Neutron Star Grand Challenge project, are still in progress [NASA]. The multi-institutional cooperation is now being continued [NumRel] with the recently founded Max-Planck-Institut für Gravitationsphysik Potsdam, also called the Albert-Einstein-Institute (AEI). One contribution of AEI to numerical relativity is the development of the Cactus simulation code [Cactus].

The efforts in numerical relativity culminated in extensive simulation runs performed in June 1999, when a new supercomputer at NCSA prior to official release was dedicated exclusively to numerical relativity simulations. The researchers could utilize this 256 processor SGI Origin 2000 for two weeks [NCSA], to perform previously undone computations of the first large-scale full 3D collision of black holes, of binary systems of neutron stars (i.e. doing general relativistic hydrodynamics) and collapsing gravitational waves [Allen, Butler]. The simulations consumed 140.000 CPU hours and output-ed about 1 terabyte of data.

In contrast to analytical solutions, whose content may be analyzed on a sheet of paper, numerical solutions need to be visualized to convey the results to scientists or to provide a notion of what happens to non-scientists. At this point, advanced visualization techniques, like those developed and implemented at the Konrad-Zuse-Zentrum für Informationstechnik Berlin (ZIB) come in [Amira]. By employing these methods it was possible to create stunning visualizations of the huge set of simulation data [MPG].

Nowadays, the gap between science and art is very large. However, this has not always been the case. In the old Greek culture, "astronomy" has been regarded as one of the basic human "arts", beside geometry, architecture, theater, sculpture, painting, music and mathematics. With the steady development of scientific methods, the gap between "fine arts" and the "arts" nowadays called "science" grew larger and larger. However, this contrast is only partially caused by a principal gap: When the results of mathematical calculations can be visualized, rational and visual understanding come again very close to each other.

2 Colliding Black Holes

Visualization of data from numerical relativity simulations is not a straightforward task. The fundamental objects describing the gravitational field are four-dimensional tensor fields of high rank. For doing numerical computations, we need to work with the numerous (up to 256!) components of a tensor field. However, these components depend on the coordinates in which they are written (i.e. which have been chosen for numerical purposes) and hence have no direct physical meaning.

Furthermore, the interpretation of these tensor fields is not straightforward: pure gravity is nothing more than curved "vacuum" and a black hole is "empty space" everywhere, so there is "nothing" to see – as compared to fluid dynamics – no particles, no gas, nothing. But this empty space is curved, according to Einstein's theory, and techniques to understand this curved space must be developed.

2.1 Horizon Information

Some way to look at curved space is to consider the "event horizon". It can be understood as the "surface of no return". Once inside a black hole horizon, nothing – not even light – can ever escape. This is the origin of the name "black hole".

For "usual" static black holes, the horizon is a sphere. But in general, e.g. during mergers of black holes, the horizon will acquire a much more complicated shape. Initially, the black holes carry their own horizons which then merge to form a larger one. This newly formed horizon will become spherical somewhat later, but at first it might be highly irregular due to the extreme gravitational forces acting between the black holes.

In order to locate the event horizon, the entire evolution of the space-time must be known – in numerical simulations this involves a huge amount of data. A concept alternative to the event horizon is the "apparent horizon". It is located always inside or at the event horizon. The apparent horizon can be located in individual time steps, which is much easier to do numerically.

2.2 The True Shape of the Apparent Horizon

To get an intuition what is really happening at the apparent horizon, it is useful to look at the Gaussian curvature of spacetime on this surface. The coordinate space, where the computed data reside, usually does not reflect the physical properties in space. Even if the computation points are distributed uniformly in coordinate space, their true physical distance may be completely different, ranging from zero to infinity (both extreme cases leading to numerical troubles for obvious reasons). This is similar to a uniform distribution of points in spherical coordinates – such a distribution is not uniform in Cartesian 3D coordinates. Using general relativistic coordinates, things might be even more complicated, and it is hardly possible to imagine the true shape of physical space. It might be (as an extreme example) that something looking like a sphere in coordinate space in reality is a rectangular box, and vice versa. By "reality" we mean what a physicist would measure locally, whereas the "coordinate space" are numbers which have been chosen for convenient computation. However, during the rendering process these numbers are treated as geometrical information, because depicting the true physical space is highly non-trivial and possibly even impossible within the "just" three dimensions that are available for rendering.

Nevertheless, by mapping the physical Gaussian curvature on some surface, e.g. the apparent horizon, we get an impression of the physical nature – for example although we might still see a box as a sphere, we might spot it as a box due to its colors. We chose the colors such that slight deviations from flat space or spherical shape jump into the viewer's eye. What we expect from the apparent horizon is to approach a spherical shape during the evolution. So all non-spherical perturbations shall attract the viewers attention, such that oscillations of the curvature stand out prominently. For this purpose, we normalize the curvature by the area of the surface, hence any sphere gets the same curvature value. We map this value to red. Somewhat larger values are mapped to the complementary color green to reveal strong contrast. Similarly, we map flat space (zero curvature) to yellow, the brightest color in the visible spectrum, and negative curvature values to the complementary color blue, see Fig. 1, left. With this color coding, regions of flat space and regions with spherical curvature are easy noticeable.

Additionally, by adjusting transparency appropriately, we may render large, homogeneous subvolumes transparent, thereby enabling a view to their inside, while still preserving significant features (Fig. 1, right).

2.3 The Gravitational Field

We also want to get a visual impression of the gravitational field. Fortunately, most of the 256 tensor field components carry redundant information. The so-called Riemann-tensor may be split up in a trace-free part, the Weyl tensor, and the Ricci tensor, which is then further used in the Einstein field equations. In analogy to the 3+1 split of electrodynamics into the well-known Maxwell equations, we may write the Weyl tensor as an "electric" and "magnetic" part. An appropriate linear combination of these quantities reveals five complex quantities, the so-called "Weyl curvature scalars" [Newman&Penrose].

A technique used commonly for visualizing scalar fields is volume rendering. Each value within the three-dimensional data volume is assigned some optical properties like local emission of light and transparency. According to the selected color mappings, the results are "glowing clouds", i.e. bright subvolumes indicating specifically interesting regions of the scalar field, and more transparent areas depicting large homogeneous regions.

However, while standard color maps result in compact clouds that hide inner structures, a special colormap may be chosen which assigns the "glow" property only to visually interesting, possibly equidistant levels of the scalar field. This results in tender, self-glowing skins of varying color and brightness, that enclose relevant regions like an onion (Fig. 3). We call this visualization technique "volumetric isolevel rendering".

3 Orbiting Neutron Stars

3.1 Matter Density and Atmosphere

Visualizing neutron stars is a much easier task than picturing black holes. Neutron stars possess characteristics like matter density, energy per volume per density (internal energy), etc., as known from fluid dynamics. These quantities can be understood quite easily. The density of a neutron star is extremely high and falls off very rapidly at the surface. The "atmosphere" of a neutron star, i.e. the region where the matter density is comparable to the density of air and water, as we are used to from our home planet, can be measured only in millimeters or fractions of a millimeter.

A realistic mapping of a neutron star's matter density to optical properties would reveal just a sphere. However, using computer graphics we have the ability to render physically opaque regions transparent, so we perceive the otherwise invisible high-density regions around the neutron star's core.

By employing the technique of volumetric isolevels it is possible to bring out quite clearly the details of energy fluctuations. Choosing a "glowing" color map is appropriate here, since higher values of the scalar field indicate more energy in nature, so this gives a more physically-correct impression.

Generally speaking, neutron stars are less "abstract" than black holes and can therefore be pictured more realistically. To support this impression, a photo displaying real stars was chosen as background image. So these images touch the border between scientific visualizations and realistic computer graphic renderings, although they are still quite far from photorealism.

If we look at the sequence of the neutron star merger, Fig. 4, computed in Newtonian theory, the fluctuations of the atmosphere that come out provide so many details that the eye nearly gets lost. But, since all moving shock fronts fit into a consistent visual context of a global flow, it does not look chaotic. As the rational brain is overflooded by information, rationality is given up and emotionality wins. Indeed a lot of people had the impression that his sequence "cries" for some acoustic accompaniment. Basically, this was one of our motivations to create a music-attributed video.

3.2 Additional Physical Quantities

In addition to matter properties, it is of interest to also visualize the geometrical shape of space around massive neutron stars, i.e. their gravitational field, especially the gravitational waves being emitted during collisions. The gravitational waves created by neutron star mergers are expected to be weaker than those from black hole collisions. Though the computational work to retrieve wave forms with clean physical foundation still needs further development, and the current data are of preliminary character, the first results look already promising, and give first impressions on how gravitational field quantities, often used in black hole spacetimes, might look like around neutron stars (s. Fig. 5).

4 Collapsing Gravitational Waves

In nature, black holes are formed when extreme matter concentrations occur, e.g. during supernova collapses of super-massive stars, or in collisions of neutron stars. Due to $E = mc^2$, extreme concentrations of pure energy may also lead to formations of black holes. This energy can also be gravitational energy ("empty space"). At the time when it was proven first that such configurations of spacetime can exist [Brill], only initial data could be investigated. Now, by means of numerical relativity, we are able to dynamically evolve these spacetimes, the so-called "Brill Waves", and to study critical phenomena, like the creation of a black hole during the collapse of a gravitational wave.

The big challenge for investigating such spacetimes is that there are lots of physical properties to be visualized at once. We have in-going and outgoing components of the gravitational field, and various polarizations of the gravitational radiations. These are only four quantities of the five complex Weyl Scalar fields, $\Psi_0 \ldots \Psi_4$, each of them having a reasonable physical interpretation. In practice it is impossible to display all these properties of spacetime in one image and still get a reasonable impression of what is going on.

Therefore the main task is to bring out – expected or unexpected – correlations between the various physical quantities. In particular, for "supercritical" Brill waves, which form a black hole, we want to see the correlation between the horizon curvature and the gravitational field. For the black hole formation we use the same method as for colliding black holes, i.e. we display the apparent horizon with colormapped Gaussian curvature.

The coincidence between curvature of the horizon and gravitational radiation is of special interest, since some theorem in general relativity states that all non-rotating, uncharged black holes of the same mass are identical, and hence all perturbations are radiated away. Therefore, we expect the largest intensity of gravitational radiations at areas where the curvature of the horizon is vastly varying, as can be seen in Fig. 6.

A special colormap has been chosen for this image, such that the volumetric isolevel rendering reveals symmetry properties of the scalar field Ψ_4 around the value zero. We can therefore detect the wave-like behavior of Ψ_4, as positive regions are displayed in red and negative regions in blue colors. As the scalar field values approach the most frequent value zero, they are rendered more transparent. By this technique, high-intensity regions stand out clearly and the volumetric isolevels give a good impression of the relationship with the horizon's curvature.

5 Video Production Issues

When the visualizations of the numerical simulations became available, it was evident that these sequences are interesting not only due to their scientific content, but also from an artistic point of view. Therefore, the idea was born to produce a film based on these simulations which takes into account both scientific an artistic aspects. Before every sequence starts, the scientific content and the motivations for the numerical simulation are explained.The elucidations for the sequences are produced with very simple techniques, just using a blackboard, a piece of chalk, and classical animation tricks. The sharp contrast between high-tech and simple, but charming low-tech methods is intended to give back a human face to the pictures that have been produced by a record-breaking use of a hyper-modern massively parallel supercomputer.

The aim of this film therefore is to give an intuitive impression and at the same time scientific overview about the current state of the research in the field of numerical relativity. It was a challenge to create a film sequence, which is understandable and attractive for people who have hardly any knowledge about general relativity, and which provides new and interesting information for insiders, too.

6 Acknowledgment

This work was done within the framework of a cooperative research project aiming at new techniques for visualization of general relativistic phenomena performed at ZIB and AEI.

Ed Seidel, head of the numerical relativity group at AEI, contributed lots of ideas and suggestions concerning the physical simulations and the visualization process. Paul Walker, Joan Masso, Gabrielle Allen, Tom Goodale, Gerd Lanfermann are the primary developers and maintainers of Cactus. Bernd Brügmann and Steve Brandt developed methods for constructing initial data for black hole collisions. Lars Nerger and Takahashi Ryoji performed the large-scale supercomputer simulations in June 1999. Miguel Alcubierre has implemented the numerical treatment of Brill Wave data. Wai-Mo Suen, head of the Neutron Star Grand Challenge project at the Washington University, St.Louis, and Mark Miller did the numerics and performed the simulations of neutron stars and their collapse to a black hole. Phillip Gressman provided the Newtonian Neutron star data.

The visualization system Amira [Amira] which was used to create all images presented here, has been developed by the visualization group at ZIB. We thank especially Detlev Stalling and Malte Zöckler for their significant contributions. Olaf Paetsch (ZIB) and Jeff Carpenter (NCSA) were doing hard work to provide appropriate film postprocessing including film titles. Last but not least, the music was composed by Frank Stanzl from the Music Academy Cologne.

References

[Allen] G. Allen, T. Goodale, G. Lanfermann, E. Seidel, W. Benger, H.C. Hege, A. Merzky, J. Masso, J. Shalf, *Solving Einstein's Equation on Supercomputers*, IEEE Computer, Dec. 1999, p. 52-59

[Amira] *Amira – an Advanced 3D Visualization and Volume Modeling System,* (1999) `http://amira.zib.de`

[Butler] D. Butler, *Computing 2010: from black holes to biology*, Nature, 402 (Millennium Supplement), p. C67, (2. Dec 1999)

[Brill] D. Brill *On the Positive Definite Mass of the Bondi-Weber-Wheeler Time-Symmetric Gravitational Waves*, Ann. Phys., 7, p. 466-483 (1959)

[Cactus] *The Cactus Code Server,* (1999) `http://www.cactuscode.org`

[Hahn] S. G. Hahn and R. W. Lindquist, *The Two-Body Problem in Geometrodynamics*, Ann. Phys., 29, p. 304-331 (1964)

[NASA] Grand Challenges supported by NASA, (1992-1999) `http://farside.gsfc.nasa.gov/ESS/grand-challenges.html`

[MPG] Max-Planck-Society, *The Biggest Crashes in the Universe*, Research News Release, 12.8.1999, `http://www.mpg.de/news99/news38_99.htm`.

[Newman&Penrose] E. Newman, R. Penrose, JMP (3) p. 566 (1962).

[NCSA] K. Green, *Colliding with a Supercomputing Record*, Alliance/NCSA Access magazine 03, Fall/Winter 1999, `http://access.ncsa.uiuc.edu/Features/o2krun/O2KRecord.html`

[Smarr] L. Smarr, *Spacetimes generated by computers: Black holes with gravitational radiation*, Ann. N.Y. Acad. Sci., 302, p. 569-604 (1977)

[Seidel&Suen] E. Seidel and W.-M. Suen, *Numerical Relativity*, Computational Physics, World Scientific, Singapore (1994)

[Seidel] E. Seidel, *Black Hole Coalescence and Mergers: Review, Status, and "Where are we heading?"*, Progr. Theor. Phys., in press (2000)

[NumRel] *International Numerical Relativity Group Webserver*, (1993-1999) `http://jean-luc.ncsa.uiuc.edu/`

Figures from:

SERGEY ALEKSEENKO

Figure 16 to 23

ECKART MEIBURG

Figure 1

PETROS KOUMOUTSAKOS

Transition to turbulence of a vortex ring at Re = 4000

TATYANA KRASNOPOLSKAYA

Figure 2 to 6

VALERI GORIATCHEV

Figure 2

ALBERT GYR

Figure 2

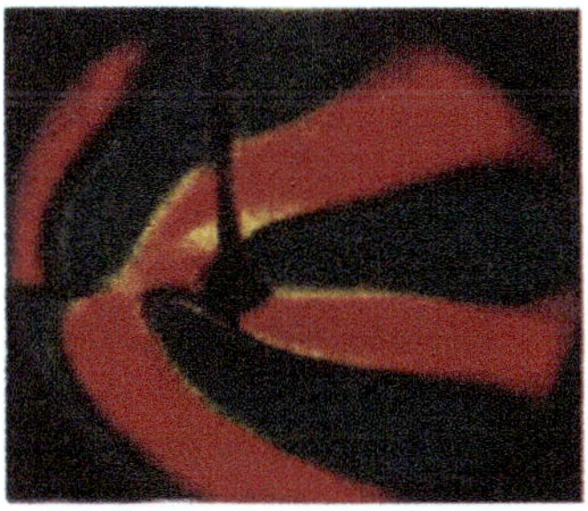

a

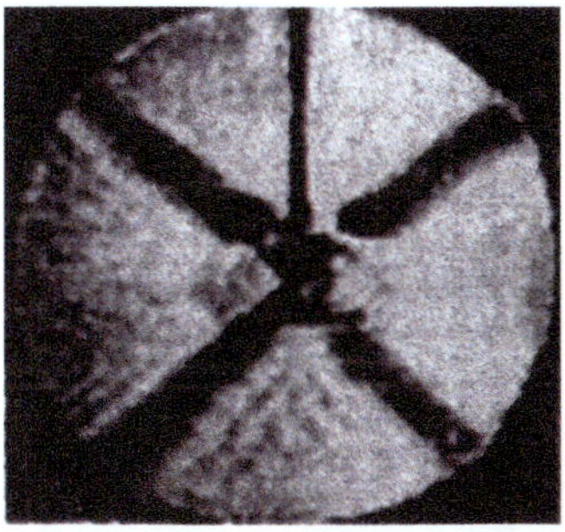

b

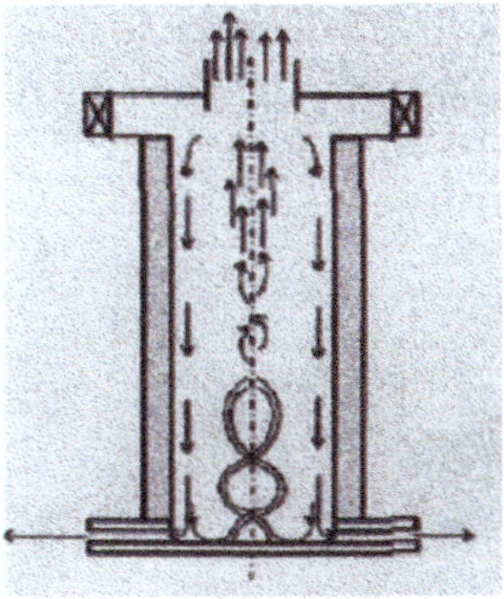

Figure 17; Ranque effect.

Figure 16; Internal waves generation during movement of a cylindrical body in stratified fluid (a) and oscillation of a body in fluid near thermodynamical critical point (b).

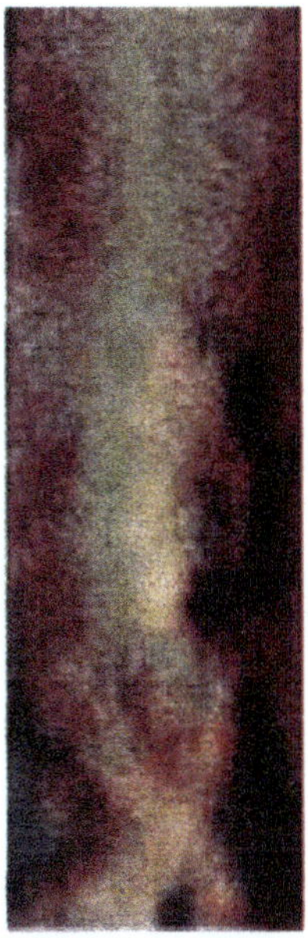

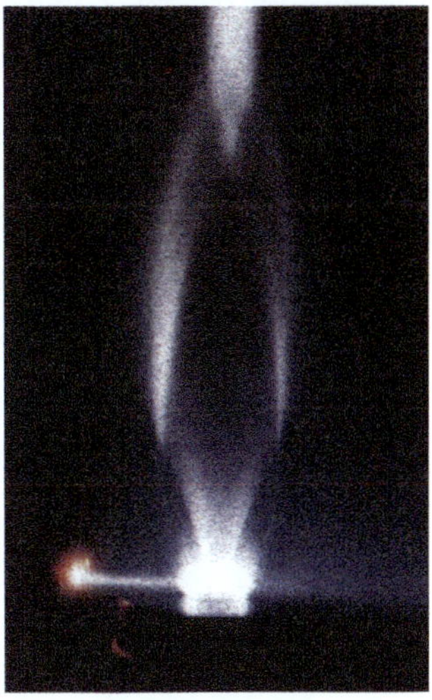

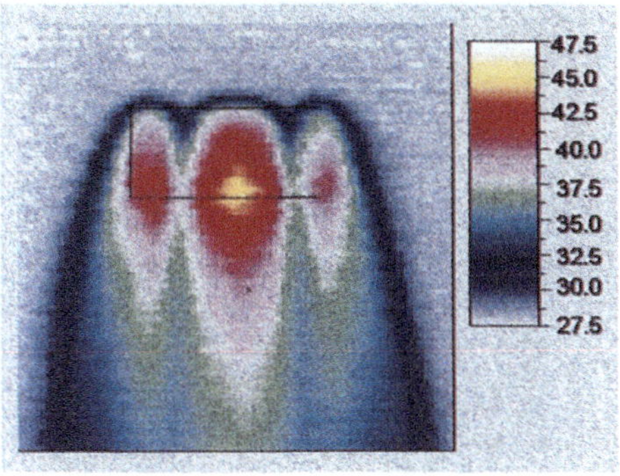

Figure 18; Formation of horseshoe structures in liquid layer with local heat source, t °C.

Figure 19; Supersonic ring jet (visualization with an electronic beam).

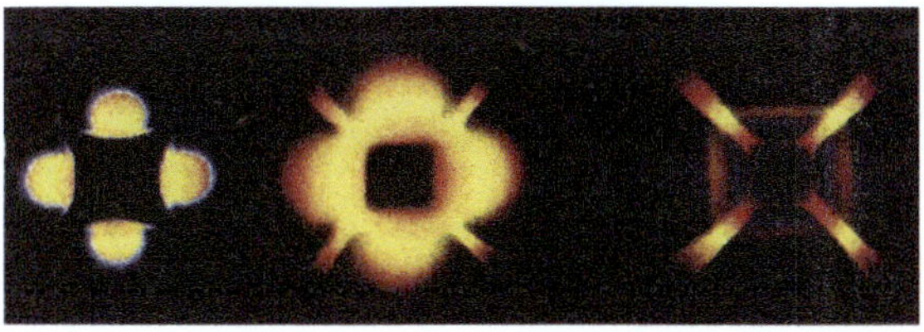

Figure 20; Flow structure behind the four-nozzle arrangement of spacecraft.

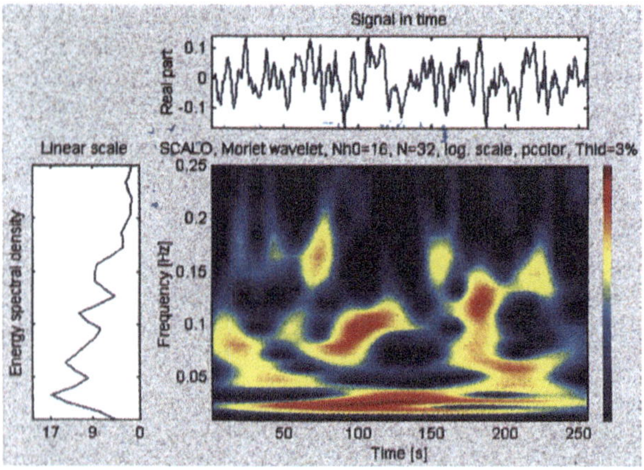

Figure 21; *Wavelet spectrum of a vertical component of velocity. Vortex flow under the rotating disk in a rectangular chamber.*

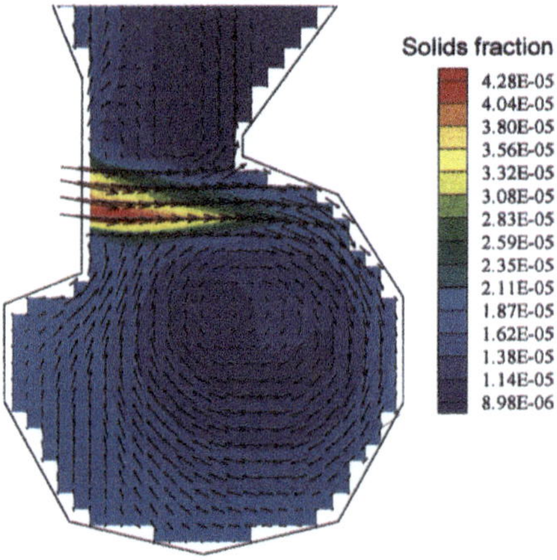

Figure 22; *Flow structure in vortex combustor (CFD). Volumetric fraction of dispersed phase and velocity vectors.*

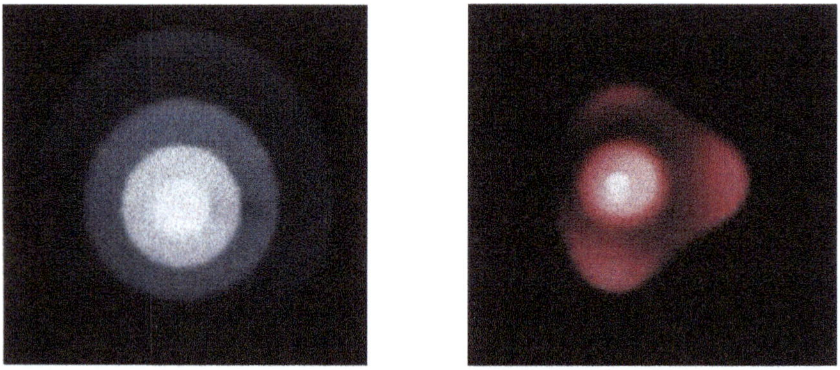

a b

Figure 23; *Spherical stratification of a glow discharge in a high-molecular gas*

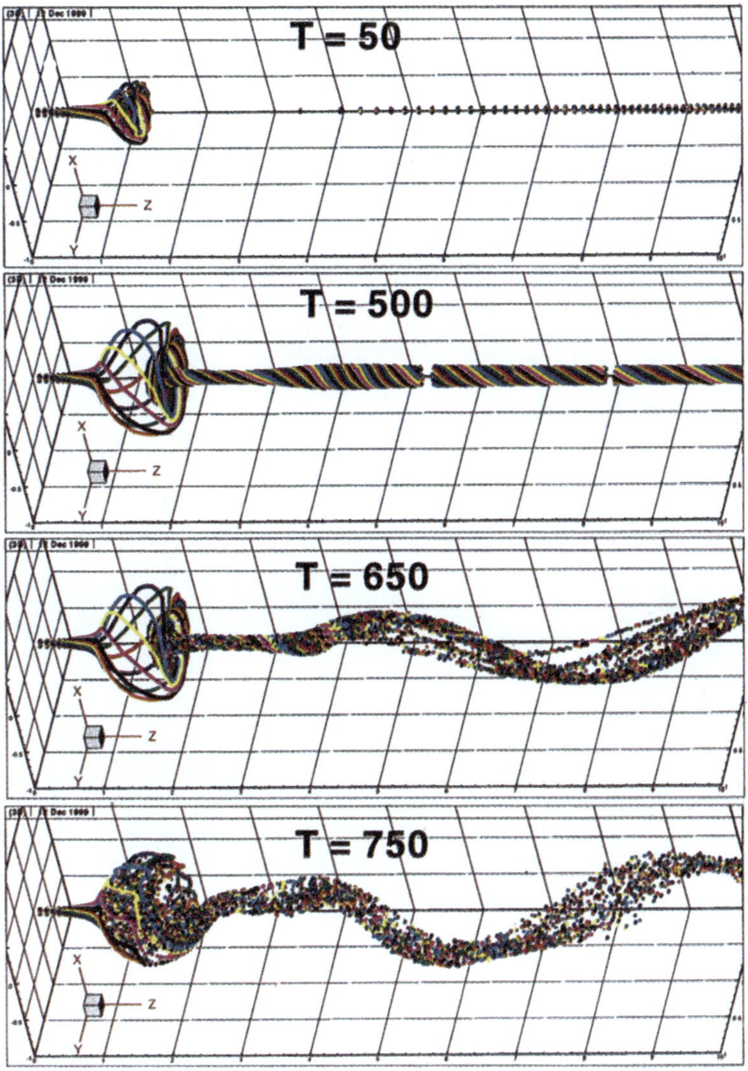

Figure 1: $Re = 200$, $\alpha = 1$, and $S = 1.095$: Streaklines for times t=50, 500, 650, and 750. The primary breakdown event is seen to be followed by the emergence of a quasisteady, axisymmetric velocity field, which in turn gives rise to a temporally periodic, helical instability.

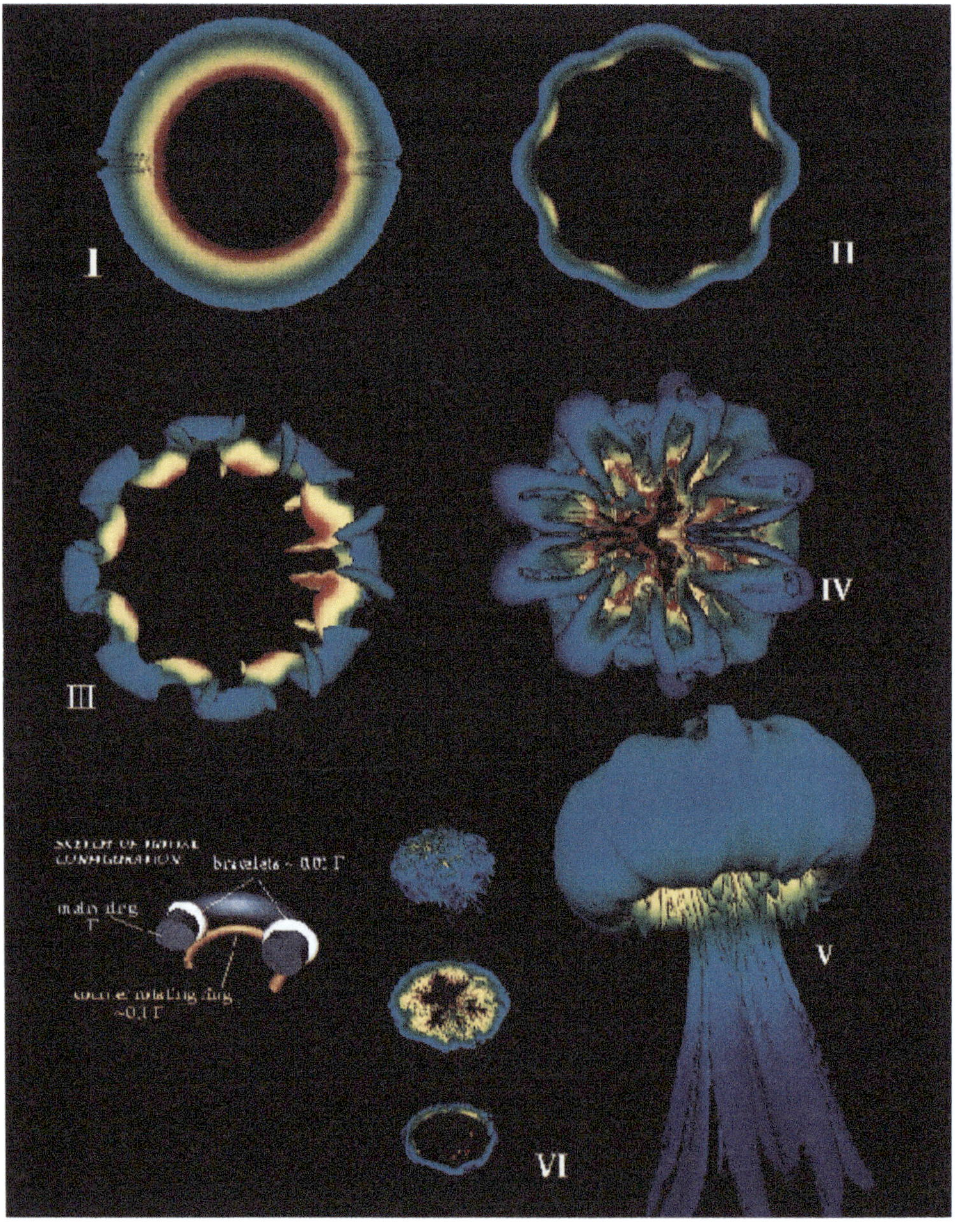

Figure 1; *Transition to turbulence of a vortex ring at Re =4000*

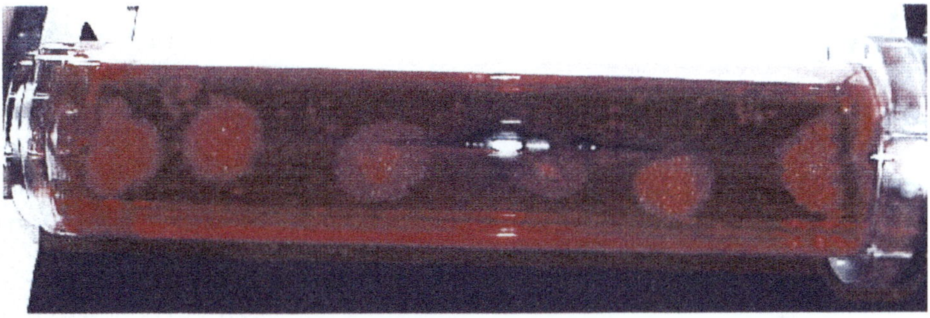

Figure 2; *Six vortices.*

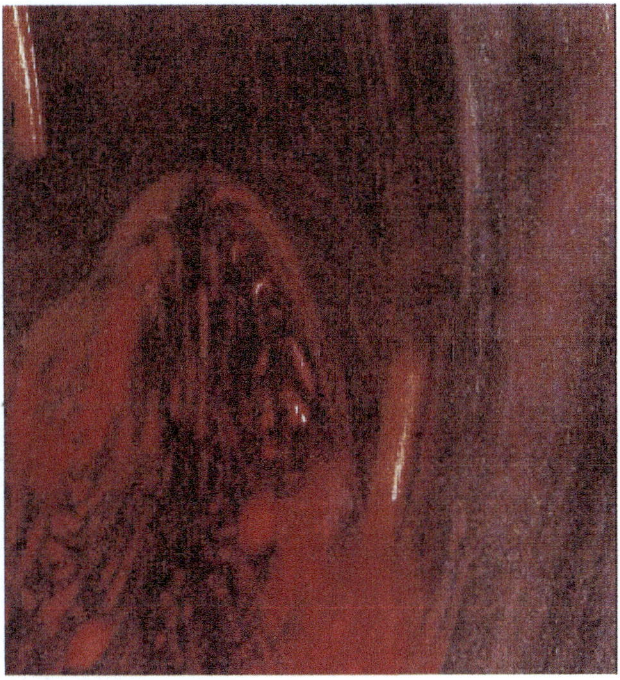

Figure 3; *"Head of a snake" pattern.*

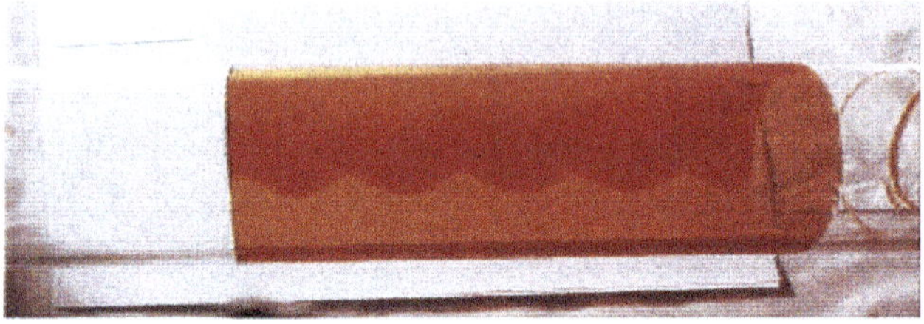

Figure 4; "Shark teeth" pattern.

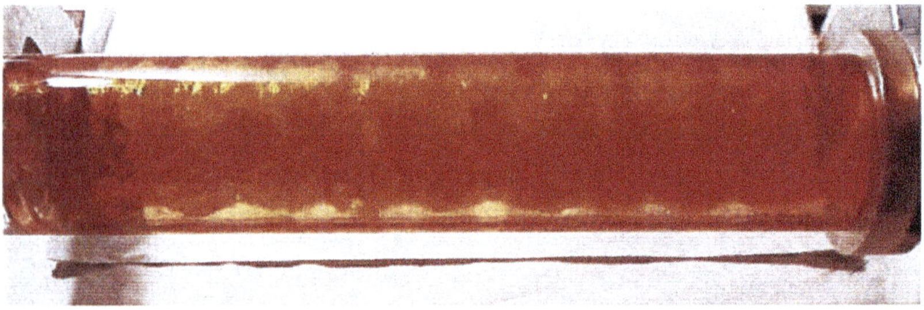

Figure 5; "Fish-like" pattern.

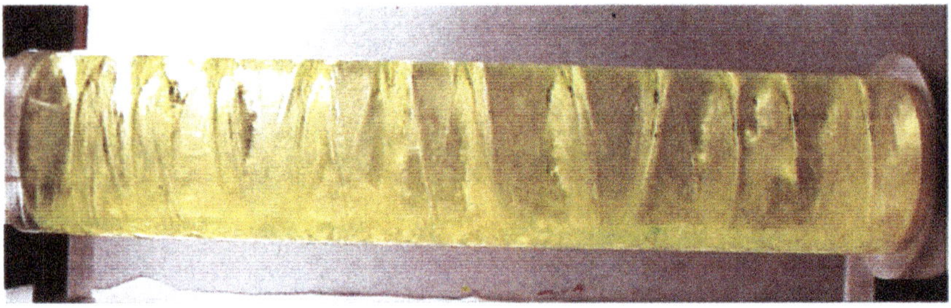

Figure 6; *"Curtains" pattern.*

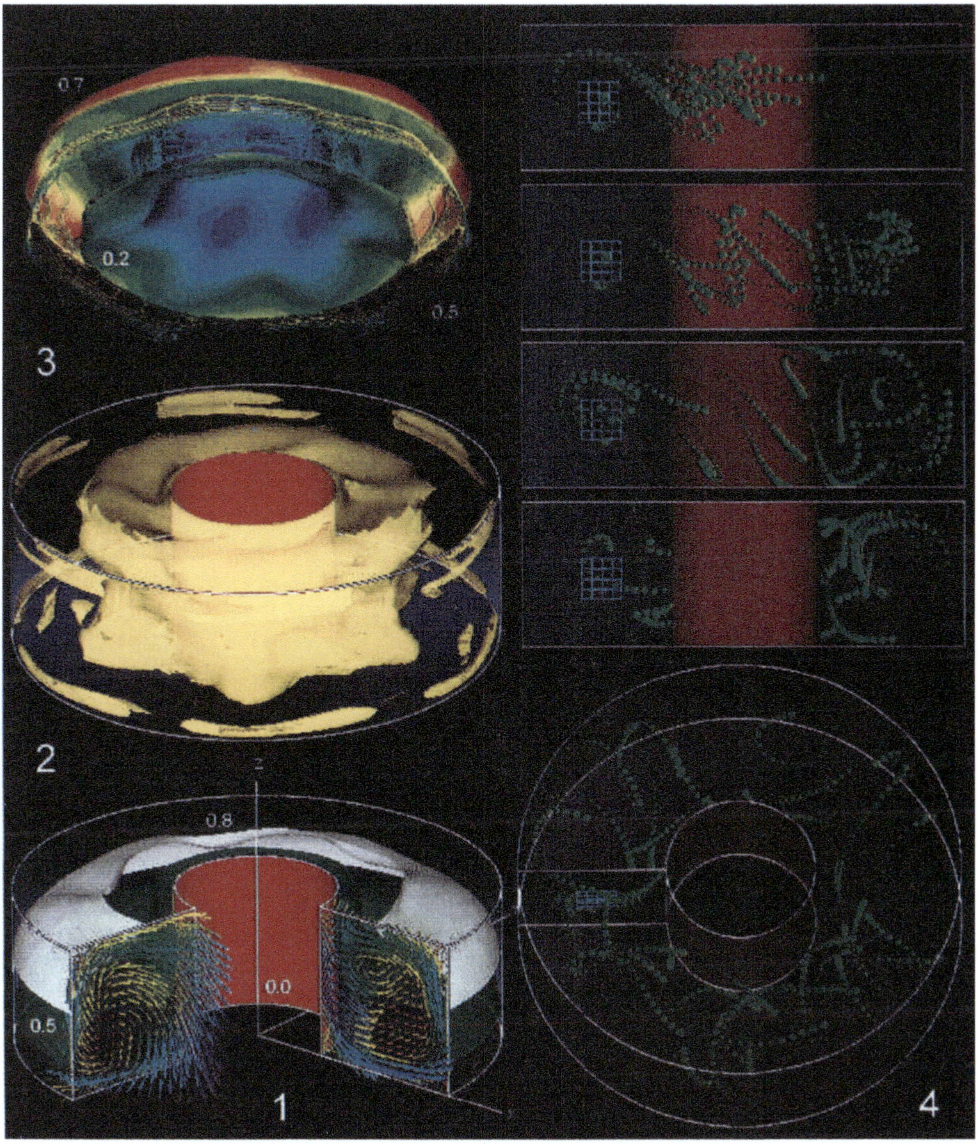

Figure 2. *Frames of convection with baroclinic waves*
(1) temperature levels of $\Theta = 0.5,\ 0.8$ *and velocity distribution in meridional sections; (2) precession of the iso-surface correspondent to* -0.05 *level of* Z *- component of velocity; (3) filtered mapping of velocity shadow contours onto temperature iso-surface; (4) path traces of diffused drops inside the annulus*

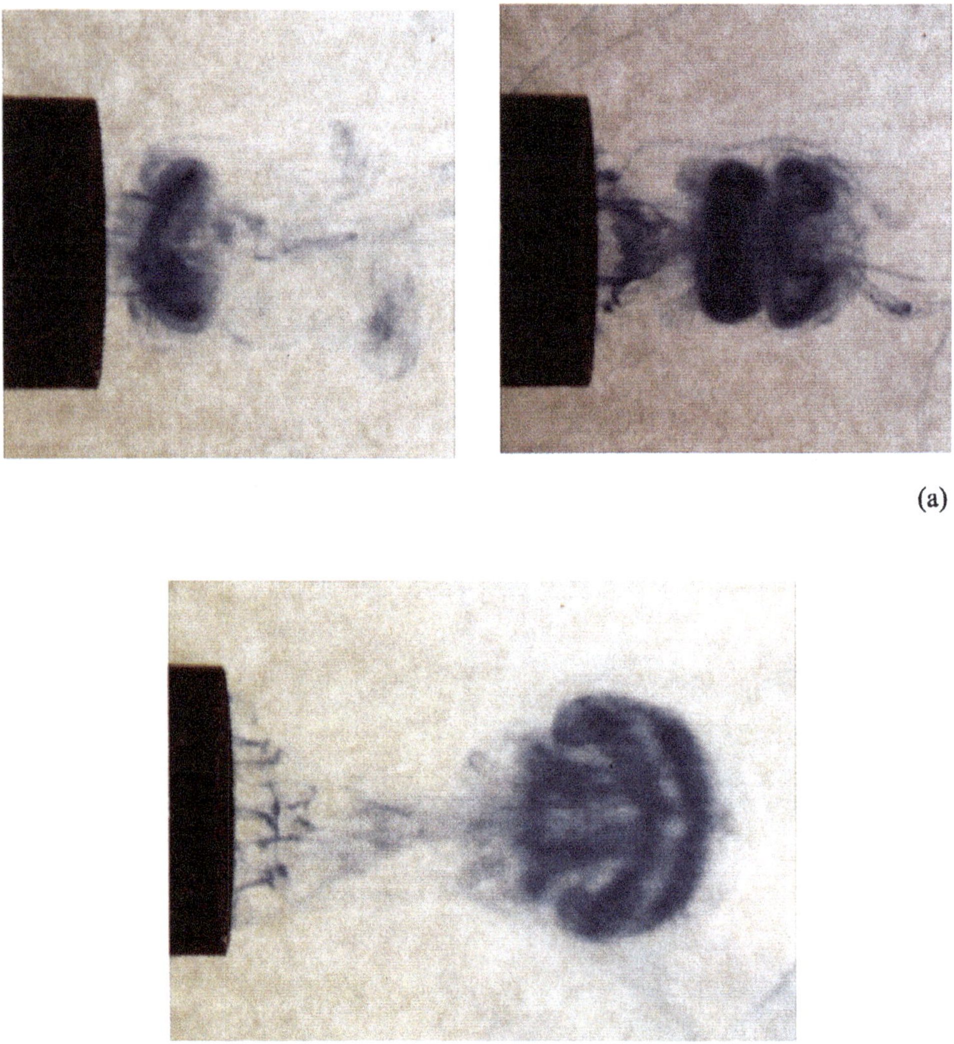

(a)

(b)

Figure 2: (a) The preceding ring is stabilizing the three-dimensional deformation of the second one. (b) A folding of an elliptical vortex ring by a 'false frog leaping' interaction.

The ordering effect of a horizontal magnetic field on liquid metal natural convective flow and its influence on heat transfer

Ulrich Burr

Eidgenössische Technische Hochschule (ETH) Zürich
Institut für Hydromechanik und Wasserwirtschaft (IHW)
ETH Hönggerberg, CH-8093 Zürich
burr@ihw.baug.ethz.ch

Abstract: Two experiments are presented where a horizontal magnetic field is imposed to a liquid metal buoyant flow. The non-isotropic character of the electromagnetic forces gives rise to the transition of three-dimensional convective vortex structures into increasingly two-dimensional ones in the sense that vortex axes become aligned with the direction of the applied magnetic field. The reduced non-linearities of local non-isotropic flow cause the formation of large-scale coherent structures which exhibit much simpler temporal dynamics and higher convective heat transport than under ordinary hydrodynamic (OHD) conditions.

Keywords: Heat transfer, magnetohydrodynamic (MHD) flow, buoyant convection.

1. INTRODUCTION

Thermal convection of liquid metals is characterised at ordinary hydrodynamic (OHD) conditions by strong time-dependent three-dimensional vortex motions. When a magnetic field $\boldsymbol{B}$ is imposed to the flow Lorenz forces $\boldsymbol{F}_L$ are introduced to the momentum balance of the flow and a magnetohydrodynamic (MHD) flow is established. For strong magnetic fields, the dynamics of vortices are strongly influenced by the electromagnetic forces and additionally to viscous dissipation, Joule's dissipation damps the convective motions. Davidson (1995) describes for the limiting case of high Reynolds numbers $\mathrm{Re} = u_0 a/\nu$ and small magnetic Reynolds numbers $Rm = \mu\sigma u_0 a$, that the component of angular momentum of vortices aligned with the magnetic field is conserved, whereas components non-aligned with the magnetic field are removed on a fast time scale. The Lorenz force elongates vortices along the field lines and thereby Joule's dissipation is continuously reduced. Here, u_0 is the characteristic velocity, a the characteristic length-scale, ν the kinematic viscosity, μ the magnetic permeability and σ the electrical conductivity. In figure 1 these principles are explained on isolated vortices evolving in both, space and time in an otherwise quiescent fluid. When the vortex axis is orientated perpendicular to the magnetic field as shown in figure 1(a), the horizontal motion of velocity $\boldsymbol{v}$ induces potential differences parallel to the roll axis. The potential gradients $\nabla\phi = \boldsymbol{v}\times\boldsymbol{B}$ left and right of the vortex centre oppose each other and therefore electric currents $\boldsymbol{j}$ close near the periphery of the vortex and a high current density is produced.

A. Gyr et al. (eds.), Science and Art Symposium 2000, 257-262.

The induced Lorenz forces $F_L = j \times B$ are opposing the vertical motions of the fluid and cause therefore strong electromagnetic damping.
For a vortex with axis aligned with the magnetic field, shown in figure 1(*b*) the circular motion in the plane perpendicular to the magnetic field induces a radial potential difference and electric currents close along magnetic field lines in the quiescent fluid zones. Within the region of swirl the Lorentz force opposes the motion whereas at the ends of the vortex the non-vanishing Lorentz force accelerates the fluid at rest and thereby propagates the vortex motion in the direction of the main vortex axis. From the longer current path the current density and the Joule's dissipation associated with it are reduced.

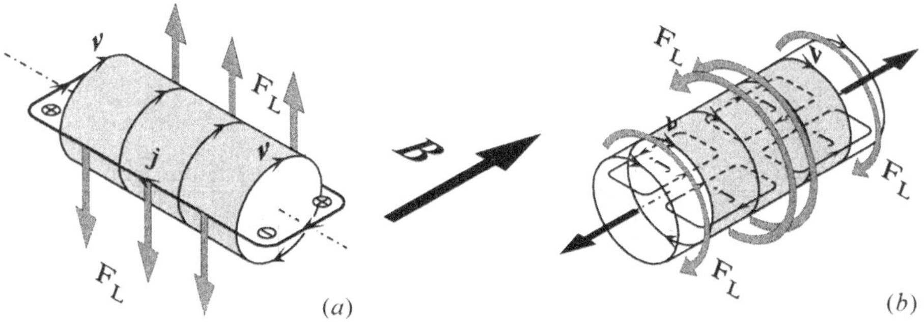

Figure1: Current density $\boldsymbol{j}$ and Lorenz forces $\boldsymbol{F}_L$ induced by the action of a horizontal magnetic field $\boldsymbol{B}$ on a vortex motion of velocity $\mathbf{v}$.

As the electromagnetic forces not only damp convective motions but also change the flow structure, integral flow properties like the convective heat flux are governed by different co-existing mechanisms. Convective velocities and with them the convective heat transfer are generally decreased by Joule's dissipation but this damping effect is balanced by the formation of two-dimensionality. The reduced non-linearities of strong non-isotropic flow cause the formation of large-scale coherent vortex structures with reduced viscous dissipation due to their larger scales. If the reduction of viscous dissipation overrides the damping by Joule's dissipation, even an improvement of the convective heat transfer becomes possible.

2. EXPERIMENTAL SETUPS

The influence of a purely horizontal magnetic field on a liquid metal (sodium-potassium $Na^{22}K^{78}$ with $\Pr \approx 0.02$) convective flow was investigated in two experimental setups. In figures 2(*a*) and 2(*b*) the geometries, the coordinates and the instrumentations are sketched. The first (figure 2*a*) is the well known Rayleigh-Bénard (RB) system, where a vertical temperature gradient ΔT is maintained opposite to the direction of gravity $\mathbf{g}$. The second (figure 2*b*) is a vertical slot (VS) with ΔT maintained in horizontal direction. Both experiments provide a rectangular fluid domain of large aspect ratio 20:10:1 with a characteristic length $h = 20mm$. The heat flux $\boldsymbol{q}$ is supplied to one wall by electrical heating and removed from the other by an evaporating water pool. The magnetic field (up to $B = 1.15T$) is produced by an superconducting magnet. The temperature difference across the layer ΔT and the magnitude of the applied magnetic field B are expressed in non-dimensional form by the Rayleigh number $Ra = \beta g \Delta T h^3 / \nu\kappa$ and the Chandrasekhar number $Q = B^2 h^2 \sigma / \rho\nu$, where β is volumetric thermal expansion coefficient, g the magnitude of the acceleration of gravity and $\kappa = \lambda / \rho c_p$ the thermal diffusivity calculated from the density ρ, the thermal conductivity λ and specific heat c_p.

ΔT is measured by thermocouples embedded in the in the heated and the cooled walls. Local, time dependent properties of the temperature field are obtained from a four-element temperature probe placed in the centreplane of the layer. The non-coplanar arrangement of the thermocouples facilitates the measurement of the local temperature T and the gradient vector ∇T of the temperature field.

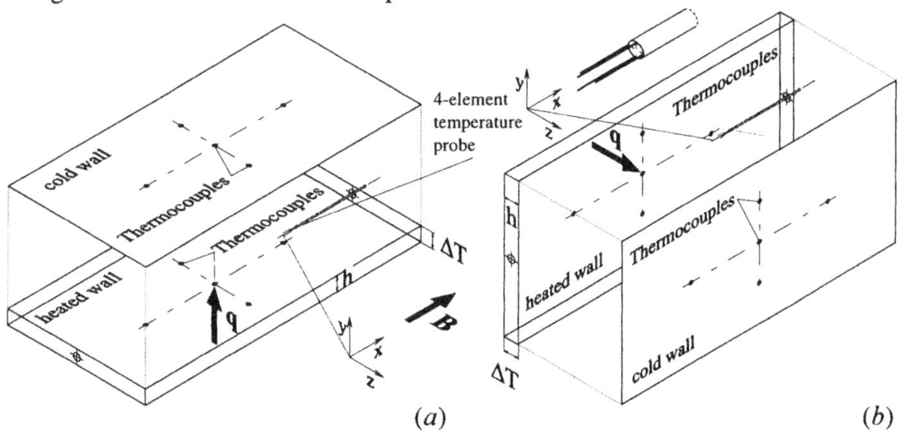

Figure 2: *Geometry, coordinates and instrumentation of the two heat transfer experiments. (a) Rayleigh-Bénard (RB) convection and (b) vertical slot (VS).*

3. RESULTS AND DISCUSSION

When the Rayleigh number exceeds a critical value of $Ra = 1708$ (see Pellew and Southwell 1940) convection in the RB system sets in as a stationary convective roll pattern with a horizontal length-scales of the order of the height of the layer h. For the very small Prandtl numbers considered here, the flow proceeds from laminar to time dependent and finally turbulent convection in a narrow range of supercritical Rayleigh numbers (see e.g. Krishnamurti 1973). In the VS configuration convective motions are present for every Rayleigh number $Ra > 0$. At small Ra the fluid at the hot wall moves upward whereas the fluid at the cold wall moves downward creating only one big vortex that is elongated along the whole experimental cell. If the velocities are increased by higher Ra the shear-layer between the upward and the downward flow becomes unstable and vortices with the length-scale of the width of the gap h are produced (see e.g. Elder 1965 or Seki et al. 1978). Thus, at high enough Ra, the flow in both configurations is characterized at OHD conditions by strong, time-dependent convective vortices.

In figure 3 time series of the fluctuating part of the non-dimensional temperature $T'/\Delta T$, recorded by the probe are shown. For both flows a constant heat flux namely $q = 2.8\times10^4 Wm^{-2}$ in the RB configuration (figure 3*a*) and $q = 4.2\times10^4 Wm^{-2}$ in the VS set-up (figure 3*b*) is applied to the layer and the magnetic field is successively increased. At OHD conditions ($Q^{1/2} = 0$), the temperature recordings are characterised by random fluctuations related to turbulent convection. For the RB system, the magnetic field exerts a significant ordering effect on the temporal dynamics of the flow. From the increase of the magnetic field the flow becomes more and more organised. It proceeds over intermittent states e.g. $Q^{1/2} = 260$ to states where the temporal dynamics are determined by few governing frequencies only. Finally, stationary flow is obtained for $Q^{1/2} = 780$. In the VS configuration such well-organised flow patterns are only observed in a narrow transition region between time dependent and stationary flow (see $Q^{1/2} = 400$).

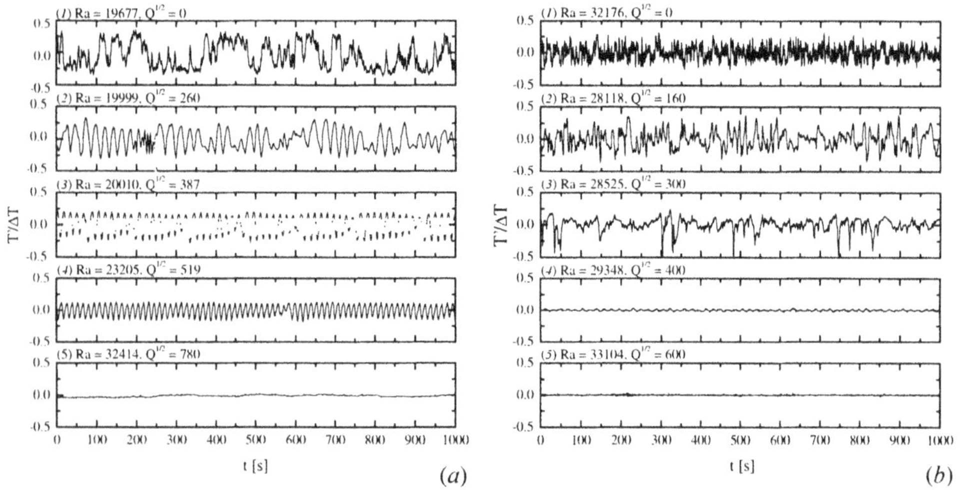

Figure 3: Time series of the non-dimensional temperature $T'/\Delta T$ recorded by the probe (a) RB convection at $q = 2.8\times10^4 Wm^{-2}$ and (b) (VS) at $q = 4.2\times10^4 Wm^{-2}$.

Nevertheless, in figures 4(*a*) and 4(*b*), it can be demonstrated on scatter plots of the fluctuating part of the local, non-dimensional temperature gradients $\partial_i T' = \nabla T'(a/\Delta T)$ recorded by the probe, that in both configurations the local isotropy properties are significantly changed by the magnetic field in a similar way. For RB convection $\partial_z T'$ is plotted versus $\partial_x T'$ in figure 4(*a*). Horizontal isotropy of the temperature fluctuations is demonstrated by the uniform distribution of values in the horizontal plane at OHD conditions $Q^{1/2} = 0$. From increasing magnetic fields the distributions become more narrow in *x*-direction and thereby show that fluctuations in the direction of the magnetic field are successively damped out. In the stationary flow regime all values collapse into a narrow spot around unity as it is seen for $Q^{1/2} = 1200$.

For the VS arrangement $\partial_y T'$ and $\partial_z T'$ are plotted versus $\partial_x T'$ in figure 4(*b*). Simular like in RB convection, the flow is more or less isotropic at OHD conditions and both distributions become narrower in x-direction from the influence of the magnetic field. As for the boundary conditions present here, the temperature field is disturbed from velocity field only and therefore, the measurements confirm, that the convective vortices become predominantly orientated in the direction of the horizontal magnetic field.

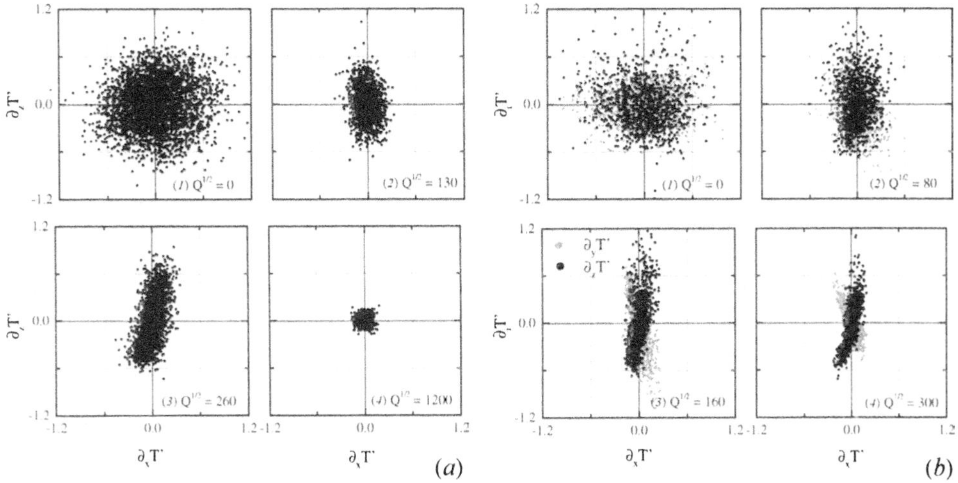

Figure 4: *Scatter plots of* $\nabla T'(h/\Delta T) = \partial_i T'$. *(a) RB convection and (b) (VS).*

In figures 5(*a*) and (*b*) the influence of the changes in the flow structure on integral flow properties is demonstrated. The convective heat transport is characterised by the Nusselt numbers $Nu = qh/\lambda\Delta T$. From Joule's dissipation, the Nusselt numbers are expected to decrease monotonous when the magnetic field is increased but such behavior is only observed at high Q. Instead, in the RB configuration a maximum in the Nusselt curve occurs before the values asymptote the state of pure heat conduction $Nu = 1$ and in the VS arrangement an even higher convective heat transport than in case of OHD flow is indicated up to $Q^{1/2} = 400$. However, one may argue that the non-constant Rayleigh numbers falsifys this important result, but generally, the convective heat transfer is increased at higher Ra.

From the fluctuating intensity of the non-dimensional temperature fluctuations $<T'^2>$ recorded by the probe it is obtained, that the fluctuating intensity of the temperature field may even be enhanced by the magnetic field in both setups. Even though above $Q^{1/2} \approx 600$ for the RB and above $Q^{1/2} \approx 400$ for the VS configuration laminar flow is indicated, significant convective heat transport is obtained nevertheless. In the regions of time-dependent flow the local isotropy properties of the flow can be estimated from the probe measurements in integral form by calculating the isotropy coefficients $A_{ij} = <\partial_i T'^2> / <\partial_j T'^2>$ At OHD conditions horizontal isotropy of RB convection is indicated by $A_{xz} \approx 1$. As the magnetic field is applied, A_{xz} immediately decreases and indicates local non-isotropy. Similar behaviour is demonstrated for the VS configuration on the coefficients A_{xy} and A_{xz}.

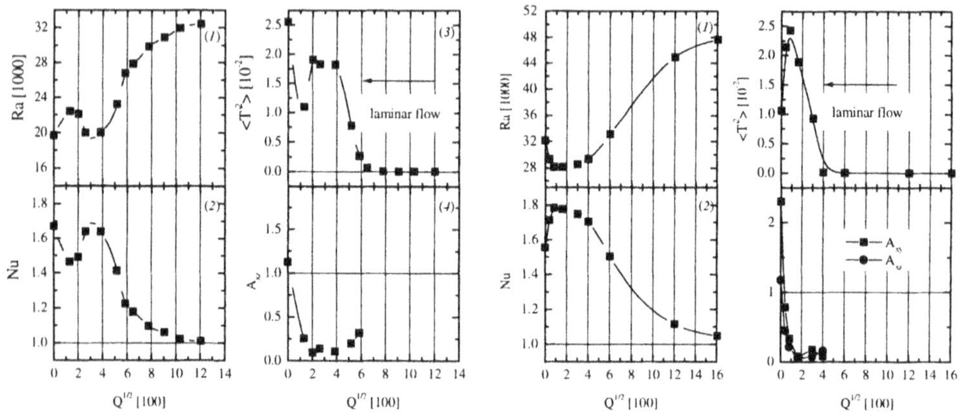

Figure 5: *Rayleigh numbers Ra, Nusselt numbers Nu, Fluctuating intensity $< T'^2 >$ and isotropy coefficients A_{ij}. (a) RB convection and (b) VS.*

4. CONCLUSIONS

Multielement temperature probes are effective tools in demonstrating the ordering effect of a horizontal magnetic field on buoyant flow fields. It is obtained that the magnetic field generates local non-isotropy in the sense that convective vortices become aligned with its direction. From the reduced non-linearities the non-organised convective flow field present at ordinary hydrodynamic conditions turns into a well organised one with increased heat transfer properties even though Joule's dissipation is present.

5. REFERENCES

Davidson, P.A. 1995 Magnetic damping of jets and vortices. *J.Fluid Mech.* **299**, 153-186.

Elder, J. W. 1965 Laminar free convection in a vertical slot. *J.Fluid Mech.* **23**, 77-98.

Krishnamurti, R. 1973 Some further studies on the transition to turbulent convection. *J.Fluid Mech.* **60**, 285-303.

Pellew, A. and Southwell, R. 1940 On maintained convective motion in a fluid heated from below. *Proc. R. Soc. (London)* **A176**, 312-343.

Seki, N., Fukasako, S. and Inaba, H. 1978 Visual observation of natural convective flow in a narrow vertical cavity. *J.Fluid Mech.* **84**, 695-704.

Knots and braids on the Sun

Renzo L. Ricca
Department of Mathematics, University College London
Gower Street, London WC1E 6BT, UK
ricca@math.ucl.ac.uk

Abstract: In this paper we show how new techniques of topological fluid mechanics and physical knot theory can be applied to estimate magnetic energy levels in solar physics. In particular, we show that magnetic energy stored in complex configurations of plasma loops present on the Sun can be quantified by geometric and topological information. These studies find important applications in the energetics of solar and stellar physics and in laboratory plasmas.
Keywords: topological magnetohydrodynamics, magnetic knots, braids.

1. SOLAR CORONAL STRUCTURES AND THE PHYSICS OF MAGNETIC KNOTS

Most of the energy emitted by the Sun that reaches us in the form of visible and invisible light and heating is generated by the intense activity of solar magnetic fields. Near the solar surface, and possibly in the entire convection zone just below the photosphere, these fields structure the plasma into coherent regions in the shape of giant arches, loops and flux tubes rooted in the solar surface and projected into the corona (see figure 1; Spruit & Roberts, 1983; Bray *et al.*, 1991). Twisted and braided magnetic loops are indeed basic structural elements of solar and stellar atmospheres. On the Sun, more than 90% of the magnetic flux outside sunspots is concentrated into intense flux tubes, which are responsible for a great part of energy emission from the solar corona. Many phenomena associated with solar activity, such as formation of sunspots, prominences and flares, are directly related to the presence of these structures. Solar loops, which are made visible through density and temperature inhomogeneities aligned with the field, have typical thickness of the order of 100–1500 km and can rise up to 400000 km in the corona, at temperatures of millions of degrees.

Direct observation of these structures reveals braided configurations with complex topology of the magnetic field. The magnetic lines of force appear to be highly tangled, and this makes it much more difficult to estimate the magnetic energy stored in the structure. In recent years, an alternative approach based on topological techniques borrowed from knot theory and applied to magnetohydrodynamics

A. Gyr et al. (eds.), Science and Art Symposium 2000, 263-268.

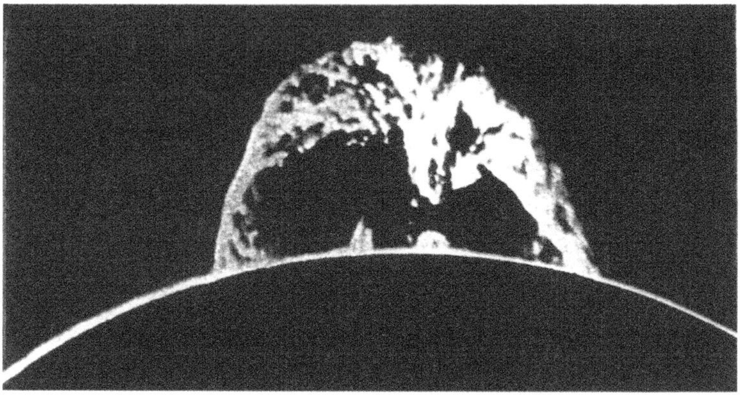

Figure 1. *Eruptive prominence as seen at the solar limb on the 13th of March, 1970. The full high of the prominence is estimated to be 370000 km and to rise at 160 km/s (from Giovanelli, 1984).*

(see Ricca & Berger, 1996; Ricca, 1998a; 1998b) has made it possible to provide new information on some important aspects of the energetics of astrophysical flows.

Mathematical models of coronal loops consider a collection of magnetic filaments, knotted and linked together to form arched braids (as in figure 2) embedded in a perfectly conducting plasma (i.e. in ideal conditions). Magnetic filaments interact with the surrounding medium through the Lorentz force and, as elastic strings, experience a tension proportional to their curvature. The curvature force is responsible for the dynamics of each filament in the medium. In ideal conditions, then, the topology of the magnetic field is 'frozen' in the fluid: this means that there are no reconnections of magnetic field lines taking place, and therefore knot and link types (and physical quantities, such as total energy and helicity) of the magnetic field distribution do not change with time. These quantities guide the magnetic relaxation of the structure towards states of minimum energy. Hence, two aspects influence the energetics of magnetic loops: a dynamical action of a curvature force (essentially related to the geometry of the magnetic field), and the type of knot and link of the magnetic configuration.

2. FROM MAGNETIC KNOTS TO BRAIDS

Knots and links are classified according to their topology. A useful measure of knot complexity is given by the minimum number of crossings $c_{\min}$ of a knot type, that is a topological invariant of the knot: its value does not change as the knot changes shape; $c_{\min}$ is determined by counting the number of crossings in the knot diagram, obtained by projecting the knot onto a plane so as that it shows the minimum

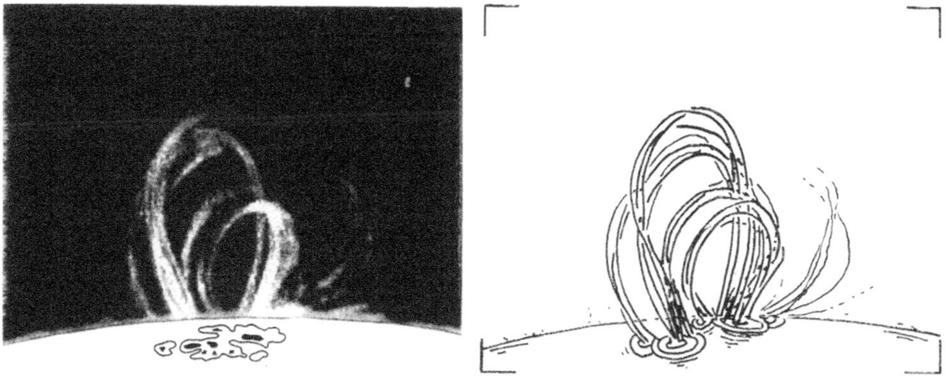

Figure 2. *Coronal loops in an active region (from Bray et al., 1991; on the left) and a sketch of the corresponding magnetic pattern (on the right). These loops are anchored to sunspots, whose locations (shown in the drawing below the picture on the left) are established with the aid of auxiliary data.*

possible number of crossings. The three knot types in the first row of figure 3, for example, are shown in their minimal projection (standard representation).

Of course a knot can be drawn to take any shape. Particularly interesting geometries are obtained when we take a standard knot and, by a series of continuous deformations, we change its shape to a closed braid with the least possible number $c_\circ$ of crossings (minimal braids), as shown by the knots in the second row of figure 3. It is known (cf. Ricca, 1998) that $c_\circ$ is bounded by

$$c_{\min} \leq c_\circ \leq c_{\min} + (c_{\min} - 1)(c_{\min} - 2) \ . \tag{1}$$

Minimal braids have interesting features: the curve goes around a centre as in an orbit-like pattern and without inflexions. An inflexion point (in isolation) is present when the intrinsic curvature of the knot vanishes: in a planar, S-shaped curve, for example, this point is visible at the change of concavity. The standard knots in the first row of figure 3 exhibit inflexional configurations, whereas their minimal braid configurations have none.

A geometric quantity useful to measure the average amount of coiling of a knot is the writhing number Wr. The writhe can be measured by the sum of the signed crossings of the knot diagram (not necessarily the minimal diagram) onto some projection plane, averaged over all projections, that is

$$Wr = < n_+(\boldsymbol{\nu}) - n_-(\boldsymbol{\nu}) > \ , \tag{2}$$

where $< \cdot >$ denotes averaging over all directions $\boldsymbol{\nu}$ of projection, and $n_\pm$ denotes the number of apparent $\pm$ crossings, from the direction of projection $\boldsymbol{\nu}$. For a nearly

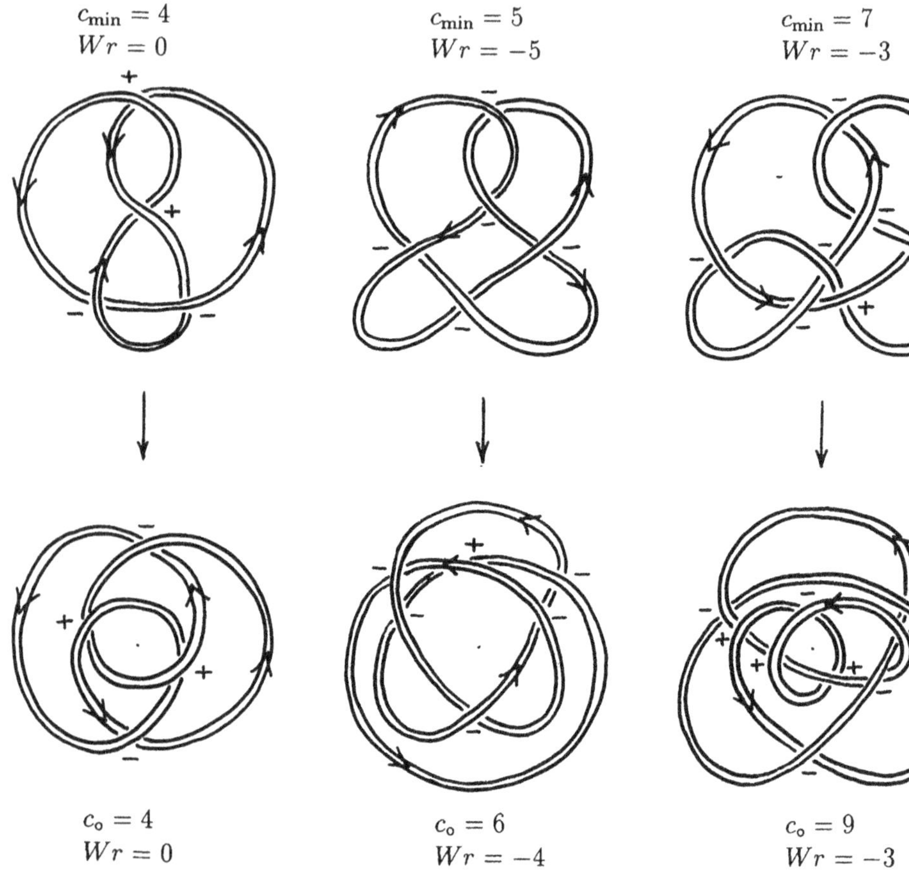

Figure 3. *Different knot types in standard representation (first row) can be transformed by a series of continuous deformations to minimal braids (second row).*

plane curve (except small indentations to allow crossings, as in the diagrams of Figure 3) the writhe can be directly estimated by the sum of the signed crossings. For a physical knot a change in writhe has physical implications and it is therefore a useful information.

Magnetic knots are dynamical objects that spontaneously tend to minimize their energy by re-arranging their shape. In a recent analysis (Ricca, 1997) it has been shown that twisted magnetic flux tubes are in disequilibrium in inflexional configurations (which are ubiquitous in astrophysical flows). Since the topology of the magnetic field is frozen and the Lorentz force drives the lines of force to an inflexion-free state, magnetic knots tend to remove inflexion points from their geometry to form closed braid configurations. Moreover, as the knot changes shape, the writhe value changes as well (remember that Wr is a geometric quantity that depends on the shape). Since in ideal conditions magnetic helicity is conserved, an

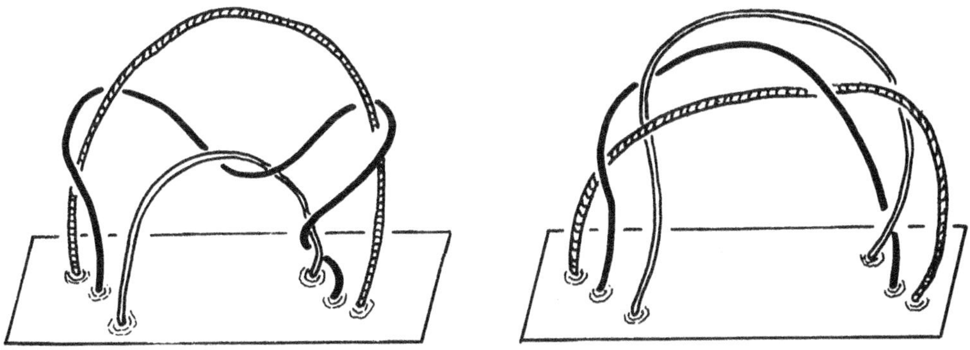

Figure 4. *A magnetic braid of 8 crossings (on the left) can relax to a minimal configuration (on the left) with* $c_o = c_{\min} = 6$.

increase in writhe is always accompanied by a decrease in the field lines twist, a mechanism that is favoured by the natural relaxation of the magnetic system.

3. MINIMAL BRAIDS AND ENERGY ESTIMATES

Is it possible to measure energy levels according to magnetic knot complexity? Recent studies on knot energy (Moffatt, 1990; Freedman & He, 1991; Berger, 1993) show that energy levels can be indeed related to crossing numbers, hence to knot complexity. Bounds on energy levels of magnetic knots are given by relations of the kind

$$E_{\min} \geq E_0 \equiv F(\Phi, V, N; c_{\min}) \ , \tag{3}$$

where $E_{\min}$ is the minimum energy level, E_0 is the ground state energy and $F(\cdot)$ gives the relationship between magnetic quantities such as total flux Φ, volume V, number of filaments N, and knot topology given by $c_{\min}$. Typically energy increases with $c_{\min}$ and therefore with knot complexity. The inequality sign, though, allows ample margin for error.

These estimates are based on calculations that provide limit values for global minima of energy, without taking into account any dynamic feature associated with the magnetic field. The competition between magnetic field distribution and the geometry of the structure may actually prevent the physical knot from reaching these limit values. We should remember that in ideal conditions magnetic knots relax their energy by moving according to the Lorentz force (a curvature force), that makes the lines of force shrinking. Since inflexional magnetic knots are forced to remove inflexions, there is a possibility that they simply cannot relax to minimal knots, if these exhibit inflexions. In this case they are forced to move to inflexion-free braids first, and then, by removing redundant crossings so as to reduce as much as possible their potential energy, to take the shape of minimal braids (that represent local minima for magnetic energy; see figure 4). Since minimal braids have $c_o \geq c_{min}$, eq. (3) suggests that the least possible energy stored in these braids ($E_{o,min}$) is likely to be higher than the theoretical limit given by E_0, i.e. $E_{o,min} \geq E_{min} \geq E_0$. Considerations based on application of the upper bound of inequality (1), then, makes us believe that the minimum amount of energy actually embodied in solar magnetic structures can be well above the theoretical limit E_0.

These considerations have important implications for astrophysical and terrestrial applications. Accurate estimates of minimum magnetic energy levels are useful to quantify the energy that can be released into heat during flares, microflares, and mass ejections from the Sun, and are also important for space applications and fusion physics in laboratory plasmas.

Acknowledgements: Financial support from the UK Particle Physics and Astronomy Research Council (PPARC) Grant no. GR/L63143 is kindly acknowledged.

REFERENCES

BERGER, M.A. 1993 Energy-crossing number relations for braided magnetic fields. *Phys. Rev. Lett.* **70**, 705–708.

BRAY, R.J., CRAM, L.E., DURRANT, C.J. & LOUGHHEAD, R.E. 1991 *Plasma Loops in the Solar Corona.* Cambridge University Press.

FREEDMAN, M.H. & HE, Z.-X. 1991 Divergence-free fields: energy and asymptotic crossing number. *Ann. Math.* **134**, 189–229.

GIOVANELLI, R. 1984 *Secrets of the Sun.* Cambridge University Press.

MOFFATT, H.K. 1990 The energy spectrum of knots and links. *Nature* **347**, 367–369.

RICCA, R.L. 1997 Evolution and inflexional instability of twisted magnetic flux tubes. *Solar Physics* **172**, 241–248.

RICCA, R.L. 1998a Applications of knot theory in fluid mechanics. In *Knot Theory* (ed. V.F.R. Jones *et al.*), pp. 321–346. Banach Center Publs. **42**, Polish Academy of Sciences, Warsaw.

RICCA, R.L. 1998b New developments in topological fluid mechanics: from Kelvin's vortex knots to magnetic knots. In *Ideal Knots* (ed. A. Stasiak *et al.*), pp. 255–273. Series on Knots and Everything **19**, World Scientific, Singapore.

RICCA, R.L. & BERGER, M.A. 1996 Topological ideas and fluid mechanics. *Phys. Today* **49** (12), 24–30.

SPRUIT, H.C. & ROBERTS, B. 1983 Magnetic flux tubes on the Sun. *Nature* **304**, 401–406.

Reconnections of elliptical vortex rings

A. Gyr & A. Müller
Institute of Hydromechanics and Water Resources Management, ETHZ
CH- 8093 Zürich-Hönggerberg
gyr@ihw.baum.ethz.ch, muller@ihw.baum.ethz.ch

Abstract: An attempt was made to produce a topological structure as predicted by Aref and Zawadzki via the reconnection of two elliptical vortex rings moving in the same direction and having a common axis. It is shown that such a reconnection is possible but so unstable that its realization is unrealistic under normal conditions.
Keywords: vortex ring, reconnection, topology

1. INTRODUCTION

As long as a fluid can be assumed to be inviscid, the topology of a flow remains unchanged. The topology of vortex lines in the ideal fluid then remains the same, while the structures of these objects, in continuos motion, can be highly distorted by the background flow. This behavior is a fundamental, intrinsic, property of the Cauchy equations, whose topological implications were studied extensively, see Arnold and Khesin, 1992, as well as Tur and Yanovski, 1993. Changes of topology start occurring when viscous effects dominate the coherence of a structure. When dissipation becomes noticeable the fluid patterns change on small time-scales compared to the evolution time scale of the flow. The changes occur through the formation and disappearance of physical connections in the fluid pattern due to vortex tubes interacting and reconnecting freely. This behavior of vortex tubes in a viscous fluid is one of the main obstacles to a description of the topology of a turbulent flow field. For this reason reconnections of vortex tubes are studied under very specific and most simple conditions to get a better insight into these complex mechanisms.

Ricca and Berger, 1996 reproduced Tait's 1898 diagrammed knot and links and discussed the possible existence of a linked pair of vortex rings as predicted by [Aref and Zawadzki, 1991, 1993, and Zawadzki & Aref, 1992] as the most simple violation of the ideal fluid topology by viscosity. Aref and Zawadzki's simulations were not the first numerical investigation of vortex ring reconnections, but all others where treating circular vortex rings, and as the authors showed, it is when the axisymmetric configuration is relaxed that significant interaction modes appear.

Zawadski and Aref (1992) investigated two identical elliptical vortex rings moving in the same direction and having a common axis. The numerical method they employed was a variation of the so-called vortex–in-cell (VIC) method originally proposed for 2D

A. Gyr et al. (eds.), Science and Art Symposium 2000, 269-274.

simulations by Christiansen (1973), and adapted by the authors to a 3D method in which every fluid particle carries a vorticity vector. The vortex interaction was characterized by the major axis of the ellipse, a; the core radius, c; the initial separation, s, of the planes containing the vortices, and the inclination angle, α, defined as the angle between the major axes of the vortex rings at interaction. All these parameters were given in grid units as the characteristic dimension. The periodic, computational box had sides of 2π and 64^3 grids. Each vortex ring had a Gaussian vorticity distribution within the core and a ratio of minor to major axis equal to 0.4, with the length of the major axis equal to 50 grid units. They found an intermediate stage with two linked vortex loops for $\alpha = 45°$, c= 3 and a s= 8. In other words, they found the violation in the topology which is caused by viscosity, however, the life time of the linked vortex ring was very short and it ended up in 'ringlets'.

The intention of the present investigation is to verify these findings experimentally.

2. THE EXPERIMENTAL APPROACH

There are a number of methods to create vortex rings, the most common one is to suddenly expel a defined amount of fluid through a sharp-edged cylindrical orifice, as shown in Fig 1. It is more tricky to produce two rings interacting in a predetermined location. Schatzle (1987) solved the problem with two pistons, but the resulting rings had no common axis of propagation. We therefore tried to produce vortex rings with a single piston arrangement with a fixed tube diameter, D, as described by Aref and Zawadzki. The parameters of such an arrangement are: the velocity of the piston, W_i, the duration of the feed T_i and the intermediate time, ΔT between the first and the second feed i= 1 and 2.

The self-induced dynamics of a single elliptic vortex ring were studied by Dhanak and de Bernardinis (1981). Based on their linearized approximation of the elliptic ring we investigated whether the mentioned five parameters together with the lengths of the half axes of the elliptical orifice, a_0 and b_0 are sufficient to calculate the situation striven for. In this linear theory the elliptical vortex ring can be thought as the first perturbation of wavenumber m= 2 of a circular ring given by

$$\underline{X} = \left(R + re^{im\theta}\right)\underline{e}_r + ze^{im\theta}\,\underline{e}_z \qquad (1)$$

where $\underline{e}_r$ and $\underline{e}_z$ are unit vectors in the radial and axial directions, θ is the azimuthal angle, R is the radius of the ring of the unperturbed ring and $|r|$, $|z| \ll R$. The elliptic ring is therefore initially of plane form given by,

$$r(t) = r_o \cos(\alpha_2 t), \quad z(t) = z_0 \sin(\alpha_2 t) \qquad (2)$$

Using the value α_2 given by Widnall and Sullivan (1973) the period of oscillation, τ, $2\pi/\alpha_2$, which depends on R and the internal structure of the vortex, its core radius c and the vorticity distribution given by A is

$$\tau(R,c,A) = \frac{8\pi^2 R^2}{\Gamma}\left[\{4(\ln c/R - A) + 0.22\}\{3(\ln c/R - A) + 2.23\}\right]^{-1/2} \qquad (3)$$

with

$$A = \frac{4\pi^2}{\Gamma^2}\int_0^c rv^2 dr \tag{4}$$

There is no axial flow in the filament, v is the swirl velocity in the core and Γ its circulation. The self-induced mean velocity V of the ring is the same as that of the unperturbed circular ring,

$$\underline{V}(R,c,A) = \frac{e_z\Gamma}{4\pi R}\left(\ln\frac{8R}{c} + A - \frac{1}{2}\right) \tag{5}$$

With these values in the (n.z) coordinates the location N_i and Z_i of the center of the developing ring and its initial conditions can be calculated as

$$N = 0.08(\alpha T)^{2/3} \quad , Z = 0.34(\alpha T)^{2/3} \quad with$$

$$\alpha = \frac{\sqrt{\pi}}{8}\frac{WD^2}{a_0 b_0^{1/2} E^{1/2}(e)}, \quad e = \sqrt{\left(a_0^2 - b_0^2\right)/a_0^2}, \tag{6}$$

where E(e) is the complete elliptical integral of the second kind.

The axes of the vortex ring are growing, and the parameters R, c, A then become

$$a = a_0 + 0.08(\alpha T)^{2/3}, \quad b = b_0 + 0.08(\alpha T)^{2/3}, \quad c = \frac{0.06\Gamma_0^2}{\alpha^2} \tag{7}$$

such that

$$\Gamma = 2.4\alpha^{4/3}T^{1/3}, \quad \overline{R}_0 = N + \frac{a_0 + b_0}{2}, \quad A = \frac{1}{2}, \quad B = \frac{c}{R}, \quad R = N + a_0 \tag{8}$$

With these initial conditions eqs. (3 and 5) can be written as

$$\tau\left(\overline{R},c,1/2\right) = \frac{79\overline{R}^2}{\Gamma}\left[(4B - 1.78)(3B + 0.73)\right]^{-1/2} \tag{9}$$

$$V\left(\overline{R},c,1/2\right) = \frac{e_z\Gamma}{4\pi\overline{R}}\left(\ln\frac{8}{B}\right) \tag{10}$$

Based on Aref and Zawadski's calculation one has to produce two elliptical vortices which are out of phase by 45°. In a coaxial arrangement as used in the experiment this configuration cannot be achieved because the vertex of the elliptical ring is the result of a standing wave on the core, and in the 45° position the wave has deformed the ring into a circular one. Therefore, an analogous configuration is investigated in which an elliptical vortex interacts with an essentially circular one which is slightly deformed in the third dimension. This can only be achieved when the second vortex has a moderately higher frequency τ_2. The desired configuration occurs when the following conditions are satisfied

$$Z_1 + \frac{(n+1)\tau_1}{8}V_1 = Z_2 + \frac{n\tau_2}{8}V_2, \qquad n = 1,2,3,... \tag{11}$$

$$\frac{(n+1)}{8}\tau_1 = \Delta T + T_2 + \frac{n}{8}\tau_2, \quad n = 1,2,3,... \tag{12}$$

Eqs. (11 and 12) contain six variables: n, ΔT, T_i and W_i, i= 1,2 in other words the system is underdetermined and a condition as given by this two equations has many solutions. New conditions can be stated to reduce these numbers, but it is not clear whether they are physically relevant. For example for vortices of the same volume of vortical fluid

$$T_1W_1 = T_2W_2 \quad \therefore \quad Z_1 = Z_2 \quad and \quad N_1 = N_2 \quad because \quad N_i, Z_i \propto (W_iT_i)^{2/3} \tag{13}$$

and eq. (11) becomes $\frac{(n+1)\tau_1}{8} = \frac{n\tau_2}{8}$, $n = 1,2,3,\ldots$ introduced into eq. (12)

$$\frac{n}{8}\tau_2\left(\frac{V_2}{V_1} - 1\right) = \Delta T + T_2 \tag{14}$$

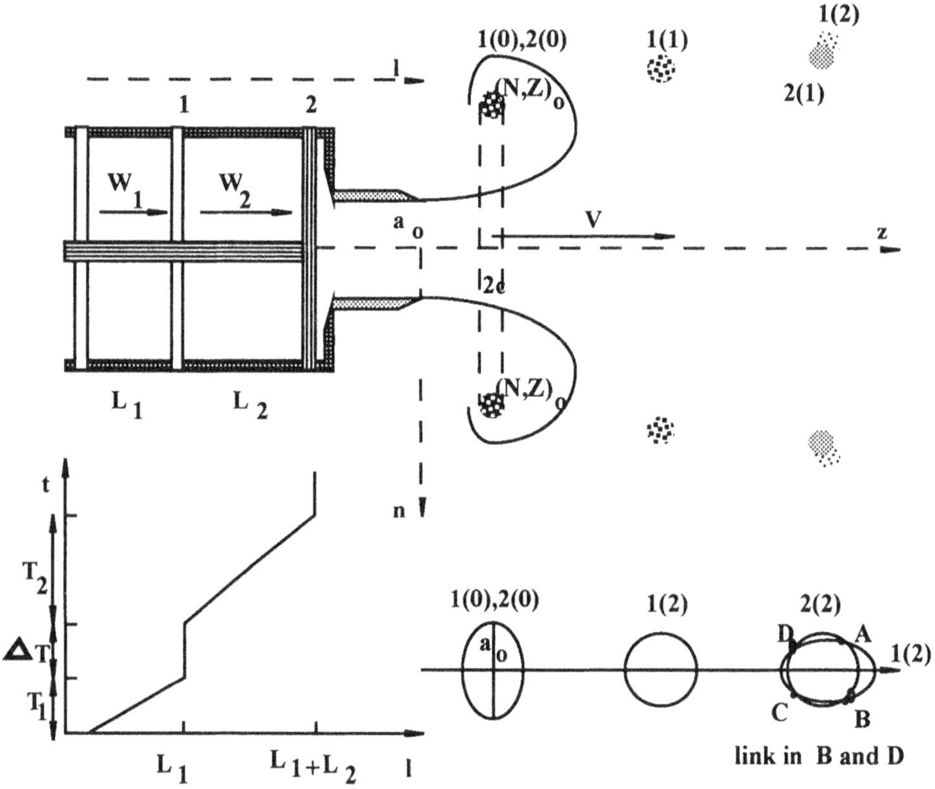

Figure 1: A sketch of the experimental set up with the definitions of the coordinate system, and a schematic of the vortex reconnections. The motion of the piston is shown in the lower part, together with a sketch of the topology at interaction.

For a given n this is a relation of the two frequencies of the vortices, and the sum of the two remaining variables. Still with this very restrictive assumption the solution is a complex implicit equation which needs additional assumptions or a fixing of variables.

It is meaningful to assign a value of 1 or 2 to n. The higher value seems more appropriate because the first condition is strongly influenced by the experimental initial condition. This is the main experimental problem as the second vortex ring is influenced during its whole life time not only by its vorticity field, but also by its wake. The clean situation as used for the numerical simulation can never been achieved. Therefore these theoretical estimates are used to fix some of the variables, but the main investigation is a trial and error experiment in the vicinity of the theoretical predicted values.

3. RESULTS

The experiments failed insofar as no picture of the intended reconnection could be taken. Nevertheless, we were twice able to see a reconnection as predicted by Aref and Zawadzki but the lifetime of this interlaced vortex rings were extremely short, and the geometry seen with the eye could not be caught with the digital camera. In addition such interacting vortex rings were not reproducible, because the wake of the first vortex ring differed from one produced in an other experiment under the identical initial conditions.

We tried to portray the dynamical behavior of the two rings with video cameras but failed too because the three-dimensionality could not be resolved with the appropriate accuracy.

Nevertheless, some picture are shown to give an impression of the very difficult task, and some to show new unexpected configurations. For example a new type of false frog leaping, where the second ring instead of going through is folding the first one (Fig. 2b). Fig. 2a shows that the first ring is by far more three-dimensional than the second one a feature, which is not predicted by the theory used here. (See color plate 8,III)

An interesting self-reconnection occurs when a strongly three-dimensional deformed elliptic ring decays into two small circular vortex rings with a remaining central circular ring.

These results make clear that the behavior of elliptical rings is much more intricate than we thought. It indicates that reconnection processes are efficient in reducing complicated topologies into simple circular vortex rings.

4. REFERENCES

Aref, H. & Zawadzki, I.; Linking of vortex ring; Nature **354**, 50 – 53, 1991, or Vortex interaction as a dynamical system. In New Approaches and Concepts in Turbulence, Eds. Dracos, T. & Tsinober, A., Birkhäuser Verl., 191- 205, 1993

Arnold, V.I. & Khesin, B.A.; Topological methods in hydrodynamics' ; Ann. Rev. Fluid Mech. **24**, 145-166, 1992

Christiansen, J.P.; Numerical simulation of hydrodynamics by the method of point vortices.; J. Comp. Phys. **13**, 363-379, 1973

Ricca, R.L. & Berger, M.A.; 'Topological ideas and fluid mechanics'; Phys. Today 28-34 Dec. 1996

Schatzle, R.S. An experimental study of fusion of vortex rings. Ph. D. thesis, CALTECH 1987

Tait, P.G.; Scientific papers, Cambridge Univ. Press **1**, 1898

Tur, A.V. & Yanovski, V.V.; 'Invariants in dissipationless hydrodynamic media'; J. Fluid Mech. 248, 67 – 106, 1993

Widnall, S.E. & Sullivan, J.P. On the stability of vortex rings. Proc. Roy. Soc. A **322**, 335

Zawadzki, I. & Aref, H.; Numerical experiments on vortex ring interactions in the large.' Private communication based on a non-submitted draft for the J. Fluid Mech., 1992 most of the content was published in Aref and Zawadzki 1993.

Vorticity induced by electrostatic forces

W.Egli and E.A.Gerteisen* in Memoriam J.J.Keller
ABB Corporate Research, Switzerland, currently LHS AG Zürich
walter.egli@ch.lhsgroup.com
**Swiss Center for Scientific Computing (CSCS), Manno, Switzerland, currently DaimlerChrysler AG, R&T, Ulm, Germany*

Abstract: In a discharge configuration with high voltage corona wires the gas flow is affected by body forces which are induced by the flow of ions. One device strongly influenced by this ionic wind is the electrostatic precipitator. A simple configuration is considered in which the corona wires are parallel to the mean gas flow, located in the center plane, and bounded by two grounded plates. Vortices are formed between the wires and the plates. The mainly non-viscous effects of the gas flow can be represented by the Euler equation. For a specific two-dimensional situation and together with some assumptions a linearized semi-analytical solution is derived. These inherently steady flow patterns are compared with finite volume solutions of the full set of nonlinear equations. This finite volume simulation of the flow field compares well with the linearized solution at an initial state, however it tends to produce instabilities and it is not certain that a steady state solution exists.
Keywords: ionic wind, electrostatic precipitator, corona wire, gas flow.

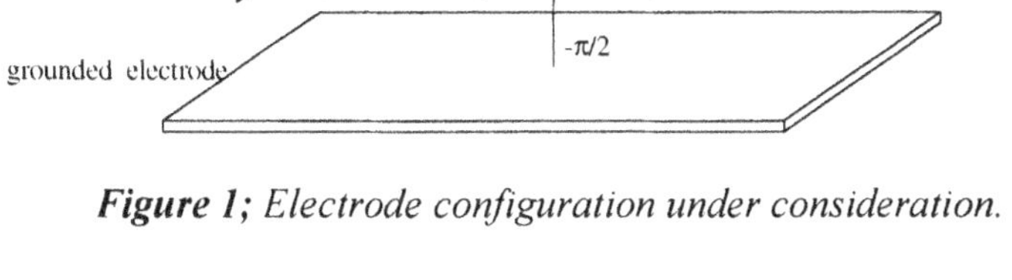

Figure 1; *Electrode configuration under consideration.*

1. SEMI-ANALYTICAL SOLUTION OF THE SECONDARY FLOW

The Euler equation describes the most important effects related to additional source-terms for momentum caused by volume forces, i.e. drift of ions from the corona electrodes to the ground plate. The investigated configuration according to figure 1 is chosen to be infinite in the main flow direction and steady state solutions are supposed

A. Gyr et al. (eds.), Science and Art Symposium 2000, 275-278.

for all dependent values. Assuming that the components of the secondary flow are small compared to the mean flow velocity v_0, typically of the order of 1 m/s, the momentum equation is linearized :

$$v_0 \frac{\partial \mathbf{u}}{\partial y} + \frac{1}{\rho} \nabla p = \frac{1}{\rho} \mathbf{F}_i \tag{1}$$

where ρ, p, $\mathbf{F}_i$ denote respectively the density, the pressure and the electrical force field through the momentum transfer of the ion flow to the gas molecules, ρ_i **E**. With the elimination of the pressure by $\Delta p = 0$ and introduction of the stream function Ψ, the flow field is described by the Poisson equation as follow:

$$\Psi_{xx} + \Psi_{yy} = \frac{y}{\rho v_0} \left[\frac{\partial F_{i1}}{\partial z} - \frac{\partial F_{i3}}{\partial x} \right] \tag{2}$$

Because the influence of the space charge on the electric field is neglected the electric potential Φ is given by the Laplace equation. For a plane wire configuration with the potential U0 at the wire the following expression holds:

$$\Phi = \frac{1}{2} U_0 \log \frac{\cosh x - \cos z}{\cosh x + \cos z} \tag{3}$$

And the electric field is given by

$$\mathbf{E} = -\nabla \Phi = -U_0 \frac{[\cos z \sinh x, 0, \sin z \cosh x]}{\sinh^2(x) + \sin^2(z)} \tag{4}$$

The continuity for the charge is given with the mobility μ of the ions by the relation with the current density **i**:

$$\nabla \cdot \mathbf{i} = \nabla \cdot (\mu \rho_i \mathbf{E}) = 0 \tag{5}$$

According to the relations (4,5) the space charge density, $\rho_i(x, y, z)$, is constant along electrical field lines. With the variable, ξ, constant along a field line and ξ = x for /z/ = π/2 the space charge density is given by

$$\rho_i(x, y, z) = \rho_i(\xi, y, z) = \frac{i(\xi, y, z)}{\mu \mathrm{E}(\xi, y, z)} \tag{6}$$

This assumption is only zero order accurate. As one can see from the solution of the characteristic equation along the field lines the ion space charge density decreases with the travel time of the ions [1,2]. As an example the current density profile f(x) on the plane grounded plate is measured from one wire and it is given by a triangular distribution, see also [4]:

$$i_1(x, y, \frac{\pi}{2}) = 0, \quad i_2(x, y, \frac{\pi}{2}) = 0, \quad i_3(x, y, \frac{\pi}{2}) = f(x) \tag{7}$$

$$f(x)=\max\left(0,\left(1-\frac{2}{\pi}|x|\right)^{c}\right),\quad c = \text{parameter.} \tag{8}$$

With the help of (4-8) the space charge density is defined:

$$\rho_i(x,y,z)=\frac{1}{\mu U_0}f(\xi)\cosh\xi \quad \text{where } \xi=\sinh^{-1}\frac{\sinh x}{\sin z} \tag{9}$$

The stream function Ψ is computed by solving numerically the Poisson equation (2). The right hand side of this equation is computed via numerical differentiation of the volume force field. For a multi-wire configuration the electric field (4) is adapted via superposition. The electric field has, see (6), an influence on the space charge. Equation (9) must also be modified. For the result presented it is assumed that influence of neighbor wires, index n, to the current density profile is neglected. The force field is then:

$$\mathbf{F}_i(x,y,z)=\rho_i\mathbf{E}=-\rho_i U_0\sum_n\frac{[\cos z\sinh(x+na),0,\sin z\cosh(x+na)]}{\sinh^2(x+na)+\sin^2 z} \tag{10}$$

The stream function is set to zero on the plates and on symmetry lines. Results for two spacing ratios in the rectangular region of one wire $[-a/2\le x\le a/2,-\pi/2\le z\le\pi/2]$ are presented in figures 2-3. In the calculation the number of neighbor wires on each side of the wire at the origin is set to 3, the range of the index n in (10) is –3 to 3.

2. FINITE VOLUME EULER SOLUTION.

The semi-discrete form of the full set of nonlinear Euler equations for the gas flow is produced from a finite volume space discretization and is solved numerically by the method of lines, i.e. the discrete space operator is relaxed by advancing in time. To retain the coupling between the pressure and momentum field for the incompressible equations, an compressibility term is added to the continuity equation [5], and local preconditioning is applied to enhance the eigenvalue characteristics of the system [6]. It is solved in pseudo time with a Runge-Kutta scheme [7] to seek a steady state solution. The numerical simulation is initiated from parallel flow in wire direction and the body forces of the semi-analytical solution are imposed. After a few cycles of iteration, when the magnitude of the secondary flow components is still small, the flow field is very much similar to the linearized solution, see figure 2, picture 1 and 2 of figure 4. However, after some additional cycles, the cross-flow magnitude is increasing and the form is changed from the butterfly propellers to round shaped ellipses.

The corresponding pressure distribution develops lens forms. These typically indicate the occurrence of instability and steady state converged solutions could not be achieved. Although the present computation were performed within an artificial time domain, the flow and the pressure patterns are regarded to be representative for the development of weak cross-flows, to strong ones, to instabilities as they may typically be induced in gas flows driven by electrostatic body forces. In the figure 4 the evolution of the streamlines is plotted starting from upper left. It can be expected that the instability formations depend on the geometric shape as well as on the strength of the exciting forces. A similar configuration was investigated earlier [3] where a steady state situation was found between the electric forces and the flow vorticity momentum.

3. CONCLUSIONS

The flow patterns from the analysis with the linearized flow equation shows counter rotating vortices in a wire plate configuration. These are analogous to the initial phase of the finite volume simulation. For higher magnitudes of the secondary flow the vortices inflate and develop more round shaped elliptic forms. A steady state condition could not be found in the nonlinear simulation. A time dependent permutation analysis and the investigation of higher order spatial permutations would help to clarify the behavior. Beside that, advanced space charge distribution models should be implemented in the semi-analytical method.

4. FINITE VOLUME EULER SOLUTION

Special thanks are due to Jakob J. Keller who developed the analytical formulation presented in this paper. He always had a great interest in the beauty of graphical presentations combined with the clarity of analytical approaches. Many thanks are due to Ulrich Kogelschatz for proof-reading and supporting this contribution.

5. REFERENCES

[1] J.E.Jones, J.Phys.D:Appl.Phys.27(19994) 1835-1847

[2] W.Egli, U.Kogelschatz, E.A.Gerteisen and R.Gruber, 3D computation of Corona, Ion Induced Secondary Flows and Particle Motion in Technical ESP Configurations, J.of Electrostatics 40&41(1997) 425-430

[3] E.A.Gerteisen, W.Egli and B.Eliasson, Vorticity Formation in an Electrostatic Precipitator, Helvetica Physica Acta 67(1994) 769-770

[4] E.Warburg, Ueber die Spitzenentladung, Wiedemanns Annalen der Physik und Chemie 67 (1899), 69 – 83

[5] A.J.Chorin, A numerical method for solving incompressible viscous flow problems, J.Comp. Phys. 2, (1967), 12-26

[6] A.Rizzi and Lars-Erik Eriksson, Computation of inviscid incompressible flow with rotation, J.Fluid Mech. 153 (1985), 275-312

[7] A.Jameson, W.Schmidt and E.Turkel, Numerical Solution of the Euler Equations by Finite Volume Methods Using Runge-Kutta Time Stepping Schemes, AIAA-Paper 81-1259, 1981

Simulation of Flow over Two Unequal Square Cylinders

***A.Y.W. Chiu*[1a] *and N.W.M. Ko*[1b]**
[1]*Department of Mechanical Engineering, The University of Hong Kong, Pokfulam Road, Hong Kong, People's Republic of China.*
[a]*alex.chiu@graduate.hku.hk*
[b]*nwmko@hkucc.hku.hk*

Abstract: A two dimensional hybrid discrete vortex method is used to simulate flow over two unequal square cylinders in side-by-side arrangements at a Reynolds number $Re_D = 550$. The large cylinder face width is twice that of the small cylinder, giving a ratio of $D/d = 2$. The normalized transverse spacing of the cylinder centres varied between $0.875 \leq T/D \leq 8$. Independent vortex streets are observed in the wake of the two unequal square cylinders at $T/D = 8$. Proximity interference commences at $T/D = 4$ with interactions between the two vortex streets. The interaction results in a synchronized vortex shedding at $T/D = 2.75$. At $T/D = 1.25$, the gap flow is predominantly biased towards the small cylinder, due to the shear layer interactions in the small cylinder wake. At $T/D = 0.875$, the gap flow is weak and vortices are shed from the outer sides of the two square cylinders, forming a single vortex street.
Keyword: bistable flow, vortex interaction, hybrid discrete vortex method.

1. INTRODUCTION

Flow over multiple bluff bodies of equal diameters, such as circular cylinders, square cylinders and flat plates, in various arrangements has gain much attention [Zdravkovich, 1977; Kim and Durbin, 1988; Ingham, 1990; Miau, Wang and Chou, 1996]. Proximity interference results in asymmetric flow patterns, or biased gap flow, leading to different aerodynamic properties [Zdravkovich, 1977]. Higher vortex shedding frequency, higher drag and outward lift are generally associated with the small wake on the biased side. At critical spacing, the gap flow flips intermittently from one cylinder to the other, revealing the bistable natural of the flow [Kim and Durbin, 1988].

Vortex and shear layer interactions are associated with the biased flow over two circular or square cylinders [Ng, Cheng and Ko, 1997; Chiu and Ko, 1999]. Although small bluff bodies of different shapes have been used as a means of control or suppression of vortex shedding from large circular or square cylinder [Sakamoto, Tan and Haniu, 1991; Igarashi, 1997; Prasad and Williamson, 1997], the flow interactions

A. Gyr et al. (eds.), Science and Art Symposium 2000, 279-284.

of two unequal cylinders in various arrangements have not been fully addressed. The present study aimed to provide a comprehensive view on the flow interactions in the proximity interference regime of two unequal square cylinders. The mechanism of bistable flow is also investigated.

2. NUMERICAL METHOD

Vortex methods have been widely used in the studies of incompressible flows. The vorticity formulation of the Navier-Stokes equation is [Batchelor, 1967],

$$\frac{D\hat{\omega}}{Dt} - (\hat{\omega} \cdot \nabla)\hat{u} = \nu \nabla^2 \hat{\omega} , \tag{1}$$

where $\frac{D}{Dt} \equiv \frac{\partial}{\partial t} + \hat{u} \cdot \nabla$ is the material derivative operator. The vortex method by solving Eqn 1 for an inviscid fluid, involves the discretization of the vorticity field, using vortex particles or vortex blobs, which are convected under the Biot-Savart induction. A fast multipole method is used to speed up the calculation [Ng, 1996].

The viscous diffusion term on the RHS of Eqn 1 is evaluated using the diffusion velocity method [Ogami and Akamatsu, 1991; Clarke and Tutty, 1994]. The diffusion velocity components (u_d, v_d) are given as:

$$u_d = -\frac{\nu}{\omega}\frac{\partial \omega}{\partial x} , \qquad v_d = -\frac{\nu}{\omega}\frac{\partial \omega}{\partial y} , \tag{2}$$

and are added to the convection velocities (u, v) of the vortices to account for the effect of viscosity.

Alternatively, the diffusion term can be approximated by using the random walk technique [Chorin, 1973]. The vortex particles are displaced randomly in the x and y directions with a mean and standard deviation of 0 and $\sqrt{2\nu\Delta t}$, respectively. This is a simple and efficient method. However, numerical noise is introduced into the calculation, leading to errors in the boundary layer development. Hence, a hybrid technique is adopted in the present study with diffusion velocity applied to vortices near the solid boundaries, while random walk is used further downstream.

The above algorithm is known as the operator splitting technique in which the convection and diffusion of vorticity are calculated sequentially.

The face width ratio D/d between the large and small cylinder is 2. The Reynolds number Re_D, based on the large cylinder face width D, is 550. The two cylinders are placed in side-by-side arrangement, perpendicular to the flow with the transverse spacing of the cylinder centres at $0.875 \leq T/D \leq 8$. All physical dimensions are normalized by the large cylinder face width D unless specified otherwise.

3. RESULTS AND DISCUSSION

3.1 T/D = 8

At $T/D = 8$, two independent vortex streets are observed in the wake with no particular phase relationship (Figure 1). This corresponds to the region of no interference as defined for two circular cylinders [Zdravkovich, 1984]. The normalized vortex shedding frequencies of the large St_D and small St_d cylinders, based on the corresponding face widths D and d, are 0.15 and 0.13 respectively. These agree with the experimental values of 0.135 at $Re = 500$ and of 0.148 at $Re = 250$ for a single square cylinder [Norberg, 1993; Okajima, 1982].

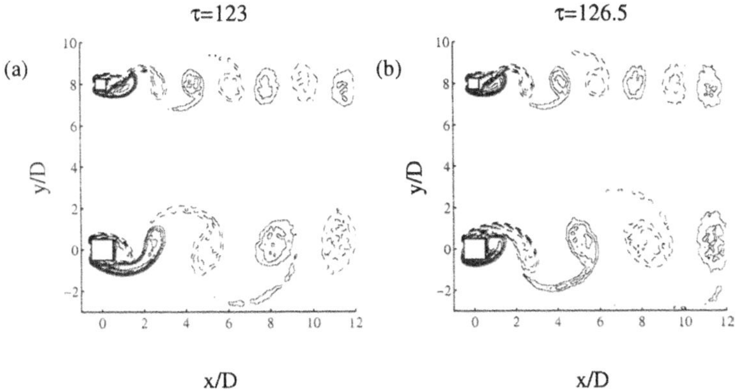

Figure 1. Independent vortex streets of two unequal square cylinders at $T/D = 8$. Normalized time τ: a) 123; b) 126.5. ——: Positive vorticity; - - - : negative vorticity. Contour level: arbitrary.

3.2 $2.75 \leq T/D \leq 4$

Wake interference commences at $T/D = 4$. The small cylinder vortex street is distorted by the mutual induction of the neighbouring vortex street. Two opposite signed small cylinder vortices are occasionally entrained between two large cylinder vortices at $x/D > 6$. This is defined as the amalgamated vortex entrainment. This was not observed for two equal square cylinders [Chiu and Ko, 1999]. No coupling or phase locking between the vortex streets is observed.

At $T/D = 2.75$, the amalgamated vortex entrainment takes place at $x/D \approx 4$, resulting in a synchronized vortex shedding from the two cylinders (Figure 2). The amalgamated vortex entrainment is triggered by the large cylinder gap vortex. The normalized vortex shedding frequencies of the large and small cylinders are $St_D = St_d = 0.143$.

3.3 T/D = 2

Due to the prominent near wake flow interations at $T/D = 2$, the vortex streets cannot be clearly identified. Sychronized vortex shedding is associated with

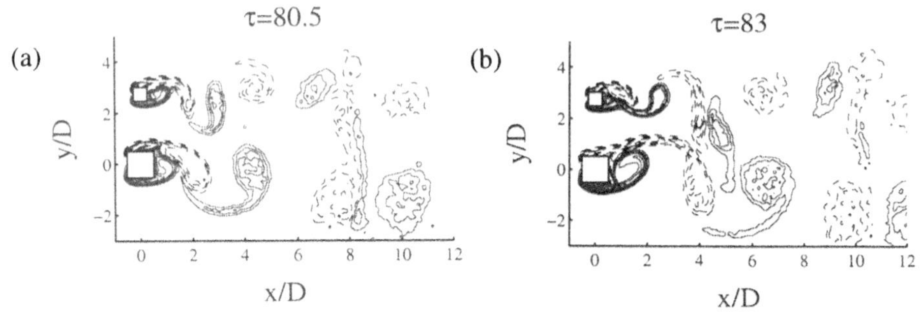

Figure 2. Amalgamated vortex entrainment process at $T/D = 2.75$. Normalized time τ: a) 80.5; b) 83. ——: Positive vorticity; - - -: negative vorticity. Contour level: arbitrary.

series of vortex interactions. Amalgamation/pairing, dipole and cycle processes as identified in the flow over two equal circular cylinders [Ng et al., 1997], are observed. Amalgamated vortex entrainment similar to that at $T/D = 2.75$ is also observed.

3.4 T/D = 1.25

At $T/D = 1.25$, the gap flow is predominantly biased towards the small cylinder, due to the successive shear layer interactions in the small cylinder wake (Figure 3). A small wake is formed on the small cylinder side with lower drag and vice versa. This shear layer interaction was defined as the coupling/entrainment process for two equal square cylinders [Chiu and Ko, 1999]. The two gap shear layers are entrained by the small cylinder outer shear layer, forming the biased gap flow. Occasionally, a coupling/entrainment process in the large cylinder wake flips the gap flow towards the large cylinder for a short period of time, indicating the bistable nature of the flow.

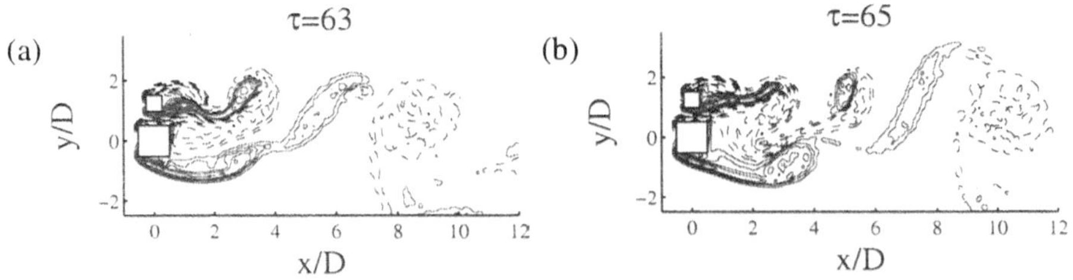

Figure 3. Upward biased gap flow at $T/D = 1.25$. Normalized time τ: a) 63; b) 65. ——: Positive vorticity; - - -: negative vorticity. Contour level: arbitrary.

3.4 T/D = 0.875

At $T/D = 0.875$, the gap flow is weak and vortices are mainly formed from the outer shear layers of the two cylinders, resulting in a single vortex street (Figure 4).

The normalized vortex shedding frequency of $St = 0.2$, based on the total frontal width (D+d), is comparable to that of two equal square cylinders at $T/D = 1.125$ and that of a sharp edged rectangular cylinders with breath to diameter ratio B/D of 0.5 [Igarashi, 1985].

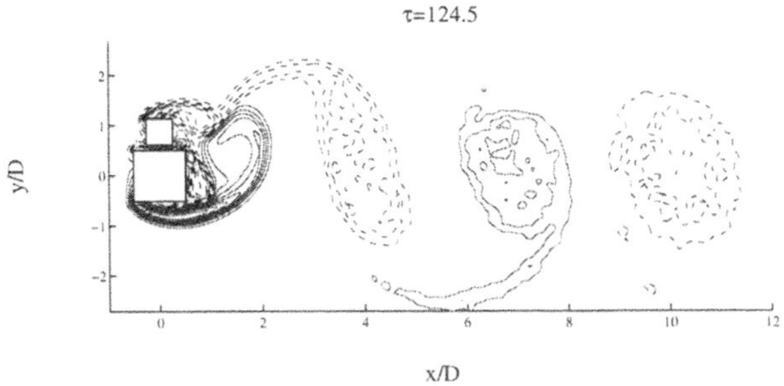

Figure 4. Single vortex street of two unequal square cylinders at $T/D = 0.875$. Normalized time $\tau = 124.5$. ——: Positive vorticity; – – –: negative vorticity. Contour level: arbitrary.

4. CONCLUSION

The proximity interference of two unequal square cylinders in side-by-side arrangement is simulated using a hybrid discrete vortex method. Sychronized vortex shedding at $2 \leq T/D \leq 4$ is associated with the amalgamated vortex entrainment process. The coupling/entrainment process at $T/D = 1.25$, similar to that of two equal square cylinders, results in a bistable flow. The gap flow is predominantly biased towards the small cylinder in the side-by-side arrangement.

REFERENCES

[Batchelor, 1967] Batchelor, G. K, *An Introduction to Fluid Dynamics*, Cambridge University Press.

[Chiu and Ko, 1999] Chiu, A. Y. W. and Ko, N. W. M.; "Numerical simulation of the proximity interference of two square cylinders"; In: *Proceedings of FEDSM'99, 1999 Joint ASME/JSME Fluids Engineering Engineering Conference, San Francisco, California*, FEDSM99-6812.

[Chorin, 1973] Chorin, A. J.; "Numerical study of slightly viscous flow"; In: *Journal of Fluid Mechanics*, 57(4), pp 785–796.

[Clarke and Tutty, 1994] Clarke, N. R. and Tutty, O. R.; "Construction and validation of a discrete vortex method for the two dimensional incompressible Navier-Stokes equations"; In: *Computers and Fluids*,23(6), pp 751–783.

[Igarashi, 1985] Igarashi, T.; "Characteristics of the flow around rectangular cylinders"; In: *Bulletin of JSME*, 242, pp 1690–1696.

[Igarashi, 1997] Igarashi, T.; "Drag reduction of a square prism by flow control using a small rod"; In: *Journal of Wind Engineering and Industrial Aerodynamics*, 69–71, pp 141–153.

[Ingham, 1990] Ingham; "Steady two dimensional flow through a row of normal flat plates"; In: *Journal of Fluid Mechanics*, 210, pp 281–302.

[Kim and Durbin, 1988] Kim, H. J. and Durbin, P. A.; "Investigation of the flow between a pair of circular cylinders in the flopping regime"; In: *Journal of Fluid Mechanics*, 196, pp 431–448.

[Miau, Wang and Chou, 1996] Miau, J. J., Wang, H. B. and Chou, J. H.; "Flopping phenomenon of flow behind two plates placed side by side normal to the flow direction"; In: *Fluid Dynamics Research*, 17, pp 311–328.

[Ng, 1996] Ng, C. W.; *Interactions of vortices from two circular cylinders in bistable flow regime*, PhD thesis, The University of Hong Kong.

[Ng, Cheng and Ko, 1997] Ng, C. W., Cheng, V. S. Y. and Ko, N. W. M.; "Numerical study of vortex interactions behind two circular cylinders in bistable flow regime"; In: *Fluid Dynamics Research*, 19, pp 379–409.

[Norberg, 1993] Norberg, C.; "Flow around rectangular cylinders: pressure forces and wake frequencies"; In: *Journal of Wind Engineering and Industrial Aerodynamics*, 49, pp 187–196.

[Ogami and Akamatsu, 1991] Ogami, Y. and Akamatsu, T.; "Viscous flow simulation using the discrete vortex model– the diffusion velocity method"; In: *Computers and Fluids*, 19(3/4), pp 433–441.

[Okajima, 1982] Okajima, A.; "Strouhal numbers of rectangular cylinders"; In: *Journal of Fluid Mechanics*, 123, pp 379–398.

[Prasad and Williamson, 1997] Prasad, A. and Williamson, C. H. K. 1997; "A method for the reduction of bluff body drag", In: *Journal of Wind Engineering and Industrial Aerodynamics*, 69–71, pp 155–167.

[Sakamoto, Tan and Haniu, 1991] Sakamoto, H., Tan, K. and Haniu, H.; "An optimum suppression of fluid forces acting by controlling a shear layer separated from a square prism"; In: *Transactions of ASME, Journal of Fluids Engineering*, 113, pp 183–189.

[Zdravkovich, 1977] Zdravkovich, M. M.; "Review of flow interference between two circular cylinders in various arrangements". In: *Transactions of the ASME, Journal of Fluids Engineering, Series I*, 99, pp 618–633.

[Zdravkovich, 1984] Zdravkovich, M. M.; "Classification of flow induced oscillations of two parallel circular cylinders in various arrangements"; In: *ASME Winter Annual Meeting, Symposium on Flow Induced Vibrations*, Paidoussus, M. P. et al., ed., New York: American Society of Mechanical Engineers, 2, pp 1–18.

List of Addresses of the Participants

Prof. R. J. Adrian Dept. of Theor. & Appl. Mech. University of Illinois at Urbana 216 Talbot Laboratory 104 South Wright St. **Urbana Il. 61801-2983,** USA r-adrian@uiuc.edu	Prof. Gido Buresti Dipt. di Ingegneria Aerospaziale Universita di Pisa Via Diotisalvi 2 **I-56126 Pisa**, Italia d4376@docenti.ing.unipi.it
Prof.S.V. Alekseenko, Director Inst. of Thermophysics Siberian Branch of RAS Lavrentyev Ave, 1 **630090 Novosibirsk**, Russia aleks@itp.nsc.ru	Dr. Rene Delfos Delft University of Technology Lab. for Aero- and Hydrodynamics Rotterdamseweg 145 **NL- 2628 AL Delft,** The Netherlands r.delfos@wbmt.tudelft.nl
Prof. S.P. Bardakhanov Inst. Theor. & Appl. Mech. Siberian Branch of Russian A.S. **630090 Novosibirsk**, Russia bard@itam.nsc.ru	Prof. H. Eckelmann Max-Planck Inst. f. Strömungsforschung Bunsebnstrasse 10 **D-37073 Göttingen,** Germany helmut.eckelmann@gwdg.de
Dr. Werner Benger Konrad-Zuse-Zentrum f. Informa- tionstechnik (ZIB Berlin) Takustrasse 7 **D-14195 Berlin-Dahlem**, Germany benger@zib.de	Dr. Walter Egli ABB Corporate Research Ltd. LHS AG Zürich **CH-5405 Baden Dättwil** Switzerland walter.egli@ch.abb.com
Pavel V.Bulychev Depart. General Physics Perm State University 15, Bukirev St. **614600 Perm, Russia** Gennady.Putin@psu.ru	Prof. Dr. Marie Farge CNRS Lab. de Météo. Dyn. Ecole Normale Supérieur 24, rue Lhomond **F-76231 Paris Cedex 05,** France farge@lmd.ens.fr
Prof. Tony Burden Dept. of Mechanics KTH **S- 10044 Stockholm,** Sweden burden@pop.mech.kth.se	

Prof. Adam Finchman LEGI-UJF-CNRS Coriolis 21 av. Martyrs **F-38000 Grenoble,** France fincham@hmg.inpg.fr	Dr. Wolfram Hage DLR (German Aerospace Center) Inst. Propulsion Techn. Turbulence Research Müller-Breslau-Strasse 8 **D-10623 Berlin**, Germany wolfrasm.hage@dlr.de
Prof. Dr. Bernhard Gampert Universität GH Essen FB 12 Angew. Mechanik Schützenbahn 70 **D- 45127 Essen,** Germany bernhard.gampert@uni-essen.de	Prof. C.M. Ho Mech. & Aerosp. Dept. UCLA **Los Angeles,** CA 90024-1597 USA chihming@ucla.edu
Dr. E.A. Gerteisen CSCS (Swiss Center for Scientific Computing **CH- 6928 Manno,** Switzerland egertei@cscs.ch	Prof. Dr. Lindy Joubert Faculty of Architecture, Building and Planing University of Melbourne **Parkville, Victoria 3052**, Australia l.joubert@architecture.unimelb.edu.au
Prof. Valery D. Goryachev Head of Dept. for Mathematics Tver State Techn. University A. Nikitina 22 **170026 Tver**, Russia valery@tversu.ru	Prof. W. Kinzelbach Inst. F. Hydromechanik u. Wasserwirtschaft, ETHZ 8093 Zürich, Switzerland kinzelbach@ihw.baug.ethz.ch
Dr. A. Gyr Inst. F. Hydromechanik u. Wasserwirtschaft, ETHZ **8093 Zürich**, Switzerland gyr@ihw.baug.ethz.ch	Prof. Norman W.M. Ko Dept.of Mech.Eng. The University of Hong Kong Pokfulam Road **Hong Kong,** China nwmko@hkucc.hku.hk
Dr. Maria Haase Inst. für Computeranwendungen Universität Stuttgart Pfaffenwaldring 27 **D-70569 Stuttgart,** Germany mh@ica.uni-stuttgart.de	Dr. H. Kocherscheidt Ruhr-Universität Bochum Universitätsstr. 150 Lehrst. F. Strömungsmechanik **D-44780 Bochum,** Germany Kocherscheidt@lstm.ruhr-uni-bochum.de

Dr. Georg E. Koppenwallner Himmelsstieg 1 leweke@marius.univ-mrs.fr, Germany Fax-0049 551 7989512	Prof. Thomas Leweke Inst. de Rech. sur les Phénomènes Hors Equilibre CNRS/ Univ, Aix-Marseille I&II 12, Av. Général Leclerc **F- 13003 Marseille,** France leweke@marius.univ-mrs.fr
Dr. Lyudmyla Koryenna Inst. of Hydromechanics NASU 8/4 Zheliabov Str. **Kiev 252180**, Ukraine lkor@rdfm.freenet.kiev.ua	Dr. K.W. Lo Maunsell Environmental Management. Cons. Ltd Gran Central Plaza, Shatin **Hong Kong**,China kwlo@maunsell.com.hk
Prof.. P. Koumoutsakos Inst. f. Fluiddynamik ETHZ **CH- 8092 Zürich,** Switzerland petros@eniac.ethz.ch	Herr Matthias Machacek Institute of Fluiddynamics, ETH Zurich **CH-8092 Zürich,** Switzerland machacek@ifd.mavt.ethz.ch
f. T. S. Krasnopolskaya Ukrainian Research Inst. of the Environment and Resources 13, Chokolivsky Blvd **Kiev 252180,** Ukraine	Prof. E. Meiburg Dept. of Aerospace Eng. University of Southern California University Park **Los Angeles, CA 90089-1191**, USA eckart@spock.usc.edu
Mr. Victor Lai 10 Lo Ping Road New Territories **Hong Kong**, China vlai@ied.edu.hk	Prof. Dr. W. Merzkirch Lehrstuhl f. Strömungslehre Universität Essen Postfach 10 37 64 **D-45117 Essen**, Deutschland wolfgang.merzkirch@uni-essen.de
Mrs. M. Lansak Inst. de Rech. sur les Phénomènes Hors Equilibre Univ, Aix-Marseille I&II 12, Av. Général Leclerc **F- 13003 Marseille**	Prof. Iraj Mortazavi Math. Appliquees de Bordeaux Universite Bordeaux I 351, Cours de La Liberation **33405 Talence Cedex**, France mortaz@math.u-bordeaux.fr

Dr. A. Müller Inst. F. Hydromechanik u. Wasserwirtschaft ETHZ **8093 Zürich**, Switzerland muller@baug.ethz.ch	Dr. L. Schouveiler Inst. de Rech. sur les Phénomènes Hors Equilibre CNRS/ Univ, Aix-Marseille I&II 12, Av. Général Leclerc **F- 13003 Marseille,** France
Dr. I. Neuweiler Inst. F. Hydromechanik u. Wasserwirtschaft ETHZ **8093 Zürich**, Switzerland neuweiler@ihw.baug.ethz.ch	Prof. Dr. R.M.C. So Dept. Mech. Engineering The Hong-Kong Polytech. Univ Hung Hom, **Kowloon** Hong Kong, China mmyliu@polyu.edu.hk
Dr. P. Petitjeans Lab. de Physique & Méc. des Mil. Hét. URA CNRS 857 ES de Phys. & de Chimie Indust. 10, rue Vauquelin **F-75005 Paris,** France	Prof. Ryuji Takaki Tokyo University of Agriculture and Technology Koganei **Tokyo 184-8588**, Japan Takaki@cc.tuat.ac.jp
Prof A. Pollard Queens University/Mech. Eng. **Kingston, Ontario K7L 3N6**, Canada pollard@conn.me.queensu.ca	Prof. A. Tamir Dept Chem. Engineering Ben Gurion Uni. of the Negev P.O.B. 653 **Beer-Sheva 84105**, Israel atamir@bgumail.bgu.ac.il
Prof. Renzo Ricca Dept. of Mathematics University College London Gower Street **London WC1E 6BT,** UK ricca@math.ucl.ac.uk	Mr. C.C.K. Tang Dept. Mechanical Engin. University of Hong Kong Pokfulam **Hong Kong,** China
Prof. T. Roesgen Inst. f. Fluiddynamik ETHZ **CH- 8092 Zürich,** Switzerland Roesgen@ifd.mavt.ethz.ch	Prof. A. Tsinober Faculty of Engineering Tel-Aviv University **IL-69978 Tel-Aviv,** Israel tsinober@eng.tau.ac.il

Mrs. Jacobiene van der Hoeven Lab. Aero- and Hydrodynamica TU Delft Rotterdamseweg 145 **NL-2628 AL Delft**, The Netherlands j.g.th.vanderhoeven@wbmt.tudelft.nl
Prof. Dr. Jose- Eduardo Wesfreid Lab. de Physique & Méc. des Mil. Hét. URA CNRS 857 ES de Phys. & de Chimie Indust. 10, rue Vauquelin **F-75005 Paris,** France wesfreid@pmmh.espci.fr
Prof. Dr. N. Yurchenko Inst. Hydromechanics NASU 8/4 Zheliabov Str. **Kiev,** Ukraine yur@rdfm.freenet.kiev.ua
Prof. N. Zabusky RUTGERS, State Univ. of N.J. Dept. Mech. & Aerosp. Engin. 98 Brett Road **Piscataway, NJ 08854-8058**, USA nzabusky@caip.rutgers.edu

Figures from:

WERNER BENGER

Figure 1 to 6 on four plates

WALTER EGLI

Figure 2 to 4 on two plates

HEINZ SCHMID

Celts one plate

NADIA GYR

The secrets of dishwater. One plate

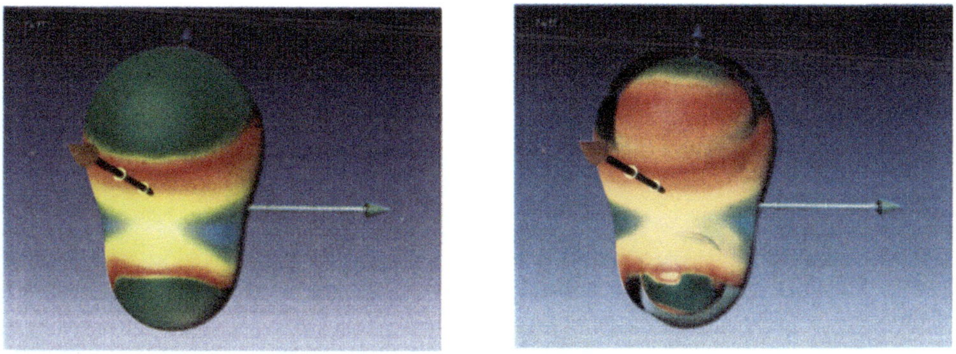

Figure 1: *Apparent horizon of two merged black holes, with color-encoded Gaussian curvatures. By assigning transparency to large homogeneous regions (right image), the inner is made visible.*

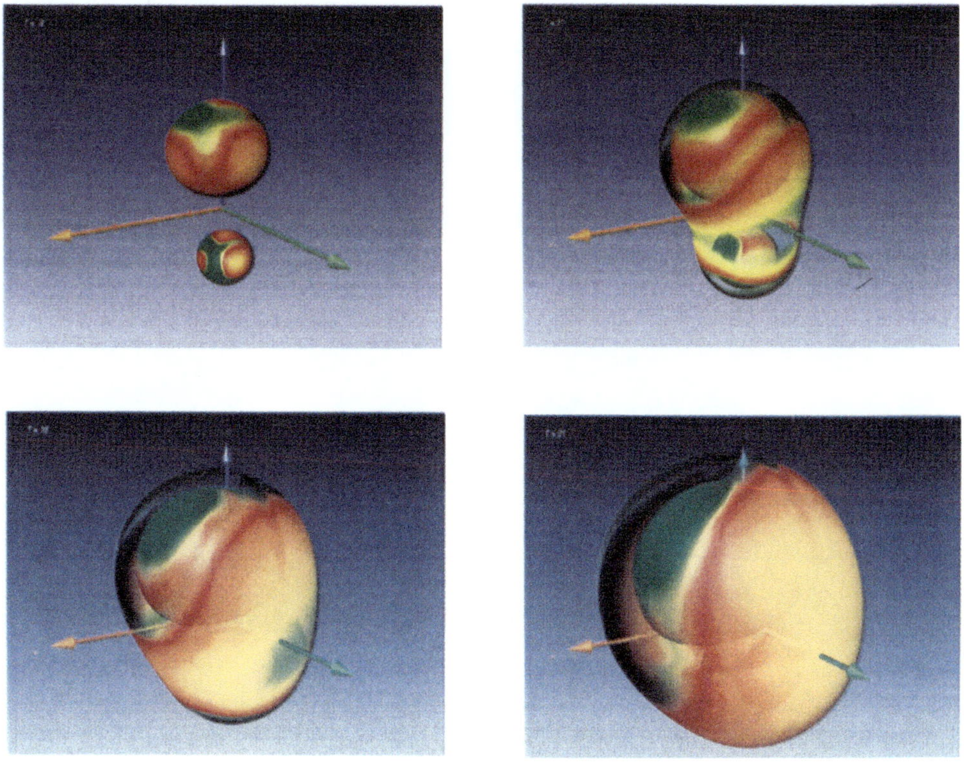

Figure 2: *Evolution sequence showing the apparent horizons of two merging black holes. An outer apparent horizon pops up at a certain time.*

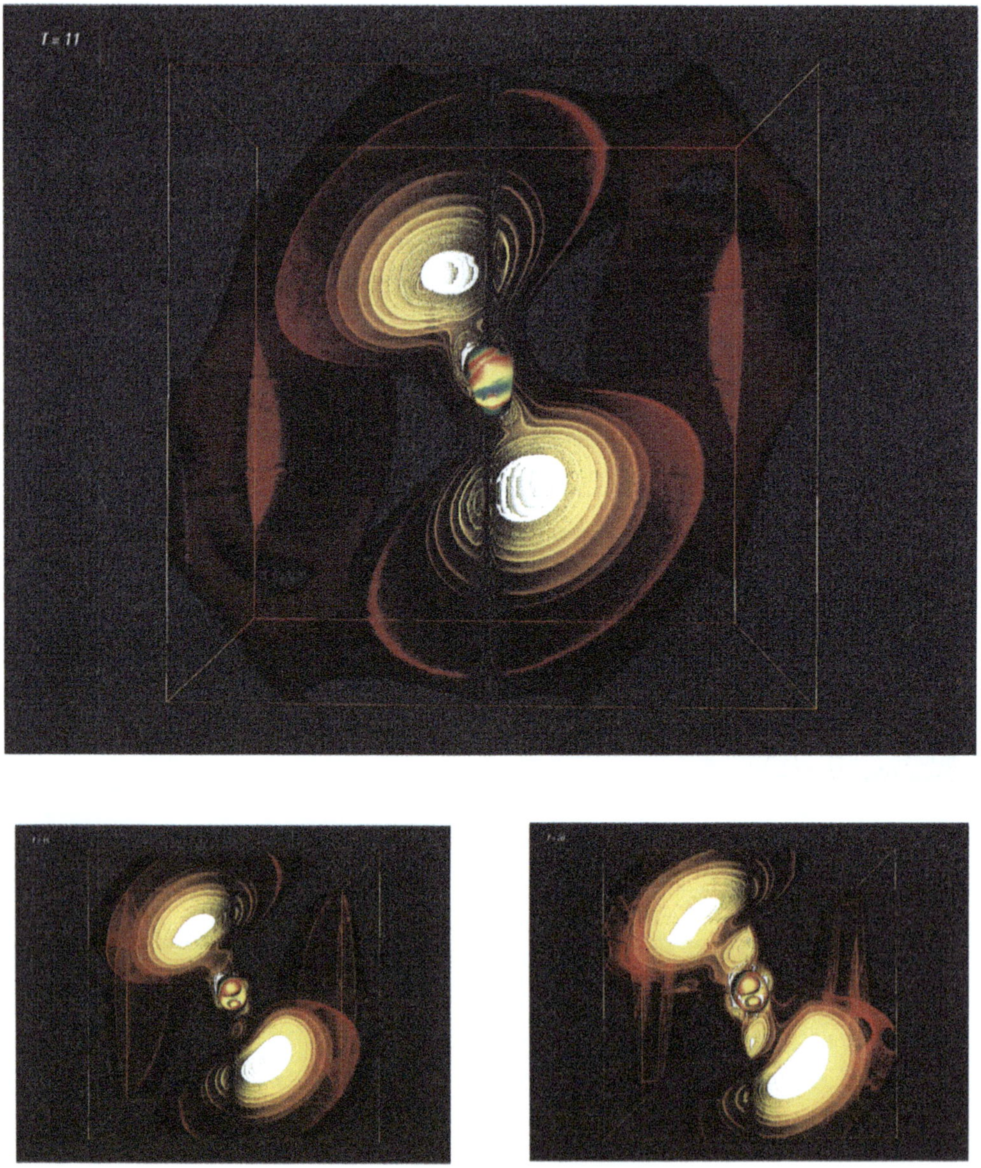

Figure 3: *A burst of gravitational energy emerging from two colliding black holes. The onion-like filaments indicate different levels of intensity of the gravitational field.*

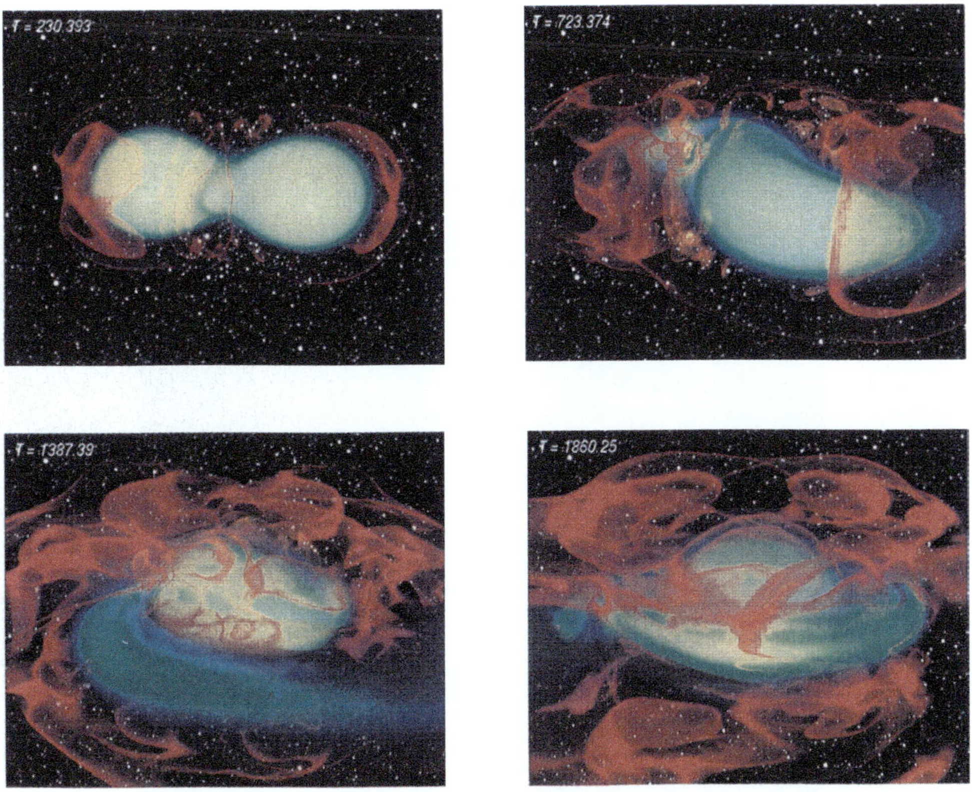

Figure 4: *Merging of two neutron stars computed using Newtonian theory. Each star has about the mass of the sun. Depicted are matter density (green) and internal energy (red).*

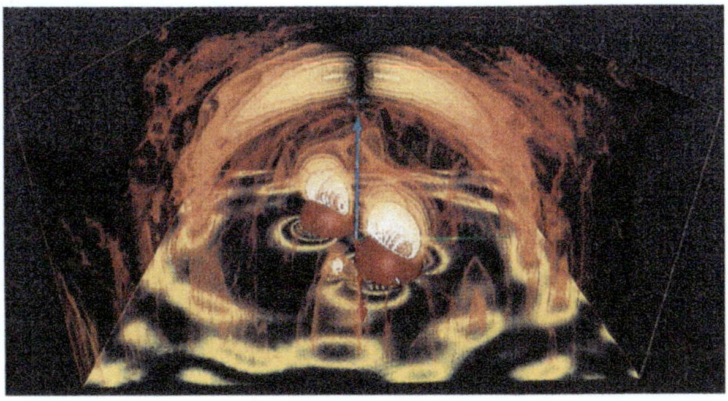

Figure 5: *Gravitational waves emitted in the spacetime of orbiting neutron stars. These results from a general relativistic hydrodynamic simulations are still preliminary.*

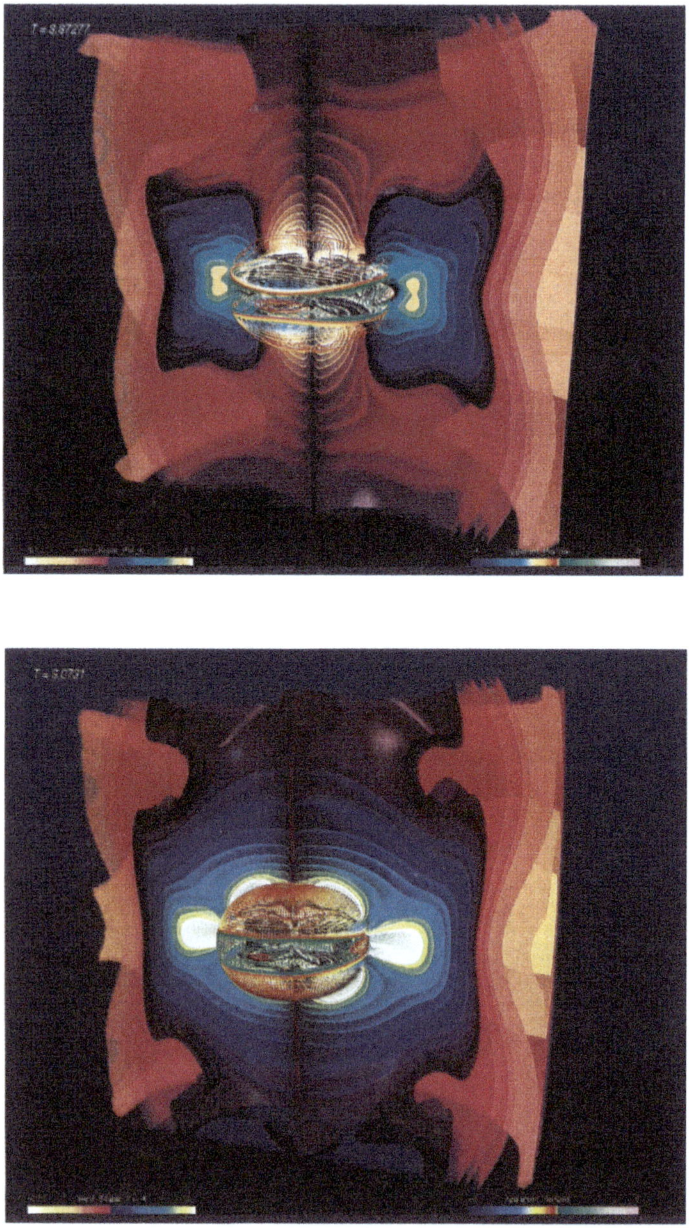

Figure 6: *A collapsing gravitational wave with the apparent horizon of the new black hole, formed during the collapse of a gravitational wave; seen as cuts along the xz- and yz-planes.*

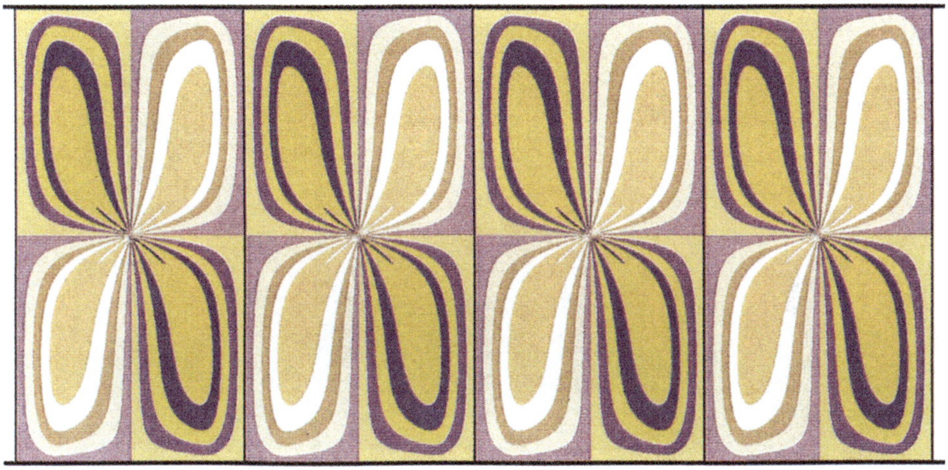

Figure 2; Stream function in a multi–wire configuration. Contour lines (texture mapped) of the stream function Ψ in xz-plane , solution of equation (2) for a wire distance of $a = \pi/2$. The gap width, measured form plate to plate, is π.

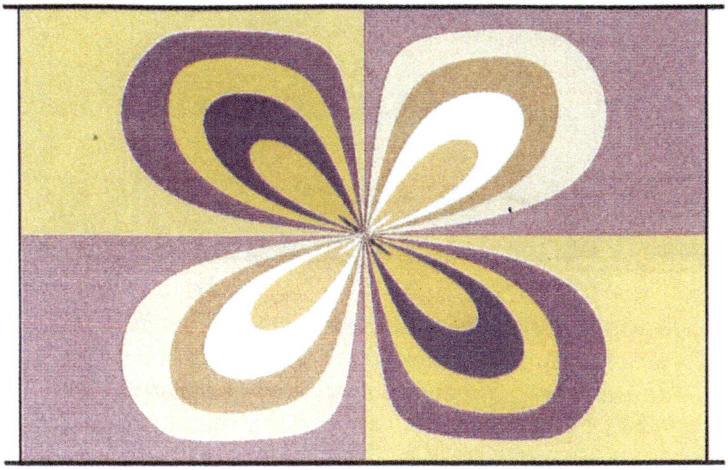

Figure 3; Stream function Ψ in xz-plane, solution of equation (2) for a wire distance of $a = 3/2\pi$. Contour lines (texture mapped) are presented. Same gap width as in figure 2.

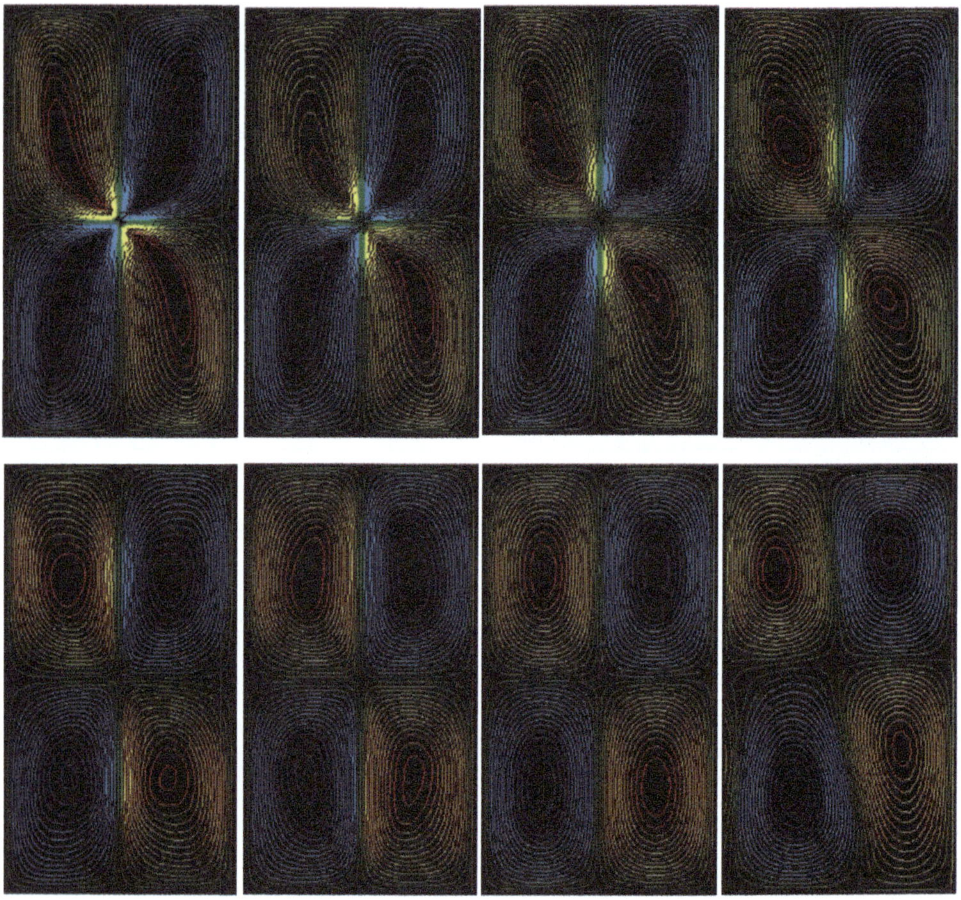

5 6 7 8

Figure 4; *Contour lines of integrated stream functions from the Euler solution of the finite volume solver. Flow patterns for different stages during the iteration. The numbers at the pictures represents an artificial time. The wire distance is set to a = π/2, same as in figure 2.*

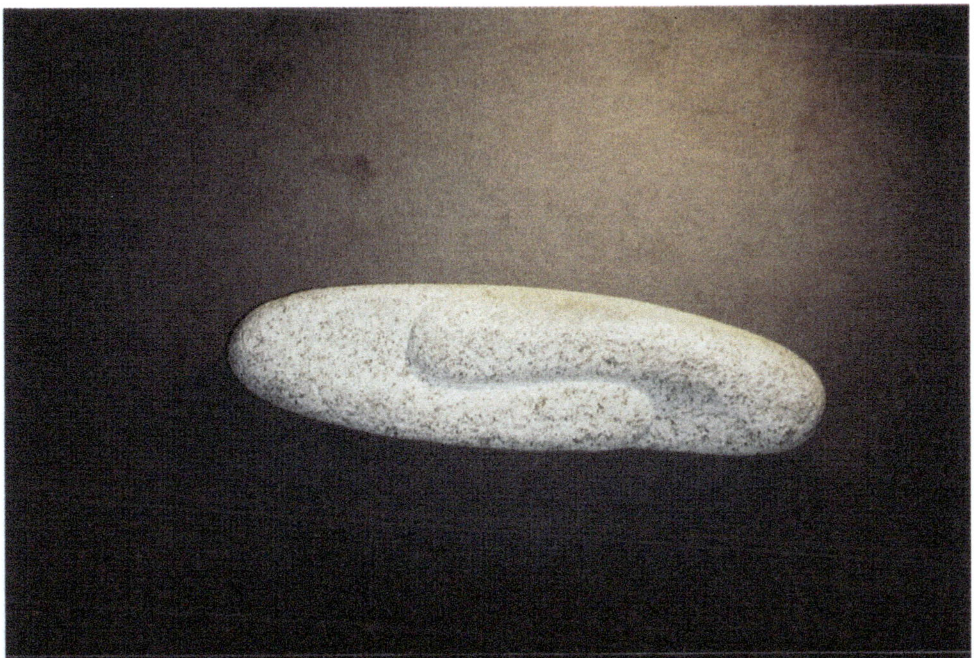

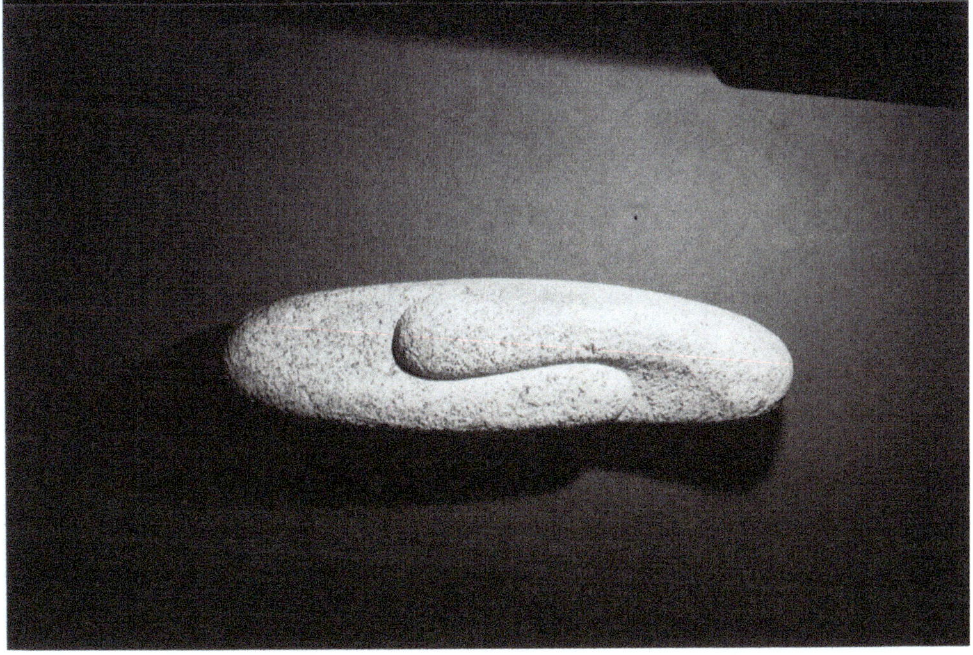

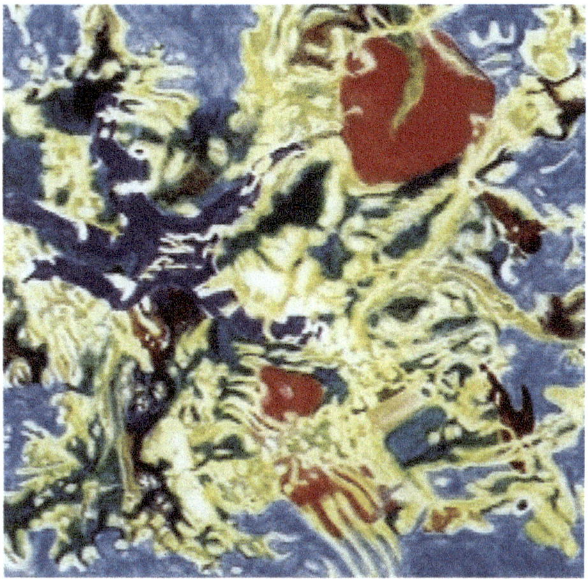

Nadia Gyr

Catalogue of the Art Exhibition

Walking on Air

In the Frame of SCART 2000
At the ETH, Zürich
From February 9th to March 3rd 2000

Walking on Air

Nadia Gyr
Bilderdijkstraat 167
NL- 1053 KP Amsterdam, The Netherlands
n@dia-gyr.demon.nl

I had my first experience with a computer as a child at an institute of the ETH. I played a computer game, of course a moon landing program and the task was to manoeuvre the orbiter during the last 10 seconds before the landing by decelerating or accelerating it with rockets loaded with a given amount of fuel. The initial condition was an orbiter at a given distance from the moon flying at a known velocity. A too rapid descent or a free fall after running out of fuel led to a crash. This was what mainly happened and showed that our feelings are not appropriate to solve the adequate differential equation. Later I saw the real landing on the moon on TV, and NASA showed us similar computer rooms but larger. We saw the first men on the moon as little figures jumping around in cute suits. The main difference to me was that the game was played without any pictures. The display showed the velocity of the orbiter, the distance from the moon, and if one wished the remaining amount of fuel.

As a child I could not guide the orbiter to a smooth landing, especially not with all the grown ups behind me who gave the essential information on what had to be done much to late, and still the game was more exiting than the reality. This is due to the emotional ambient, the small nervous light bulbs the tape recorder like elements and the feeling of participating in such an important event as a landing on the moon.

In the meantime the technical facilities have improved, the visualization techniques achieved a professionalism which allows us to follow easily a simulation and animation of a landing on the moon. However, it is not anymore the same, and if we ask ourselves why it is so? we came to the conviction that it was the situation at which we experienced the simulation. The computer room of a university, the determination of reproduce a very actual reality all this together had an additional quality . In my opinion to integrate this emotional part, art is needed. It is a sign of the professionallity of the artist how he or she integrates the feeling with fantasy and sagacity for the details.

With the exhibition 'Walking on Air' artists submitted contributions treating the flow of air and water. The exhibits show the fascination of the artists have with scientific aspects, without being scientific illustrators.

Every exposit is presented briefly by the artist and documented by a limited number of reproductions. Since these are still pictures of a dynamic process the pictures in these proceedings can only be understood as a motivation to study their work in more detail.

SCART is giving the artists a unique opportunity to get a deeper insight into a scientific field which is fascinating, but, normally not accessible to a broader public. In turn the participating artists with their fine sensitiveness could, by their contributions to the discussion, elicit new aspects, and help scientists in finding novel approaches. The result should be 'a school of astonishment' for both sides resulting from a dialogue based on the vigor of both professions.

SCART is an experiment, and I hope it will give impulses to art and science.

Amsterdam, January 2000 Nadia Gyr

Thomas Wildner

Gerard Doustraat
1072 VJ Amsterdam
Tel. 0031 20 6766665
twildner@xs4all.nl

10000 / 1-5-9 / II, 1999

A number of simple lines and consecutive even and uneven telephone numbers from the Amsterdam telephone directory form the basis for this drawing. By writing the numbers along the lines of a grid, a drawing is produced in which abstraction transcends the symbolic and presents a simulation of a concrete reality.
The straight lines and corners generate enigmatic, complex structures which display strangely natural characteristics. These structures are redolent of complicated maps and bear a resemblance to natural, non-linear structures. It is the combination of accident and order that generates tension in this drawing.
It is a selection extracted from noise.

Born: 26.10.1966 in Essen , Germany

1997 educational trip to Mexico City The annexation of public space, Grant of the Fonds voor beeldende Kunst, Vormgeving en Bouwkunst

1995 educational trip to the south west of USA, visit to the Chinati Foundation Marfa, TX USA

1986-1990 AKI academy of art Enschede (NL)

Exhibitions

1999 De Parel, solo Amsterdam (NL) / Music and Media Composition for flute played by Ned McGowan Posthoorn church, Amsterdam (NL)

1998 Foundation ŒNieuw en Meer, Archipel Nieuwe Meer, Amsterdam (NL)

1997 Progressiev ,96 eV International forum 1997, Köln (BRD)

1996 Foundation Steyn , solo Masterplan, Haarlem (NL)

1995 Inge Romgerg Gallery, solo, Borken (BRD) / Museum Van Bommel van Dam Van, Bommel van Dam prix 1995, Venlo (NL)

1994 Gallery Erhard, in case of emergency: schwartz, Dresden (BRD)

Yves Netzhammer

yves@netzhammer.com

BLINDES WASSER
Standbild aus einer Videoarbeit

Born: 1.08.1970

1996	Award, Hochschule für Gestaltung Zürich (CH) / Digitale, Messe Köln (D)
1997	Was sich erzählen lässt, wird verbessert werden., publication, Ricco Bilger Verlag / Award, Kunst aanmoedigingsprijs amstelveen (NL) solo exhibition, Gallery Haus Schneider, Karlsruhe (D)
1998	Was man frisst, wird zum Inhalt, solo, Gallery Stähli (CH) / Gallery Stähli, Art Cologne (NL)
1999	Curators`Digest, Helmhaus, Zürich (CH) / Manor- Kunstpreis, Award and catalog, Museum zu Allerheiligen, Schaffhausen (CH) / Digital sunshine: 4 Positionen, Brandt Kleadefabrik, Odense, (DK) / Video Fictionale, Museum Ludwig, Köln (D) / Der anagrammatische Körper, curator Peter Weibel, Steirischer Herbst 99 (A) Die Zukunft, Newspaper Tages-Anzeiger, weekly Sunday magazine Nr. 47 (CH)

Gabriël Kousbroek

Media Producties
Tel / Fax 0031 20 6391684

TURBULENCE IN TRAFFIC FLOW

Turbulence as in water, or airflow is a known 'force' of nature. Road traffic seems an unlikely comparison regarding turbulence, however I've noticed certain similarities, for example; if we were to drive a car through rush-hour traffic in Paris, over a well known square like the Arc de Triomphe, as driver it would seem to us a total madhouse. But if we were to stand on top of the Arc, we'd notice, that flow of the traffic follows a similar path, as water follows, when descending a slope, evading rocks and other obstacles; The path of the least resistance. These comparative pathways interest me as an artist and a amateur physicist.

Born: 11.10.1965 in Neuelly sur Seine, France
Training: Interdisciplinary arts (graphics, sculpture and audio- visuals) at the Gerrit Rietveld akademy Amsterdam (NL)
1992-1995 Director of the Verdieping gallery I & II
1993-1997 Managing-Director of Eyegasm, video-art collective
1998-1999 Audio-visual teacher for the Hogeschool v.d.Kunsten s'Hertogenbosch
Exhibitions
1993 Verzameld Werk, Beurs van Berlage, Amsterdam (NL)
1994 Vigliat ut Quiesant, galery Aschenbach, Amsterdam (NL)
1995 60 minuten, solo, in het hoofd Made in Heaven gallery, Utrecht (NL)
1996 De zoon van, solo, Witzenhausen art space, Amsterdam (NL)
Audio-visual
1995 Suitcases, video-movie and live video-manipulation in 9 European cities
1996 Signal to Noise, feature film (3de prize Night of the short film 1997) in cooperation with Boris v.d.Meyden
1997 L'Histoire du Strawinsky, virtual representation in various theaters (NL) in cooperation with the Dutch Horn Ensemble
1998 Bio Gotchi, feature film in cooperation with Boris v.d.Mayden (NL)
1999 Pas 18, Ballet-Artdirector * National Ballet Academy Amsterdam (NL)
2000* Tu Tuage Ballet-Artdirector for Le Maison de la Musique, Nanterre-Paris F
*in cooperation with choreographer Pieter de Ruyter

Erik Nicholas Johan Hobijn

THE DANTE ORGAN AND THE AESTHETICS OF VIOLENCE

Gordon Monaham E.Hobijn 1996

The Dante Organ is shown in the reproduction whereas on the exhibition the Delusion of self molestation is shown. This is an installation consisting of 11 to 12 flame-throwers with pillars of fire 13 to 20 meters high, creating an environment dealing with the aesthetics of violence. The performance is technically based on military knowledge, and this also determines a substantial part of its artistic form and content. It is an example of an extreme level of industrial art.

The reality of the aesthetic experience develops in multiple levels that derive from the viewer's preconditioned aesthetics. The first level is that of amusement, where the audience is charmed by the visual and conceptual aspects of the piece. The public desire for fulfillment of pure amusement satisfaction is appeased.

This leads to the second level. There is still the comfortable situation of looking at these 'images' of violence, as though they are sequences in an action movie, or newsreels on television. But the 'beautiful' images of disaster-fires and exploding rockets as seen in the media are never true representations of reality, and the increasing violence of the Dante Organ undermines the media-violence principle. The sense of danger of the Dante Organ becomes real, as the intensity of the flames develop from a visual level to a reality of their own. The safety screen of media violence is dissolved by the flames.

This is the third level. The Dante Organ becomes real time media violence, as the intensity and heat turns into a dimension which is normally not possible to create out of, for example, a war field, or any other human disaster situation as represented in the media. One is forced to abandon the pure artistic aspects of the piece to deal with the real and present danger. The computer orchestration of the flames may cause the audience to view the piece as a media spectacle, but the motored essence of the oil fires plays a game of cat and mouse with their adversaries. The audience is required to take a physical decision either to step away or to accept the heat and get burned. The speed of this violence burns the time. Nothing else rests than a blistering image on the retina.

The Dante Organ is a work about the role of pure images. Any disaster has its value for the media market, and the more it is transported through public information channels, the more it changes and becomes a thing of its own. It becomes a utility for the media and a non-emotional object for everyday living. This distancing of reality, and the impossibility of involvement in it, create a dilemma: the more you see it, the less you feel it. This is not a moral question but one of the obvious mechanisms of data flow. The Dante Organ makes a choreography out of the possible experience of heat. In this sense it is a live monument coming from a generation after World Wars. A monument that does not point back to the past but operates in this time on its own specific terms.

Born: 07.11.1956 in Amsterdam, The Netherlands

Training	Gerrit Rietveld Akademie, Amsterdam MTS Patrimonium, Amsterdam
1999	Dante Organ II; Feuer Zeichnen, Bundesgartenschau Magdeburg (D) / Chemobar; 2000 Beats festival, Rotterdam (NL) / Chemobar, Next 5 minutes Amsterdam (NL) / Funeral: 2 Dante Organs, Amsterdam (NL)
1998	Chemobar; Dutch-arts festival, Maria club Berlin (D) / Organisation: Barre Tocht, festival at the Nieuwe Meer Amsterdam (NL) / Dante Organ II: Int. medusa arts festival Poland, Gezisko (PL) / European media Festival Osnabruck (D)
1996	Dante Organ II; Feuer und Flamme Dresdner Kunstfest (D) / Netband; Labyrinth, Dresten (D) / Ars Electronica, Linz (D) / Under the Capricorn, Stedelijk Museum Amsterdam (NL) / Techno-Parasites: Stadt Galerie Zwartz Tirol (A)/ min, V2 Rotterdam (NL) / Presentation, Künstlerhaus Bethanien Berlin (D)/ 5 Syberconf, Madrid (S) / Lectures: HDK, Berlin (D) / Rietveld Akademie, Amsterdam (NL) / Hoogeschool voor de Kunsten, Den Boch (NL) / Video Fest, Berlin (D) / Auto-Auto Mutulatie: Geben und Nehmen, Künstlerhaus Schloss Plueschow (D) / Dood, Arti et Amicitae Amsterdam (NL)
1995	A grant; Kunstlerhaus Bethanien, Berlin (D) / Netband: Founding of the nieuw Internet artgroupe with FFFeigel, Dick verdult, Amsterdam (NL) / The Egg of the Internet, Fabriek Eindhofen (NL) / V2, DEAF, Duth Electronic Art Festival EGG. NET, Rotterdam (NL) /

Yvonne Fontijne

Amsterdam
Tel/Fax 0031 20 6168261

UPPER AND LOWER PART II

Stills from computer-animation / motion-capture; 1998 in collaboration with K. Post / choreography
Researching the processes of thinking by analysing human motion.

Born: 1967 in Amstelveen, the Netherlands

1986-1991	Hogeschool voor de kunsten, Utrecht
1995-1996	Rijksakademie van beeldende kunsten, Amsterdam
1996	Malmo-Amsterdam; Gesellschaft für Aktuelle Kunst. V., Bremen (D)
1997	Imagina; motion capture theatre Intergraph Softimage SP, Monaco (F) Diskland Snowscape; Shed im Eisenwerk, Zürich (CH) / Het eigen gezicht; Museum Beelden aan Zee, Scheveningen (NL) / New Dutch Sculptors 10; Glazen Huis Amstelpark, Amsterdam (NL)
1998	'Unlimited.nl'; De Appel, Amsterdam (NL) / Installation for entrance Unilever Research Laboratorium Vlaardingen (NL) / Traces of science in art; KNAW, Trippenhuis Amsterdam (NL) / Eurocentral; Transmission Gallery, Glasgow (GB)/ Grown in frozen Time, Shed im Eisenwerk, Zürich (CH)
1999	Aan / Uit; Dance-show K. Post, on tour / Possible Side Effects; M. Schumacher and J. Ryan, Nes theaters, Amsterdam (NL) / Natural Walk; Galerie Fons Welters, Amsterdam (NL) / Motion Library; Siggraph, Los Angeles (USA) / Galerie Wolfram Bach, Düsseldorf (D)

Zilla Leutenegger

zilla.l@bluewin.ch

FÜR ELISE

from the Press release gallery peter kilchman
In all of her works (videos, video installations, video stills, paintings on video stills) Swiss artist Zilla Leutenegger, encounters her own personality. She is onstage as lead singer of the band 'the Moles' observes eventually suspicious deeds in the staircase, dances as several selves, sits on a sofa and flirts with her double. In her videos she is to be seen as 'Little Zilla' chewing bread and butter and honey at the kitchen table, as pianist of all time torch favorites as 'For Elise', stretches her legs as prima ballerina or shouts orders as aerobics teacher 'Zilla Zack'. Thought and concept as projection of and for years, the artist works on the flood of time, the changes of environment and of the artist herself.

Born: 1968 in Zürich, Switzerland
1998 Gallery Peter Kilchmann, Zürich
1999 Diplom Studienbereich Bildende Kunst Hochschule für Gestaltung und Kunst Zürich

selected exhibitions
1997 Art'97, Videosampler Stampa, Basel (CH) / 7e Semaine Internationale de Vidéo, Genève (CH) / Gallery HOTEL, solo, Zürich (CH), in cooperation with J. Lenzlinger and B. Brogle
1998 Galerie Brigitte Weiss, Apartmentshow, Zürich (CH) / Technoculture, FRI-ART Centre d'art contemporain Kunsthalle Fribourg (CH) Stipendien-ausstellung, Helmhaus Zürich (CH) / The way I make you feel» Videosampler ll, Kunsthaus Glarus (CH) / Video- und Performancetage, Kunstmuseum Luzern (CH) / Galerie Ars Futura, solo, Zürich (in cooperation with M. Staub und I. Truniger (CH) / Galerie Peter Kilchmann, solo, Zürich (CH) / Gallery attidudes, Paris (F) / Centre culturel d'art suisse, Paris (F)
1999 Kunstszene 98/99, Helmhaus Zürich (CH) / Binz 99, Zürich (CH) / Eine Insel, Stadtgalerie Bern, exhibition from and with Markus Wetzel (CH) Art club berlin, Berlin and New York (CH;USA) / Art Forum Berlin, Messe

Berlin, Gallery Peter Kilchmann, (D) / Messe Zürich, Galerie Peter Kilchmann (CH)

Actions and Performance

Performance with Annelise Coste, Kunstmuseum Luzern (CH)
Performance with Annelise Coste, Klinik, Zürich (CH)
Performance AM PIANO, Kleines Helmhaus, Zürich (CH)

Jeroen Werner

Bilderdijkstraat 167, 1053 KP Amsterdam
TEL 0031 20 6168651
Jjwerner@xs4all.nl

ZOOM-MIRRORS
The I in the Eye.

Jeroen Werner's work distinguishes itself by the use of light and Werner's fascination with optics. Visual research into projection and reflection has led to a series of Zoom-Mirrors, some of which are shown here.
Picture of a child baffled by the image of itself reflected in a well-polished spoon during a tedious dinner-party of adults. While on the convex outside the face appears puffed up and ridiculous it is turned upside down when reflected in the hollow inside of the spoon and remains so even when in wonderment the child turns the spoon itself. "Why?" the child asks.
Jeroen Werner's Zoom-Mirrors are a way of controlling this childhood experience and of making visible what is obscured in the turning around of the spoon, bridging the baffling opposition of the puffed up and the upside down image of ourselves. The Zoom-Mirror consists of a reflective membrane which has been stretched over a closed cylinder rather like a drumhead. The motor of a vacuum cleaner is used to inflate and deflate the drum, thereby stretching the membrane from convex to concave and vice versa with an almost film-like fluidity. Deflated the membrane's image zooms in on the on-looker, inflated one sees oneself as if through a fish-eye lens. Different trajectories as well as zoom-speeds can be chosen by regulating the actions of the pump.
When the focus point of the mirror has shifted beyond the eyes of the on-looker the image turns upside down and seems to be floating three-dimensionally in the space between subject and mirror. To an extent this is how the human eye projects what is seen onto the inside of our head so that while watching ourselves in these mirrors we are also looking at the way we look. The three-dimensional aspect of the reflected image means that this perceiving of our own perception becomes an almost physical experience.
In this sense the Zoom-Mirrors are a childhood dream coming true; not only to have a spoon this big but to be able to see what actually happens when we turn the spoon around. Something still eludes us however; while we witness how the fluid mass that was our face is reshaped into the same but upside down we realize that by having been

shown we haven't necessarily understood. By being placed within the image of ourselves as well as of how we look our notion of 'I' is contained within something we perceive as 'I' as well as holding 'I'. Something the exact workings of which elude us although they are shown to us. The question emerges what our desire to understand is really asking for. When confronted with Werner's Zoom-Mirrors we, rather than 'walking on air', find ourselves standing on the shaky fringes of our understanding of ourselves.
Sicco Heyligers , London.

Born: 25.09.1960 in Amsterdam, the Netherlands

1978 -1979	Gerrit Rietveld Academie, Amsterdam
1978-1982	Rijksacademie van Beeldende Kunsten, Amsterdam

selected exhibition

1992	Op Persoonlijke Titel, Aleph Almere (NL) / Verder dan Foto, Schouten en de Vries, Haarlem (NL) / Brain Internal Affairs, Beatrix Ziekenhuis, Gorinchem (NL) / Beauty of the Last Minute, Hannema de Stuers Heino Foundation (NL)
1993	Kunstwerken, Tunnelontwerpvoor Stad Amsterdam, Stadhuis Amsterdam (NL)
1996	De Verwondering 1, Ideas of Nature, Centrum Beeldende Kunst, Leiden (NL)
1997	De Verwondering 2, Ideas of Nature ,Centrum Beeldende Kunst ZuidOost Amsterdam (NL)
1998	Spiegelreis,through the looking glass, 2,5 x 4,5, Amsterdam (NL)/ Automobiel 1, Moving Perspectives, De Veemvloer Amsterdam (NL) / Second Original,blanco als het moet, with Franky DC, Hedah, Maastricht (NL)/ Zoomspiegel, Akzo Nobel Art Collection, permanent installation, Arnhem (NL)
1999	Tien DNA-ramen in liften, permanent installation, Mondriaan Stichting en Ijsselmeer-ziekenhuis (NL) / Zoomspiegel, permanent installation, AZU/WKZ Utrecht (NL) / De Neus, Sensorische opera, in cooperation with Manel Esparbe Gasca, Theater De Balie Amsterdam (NL)

Thomas Wildner

10000 / 1-5-9 / II, 1999

Yves Netzhammer

Blindes Wasser (Videostill from a Video work)
Blind Water

Gabriël Kousbroek

Turbulence in Traffic flow

Erik Hobijn

Delusion of self molestation

Yvonne Fontijne

Still from the computer-animation: Upper and lower Part 2 / motion-capture,1998
in collaboration with K.Post / Choreography

Zilla Leutenegger

Videostill from Für Elise 1995
For Elise

Jeroen Werner

Zoom Mirror

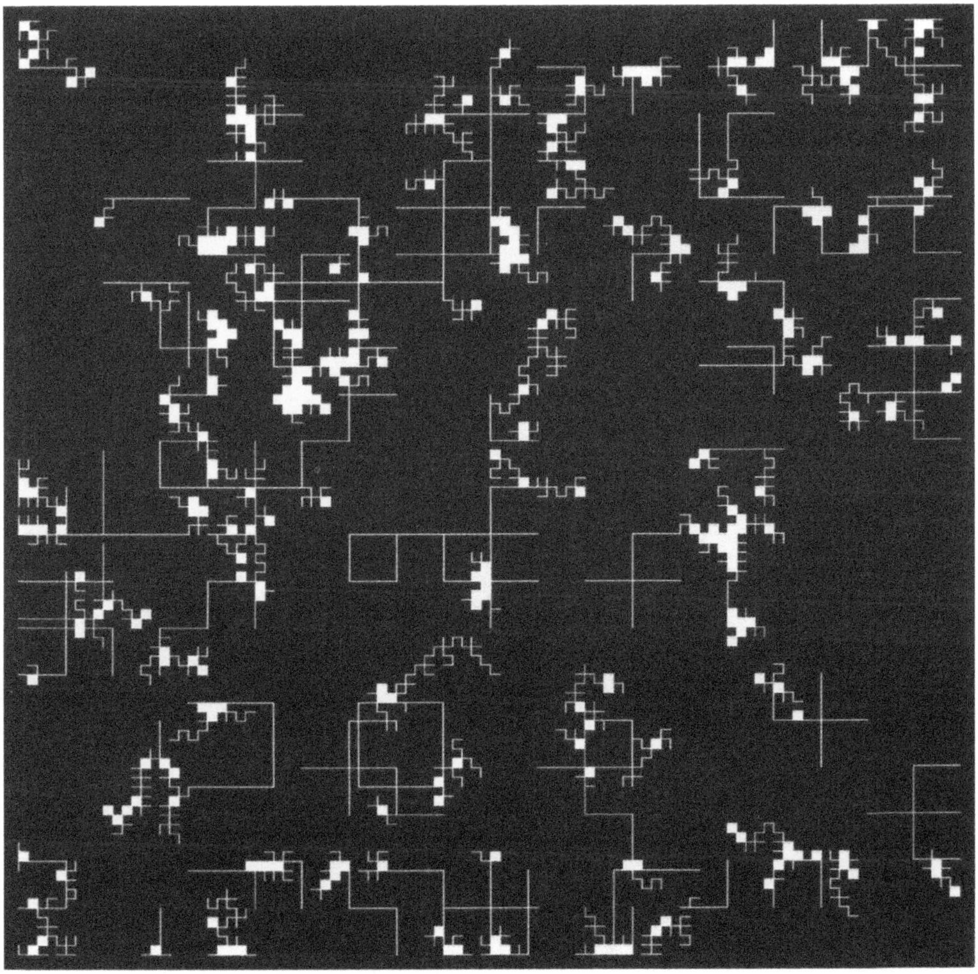

Thomas Wildner

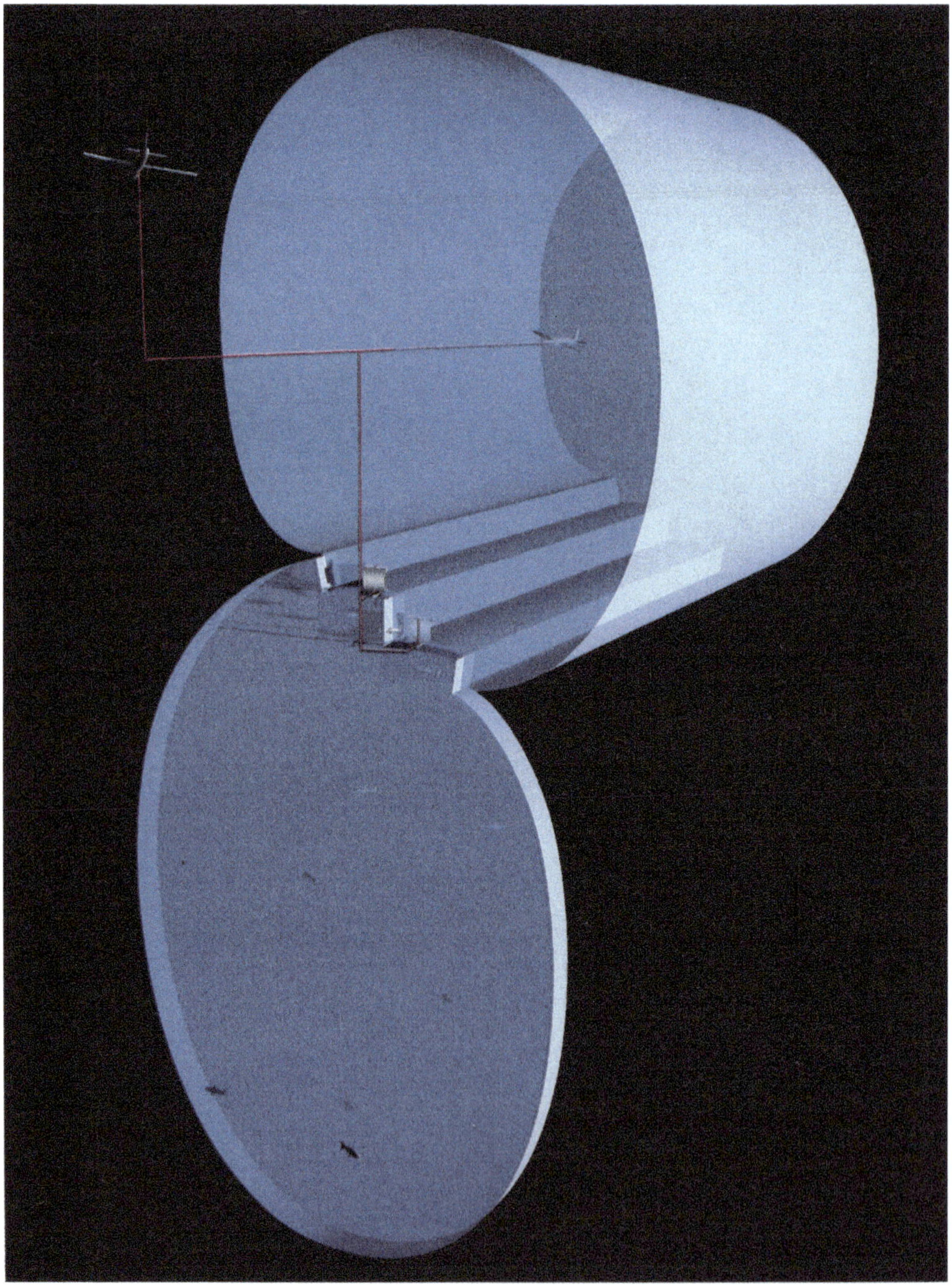

Yves Netzhammer

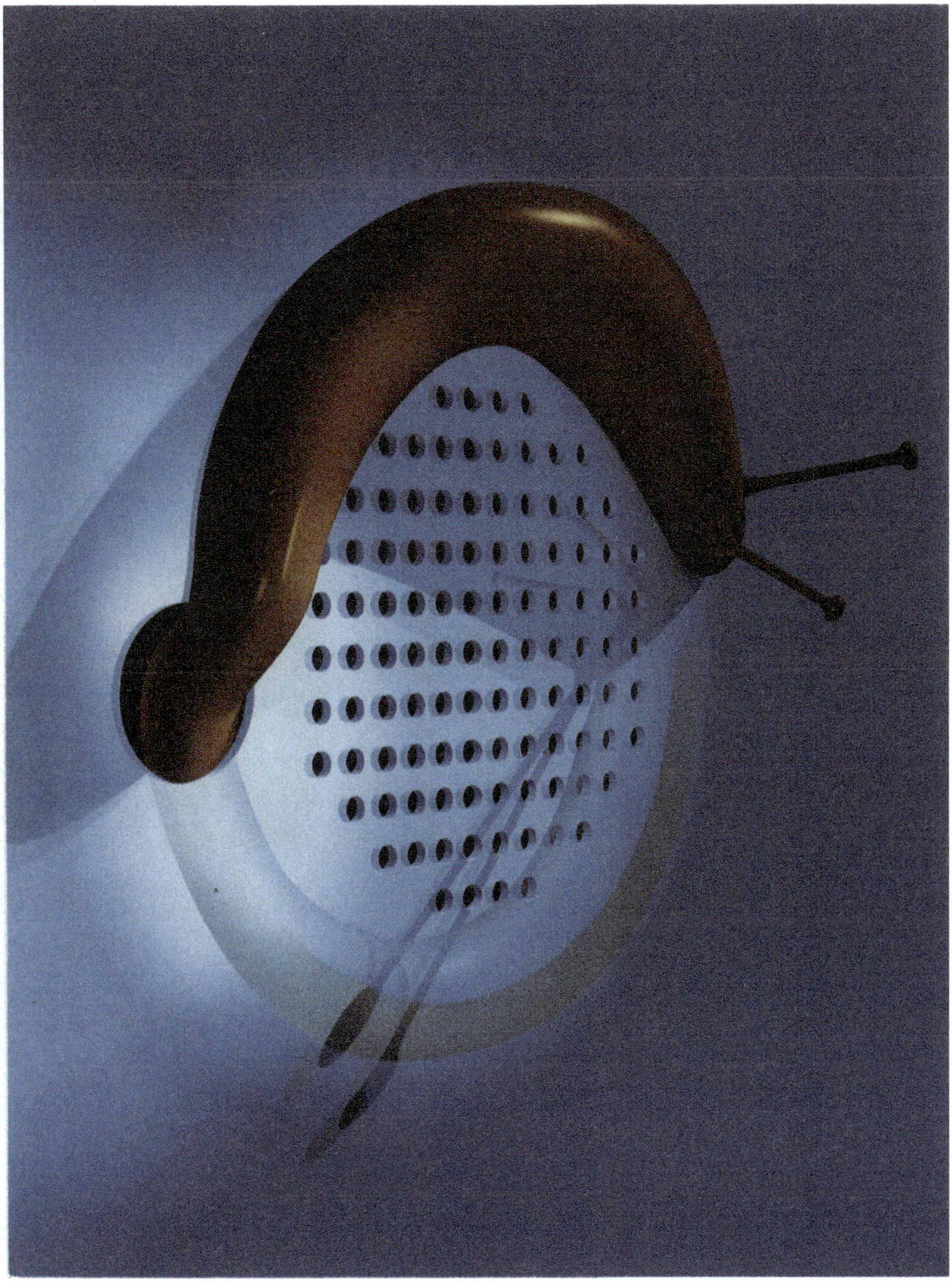

Yves Netzhammer

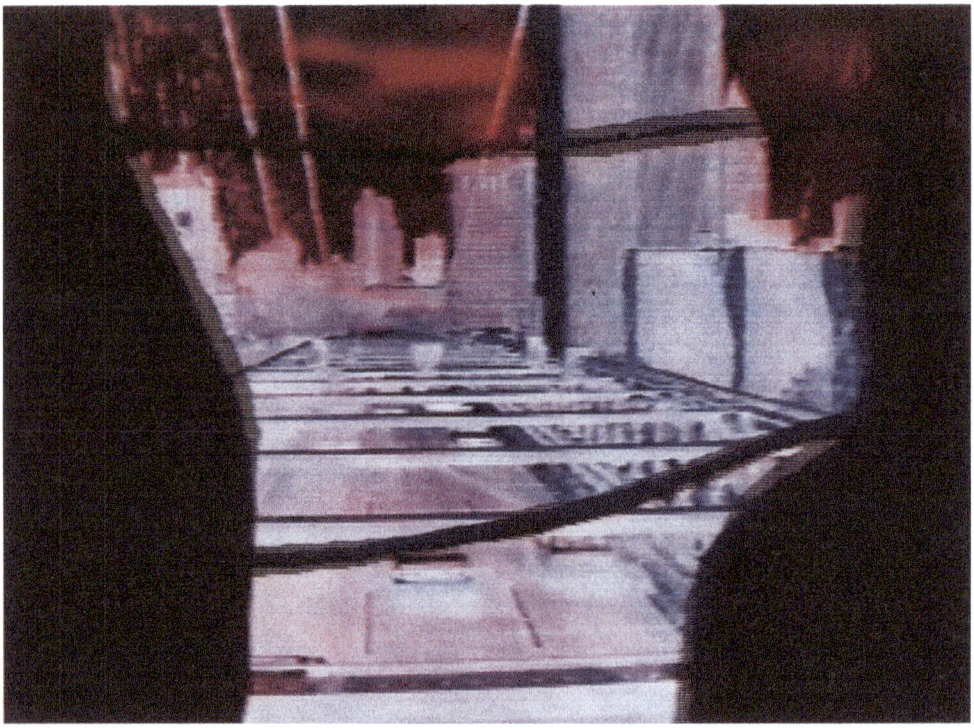

Gabriël Kousbroek

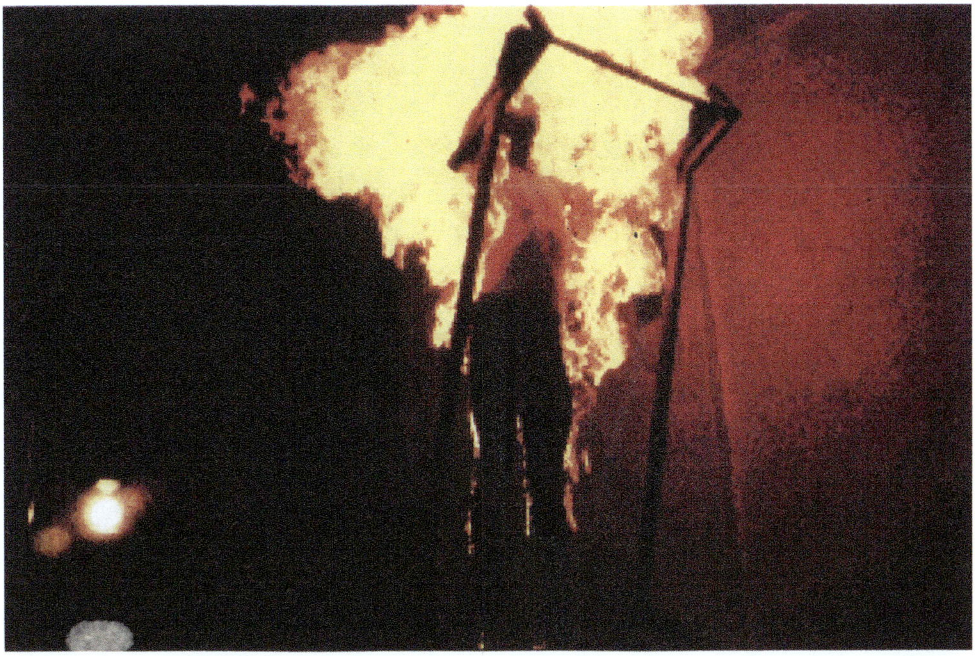

Erik Hobijn

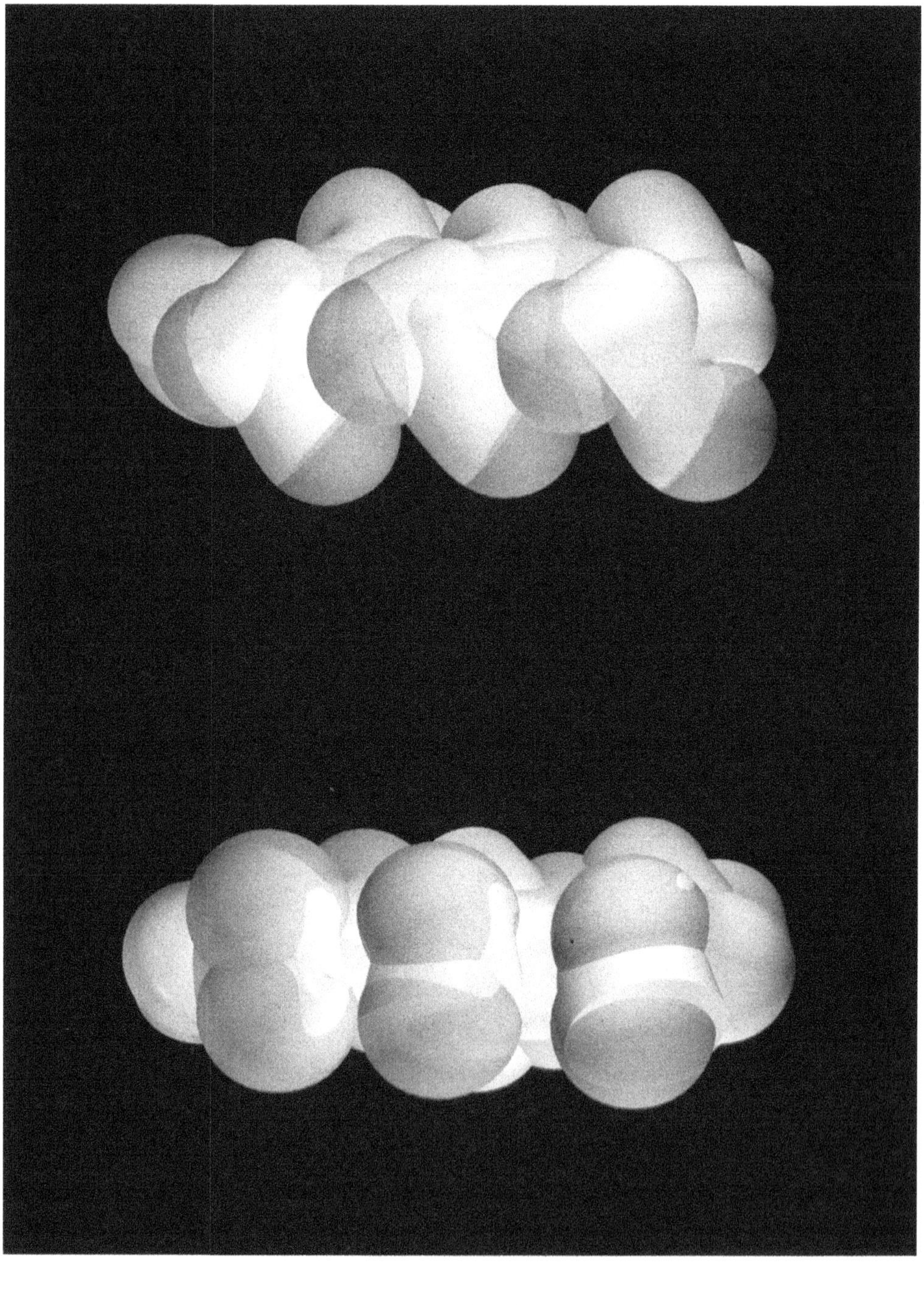

Yvonne Fontijne

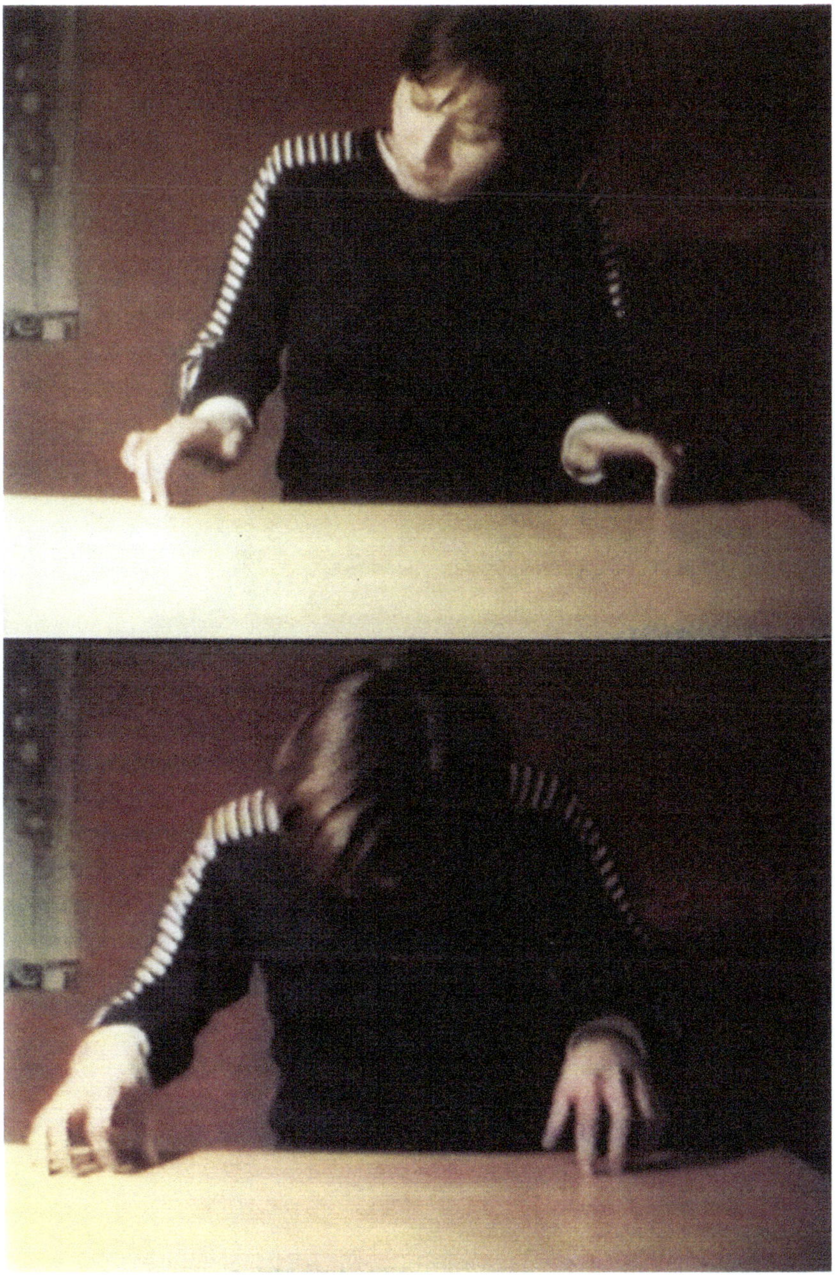

Zilla Leutenegger

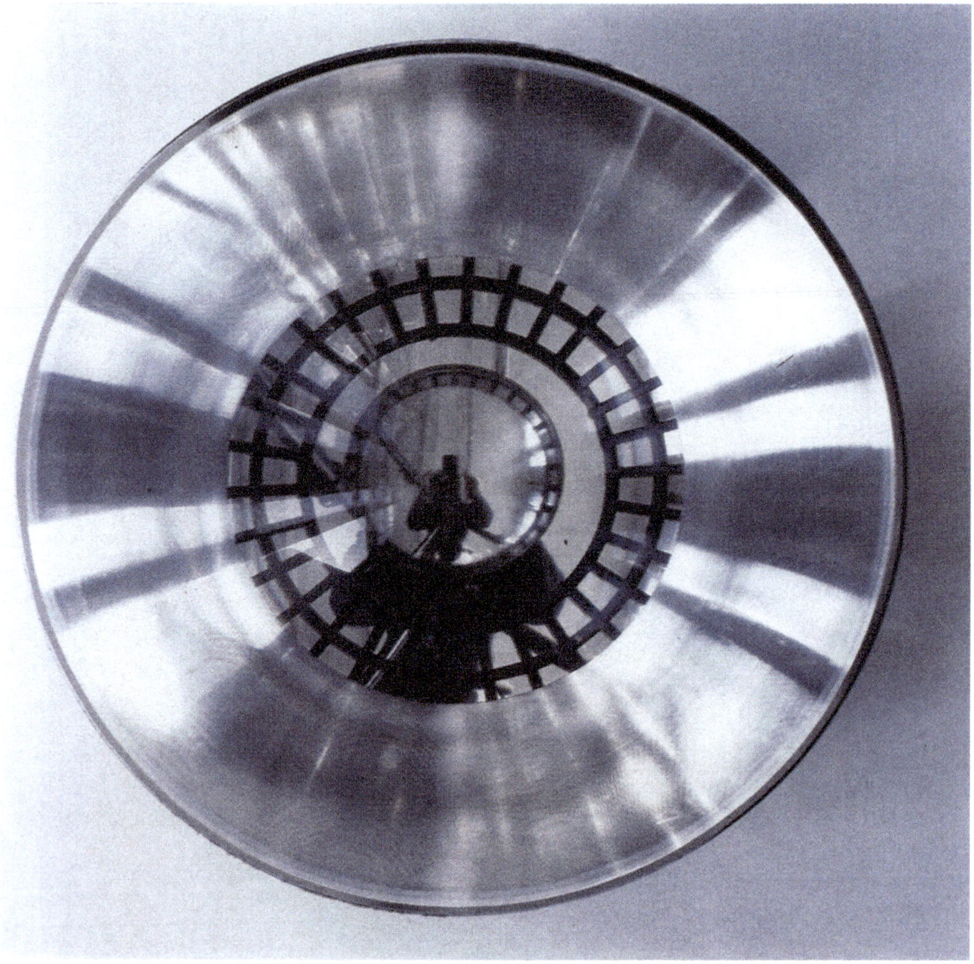

Jeroen Werner

Theo Jansen

*Koornmarkt 48^H^*
2611 EH Delft
Tel: 0031 152145297

ANIMARI

Since 1990 Theo Jansen is working on a new nature. By doing this he hopes to fathom real nature.
His nature exists of skeletons made of yellow electricity tubes. They walk by the wind. In the end Theo Jansen wants to set out his animals (Animari) on the beaches so they can live their own lives.
During the years the animals were due to evolution; they become better and better...

Born: 1948 in The Netherlands

1968	Studies Technical Science at Delft University
1975	Cuts off his study and becomes an artist
1980	UFO over Delft, Flying saucer (four meters) flies over Delft (NL) and throws the town in commotion / Film about the UFO by Otakar Votochek
1986	Start of the column 'Reflectie' in the science-part of De Volkskrant (national newspaper) / Rotterdam Computerprint, fourty meters, three meters high Central Station. (NL)
1987	The Hague Drillingmachineproject. Instead of a drill there was placed a propellor in the drillingmachine. It was hanging at the cord between two high buildings.It flew so called lissajous-figures. Recorded by openlens-photos (NL)
1992	Appearence of the book 'Zogenaamd ik' (So Called Me) Bzzt h The Hague Ideas for televisionprogram 'Watt!?' VPRO about philosophy and science (NL)
1994	Columns for Metropolis-M Artmagazin (NL) / Gets the Sandbergprize (encouranging part) by the city of Amsterdam (NL)
1996	Max Renemanprize / Appearence of the book 'Klimmen in lucht' Uitg. SUN (NL)
1999	Panorama 2000, Centraal Museum Utrecht

Martin Frei

ARTBURO
265 McKibbin St.3rd Floor
Brooklyn 11206, NewYork USA
Tel: 718 628 1962, Fax: 718 628 1129
mfmontrose@aol.com

"RIVER OF RETURN"

You leave the movies to turn back towards life, but this transition occurs unnoticed while floating in the stream, the stream of time. With this return, time provides itself with a future. Free adapted by Deleuze 1999

Born: 25.2.1966 in Winterthur,Switzerland

1984-87 School of Visual Arts Zürich,
1987-90 School of Fine Arts Luzern, 1990 Bachelor of Arts,
1990-9 School of Fine Arts Luzern Film and Video-department,
since 1997 lives and works in New York and Zürich
©USA - United Swiss Artist (Martin Frei and Christoph Draeger)

selected group exhibitions

1999 Un ga nai- bad luck, Kunsthaus Zuerich (CH) Dog days are over, Centre Culturel Swiss, Paris (F) Videolounge, freie Sicht aufs Mittelmeer, Kunsthaus Zuerich (CH)
1997 Speed, 2nd Biennal (curated by Harald Szeemann) Kwangju, Southcorea
1995 Sicherheit und Zusammenarbeit, museum for visual art, Zürich (CH)
1995 Vision-Illusion-Reality,Kunsthaus Zürich - Linz - Kunsthalle Wien (A)
1995 Jedes Haus ein Kunsthaus, museum for visual art, Zürich (CH)

selected solo exhibitions

1999 Yellow and Red Institute for Contemporary Art, Dunauyvaros, Hungary
1998 Hejdji, Luciano Fasciati, Chur (CH)
1998 KOMMERZBAU, Bernhard Schindler Gallery, Bern
1997 the discovery of the marsian sea, Bernhard Schindler Gallery , Kunst 97 ZH
1996 It's the end of the world as we love it/ simulated catastrophes, Shed im Eisenwerk, Frauenfeld (CH) with the participation of Roman Signer, Claire Durlin, Pascale Wiedemann, Urs Lehmann, Ali Durt Morimoto, Bessie Nager.

David Renggli

CAR HITS BLANKET
Taken from the HiÏ series

Born: 1974 in Switzerland

Training	in the Netherlands and Germany
lives	in Paris

Roman Signer

HOT SPRING

Born: 1938 in Appenzell Switzerland

selected exhibitions

1999	Panorama 2000, Utrecht (NL) / Bonnefanten Museum, Maastricht (NL) /Art Focus,Jerusalem (Israel)
1998	Centre d' Art de l'Espace Jules Verne, solo, Brétigny sur Orge (F) / The Living Art Museum, Reykjavik (Island) / Aktion Ballons, Kulturzentrum Kunstverein, Konstanz (D)
1997	Gallery Hauser & Wirth, solo, Zürich (CH) / The Swiss Institute, solo, New York (USA) / Skulptur, Projekte in Münster (D) / Zeitskulptur Landes-galerie, Linz (A)/ Aktion Fallendes Fass, EMPA, St.Gallen (CH)
1996	Slunkariki, solo, Isafjördur (Island) / Wunschmaschine Welterfindung, Kunsthalle Wien (A)
1995	Zeichen und Wunder, Kunsthaus, Zürich (CH) / Shift, De Appel, Amsterdam (NL)/ Station Deutschland, Künstlerhaus Bethanien, Berlin (D) / Aktion Helikopter, 20-er Haus, Wien (A)
1994	Heart of Darkness, Rijksmuseum Kröller-Möller, Otterlo (NL) / Aktion mit Kajak, Steirischer Herbst, Graz (A)

ANDREA EHRAT

Grüngasse 20
8004 Zürich
Tel / Fax 0041 1 242 80 14
Mobil 0041 763850184

LA BAIGNEUSE / THE BATHER

By diving into the water all becomes blurred, the noises gets damped. There's a fizzle in the supply-pipe. In the shop downstairs flowers become irrigated.
I live here since a year, on the top of the shop. The one with the most beautiful flowers in town, they say. It's a pity I never receive one. Hardly the ears are out of the water the eyes still misty, everything sounds so shrilly.

Born: 1971 in Zürich, Switzerland
1992-93 Stage Assistant / directory and décor
1993-1998 Film and Video at the: Hochschule für Kunst und Gestaltung Zürich

Films:
Muschel 3' Essay '93 / Auf See 8' Dok '94 / Rub Through 3' Exp. '95 / Game-street 2' Music-essay '96 / Hier, sonst nirgendwo 5' Kurzspielfilm '97 / Kirschenmund 9' Exp. '98 / La chose qui est la plus belle,. 23' Exp '99

1995-97 Rub Through; Media Festival Osnabrück (D) / Viper Luzern (CH) / Impakt Festival Utrecht (CH) Up and coming Film Festival Hannover (D) Kino Aarau (CH) Kino Xenix Zürich (CH)
1996 Game-street 2; Mediaskulptur Langenthal (CH)
1998-99 Kirschenmund 9; Centre de la vieille charité Marseille (F) The new Festival N.Y. (USA) Int. Videokunstpreis Karlsruhe (D) Agliff Festival Austin (USA) Viper Luzern (CH) Up and coming Film Festival Hannover (D)

Others
1998 UPS Morphing system, installation # 4 Klinik Zürich (CH) Les Poissons morts, Décor Theater Grütli Genf (CH)
1999 Horla, Der, Décor Postführamt Berlin (D)

Brigitt Lademann

Birchlenstrasse 10
8600 Dübendorf
Tel: 0041 1 8219502 Fax: 0041 1 8203057

APHRODITE DRANK TEA IN MY KITCHEN

Some experiences of imaginative housewives
On the floor stands a basin made of black PVC, 300x200x40 cm in size and stabilized by a metal frame. The basin is filled with water containing a bubble bath up to a level of 5cm. In it, five manual agitators with correspondingly long oscillating brushes stand on metal supports of different heights. The supports and agitators are yellow. They are connected to the power supply via automatic timers and at various times produce different clusters of bubbles witch disappear and reappear. The bubbles have a subtle hint of scent about them.
This work is linked to others which may be subsumed under the term 'inconstant volumes'. But it also underlines the importance of a playful approach to everyday things, and to the world which appear new in response to how we set the zoom function in our head. Did I not drink tea with Aphrodite in the clouds above Mount Olympus or when preparing to land on Venus? With, at least, a touch of Lem and Tarkowski

Exhibitions

1992	Kunstpreis Oekologie der AEG, (D,A,CH) catalog / Projektraum Hohlstrasse, solo, Zürich (CH)
1993	X Hoch 2, Helmhaus, Zürich (CH)
1994	'Netto, Nichts als Inhalte, Museum für Gestaltung, Basel (CH), catalog / gallery Brandstetter & Wyss, solo, Zürich (CH)
1995	Studio Leonardi V -Idea, solo, Genua (I), catalog
1996	Stadtgalerie Dübendorf (CH), solo, catalog
1997	Gallery Bob Gysin, Zürich (CH) / Die elf Sinne, Museum für Gestaltung Basel in Weil (CH)
1999	Dicke Luft, Forum Vebikus Schaffhausen (CH)

1982+84	Scholarship from the city of Zürich
1992+97	Scholarship from the state of Zürich
1992	Kunstpreis Oekologie derAEG,
1994	Guest-atelier from the city of Zürich in Genoa (I)

Nadia Gyr

Bilderdijkstraat 167
1053 KP Amsterdam
Tel / Fax 0031 20 6168261
n@dia-gyr.demon.nl
www.dia-gyr.demon.nl

THE SECRETS OF DISHWATER
In a clean kitchen one cannot swim between the salad

Born: 26.9.1961 in Milwaukee America

1980-1985	Textildesign Hochschule für Kunst und Gestaltung Zürich
since 1988	Lives and works in Amsterdam

Exhibitions

1994	Stichting Achterstraat, Hoorn (NL) / Solo, Kulturamt Prenzlauer Berg (D)
1995	solo, Stichting Alarm, Beugen (NL) / An invention, in collaboration with S.Mooibroek, Arti et Amicitiae, Amsterdam (NL) / By 57 Degrees Celsius, in collaboration with S.Mooibroek, P.A.R.K. TV Amsterdam (NL)
1996	Klotsende Lucht, solo, Marmeren Hal, Amsterdam (NL) / Park Haven, W139, Amsterdam (NL) / Safe Box K.I.Safe Dalfsen and Aa- church (NL) Kunst Halle St. Moritz (CH)
1997	SCART 2, TU Berlin (D) / catalogue 1997, Amsterdam (NL)
1998	Veni Vidi Video, in collaboration with M.Verhoeckx, Forum Claque, Baden (CH) / catalogue 1998
1999	Shootingstars, Xposition, New York (USA) / The Sharjah Art Museum, Sharjah, United Arab Emirates/ catalogue 1999

Heinz Schmid

Buchenweg 25, CH-8121 Benglen, Switzerland
Tel. 0041 1 825 47 63

KELTISCHE WACKELSTEINE
CELTS

Most of my sculptures are based on precise geometrical fundamentals. I am fascinated by the elegance comprised in pure mathematics and by the possible symbolical expression of abstract forms based on this mathematics. Usually the conceptualization starts with an idea of a figure, and later I search for the geometrical fundamentals which I need to translate my imagination into a concrete form.
The bottom side of the celt is softly curved and finely polished. With a small effort it can be turned into rotation and by tapping it lightly it starts to rock. The celt as a rocking stone is behaving in unexpected and self-willed ways. For example, an initial rotation can turn into a rocking motion which can end up as a backward rotation of the celt. Stones with this properties were admired by the Celtics as magic stones. Although one can now explain the behavior of such stones, some mystery remains.

Born: 1947

Civil engineer by profession (traffic planing and road construction) I have been working as artist for about 15 years.
Apprenticeship in modelling and sculpturing with several teachers.

selected exhibitions

	Participated in various group exhibitions
1998	Exhibition in the Gallery Vision, Zürich
1999	Exhibition in the Gallery Bally, Zürich

Theo Jansen

Animaris Currens Ventosa

Martin Frei

Videostill from The river of return

David Renggli

Videostill from Car hits Blanket

Roman Signer

Hot spring
(I-III)

Andrea Ehrat

La Baigneuse / The Bather

Brigitt Lademann

Schäume, Foam of bubbles

Nadia Gyr

The Secrets of Dishwater

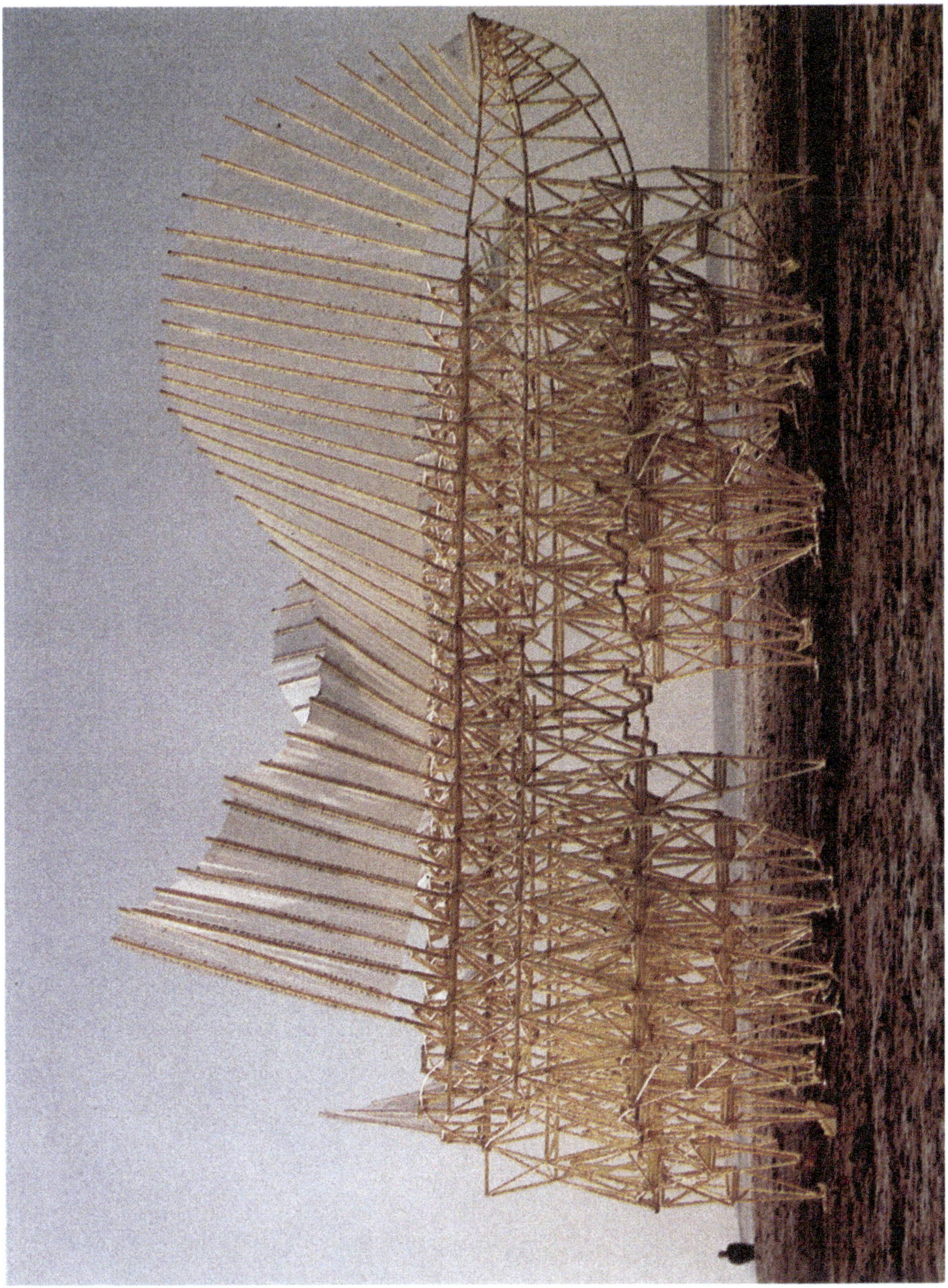

Theo Jansen

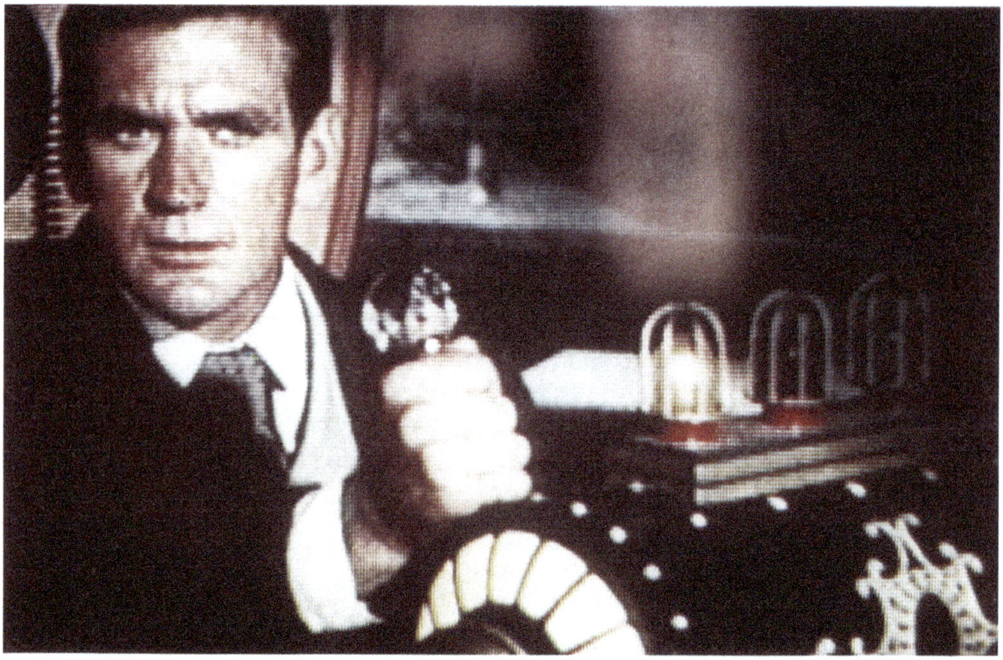

Martin Frei

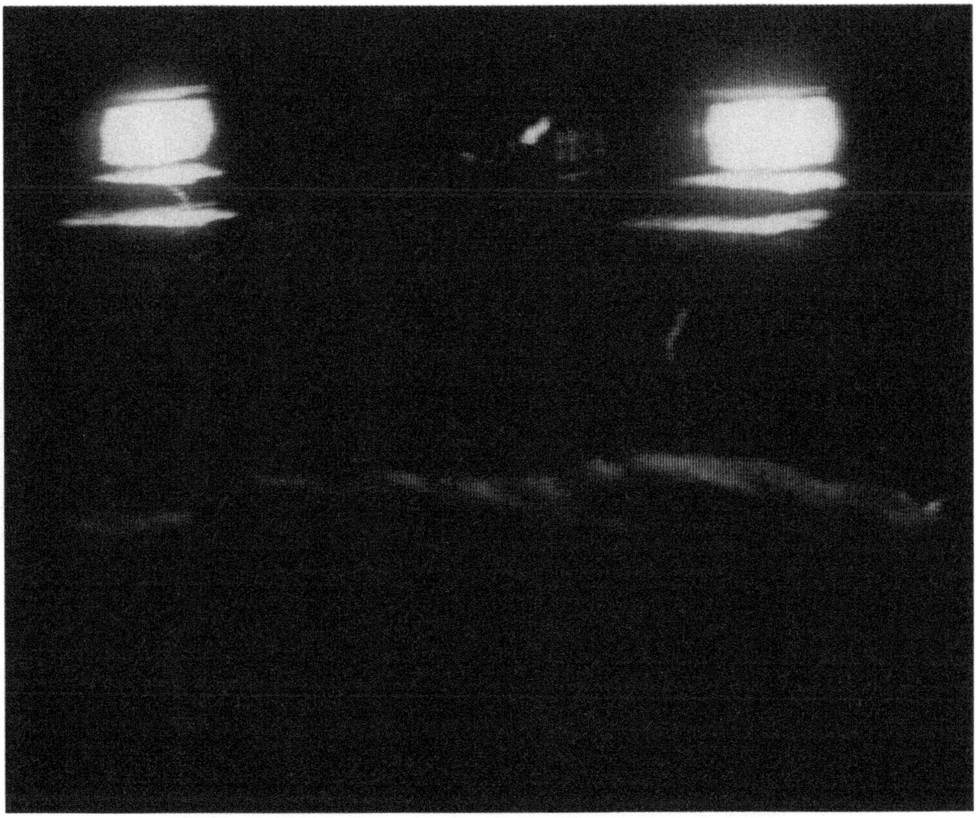

David Renggli

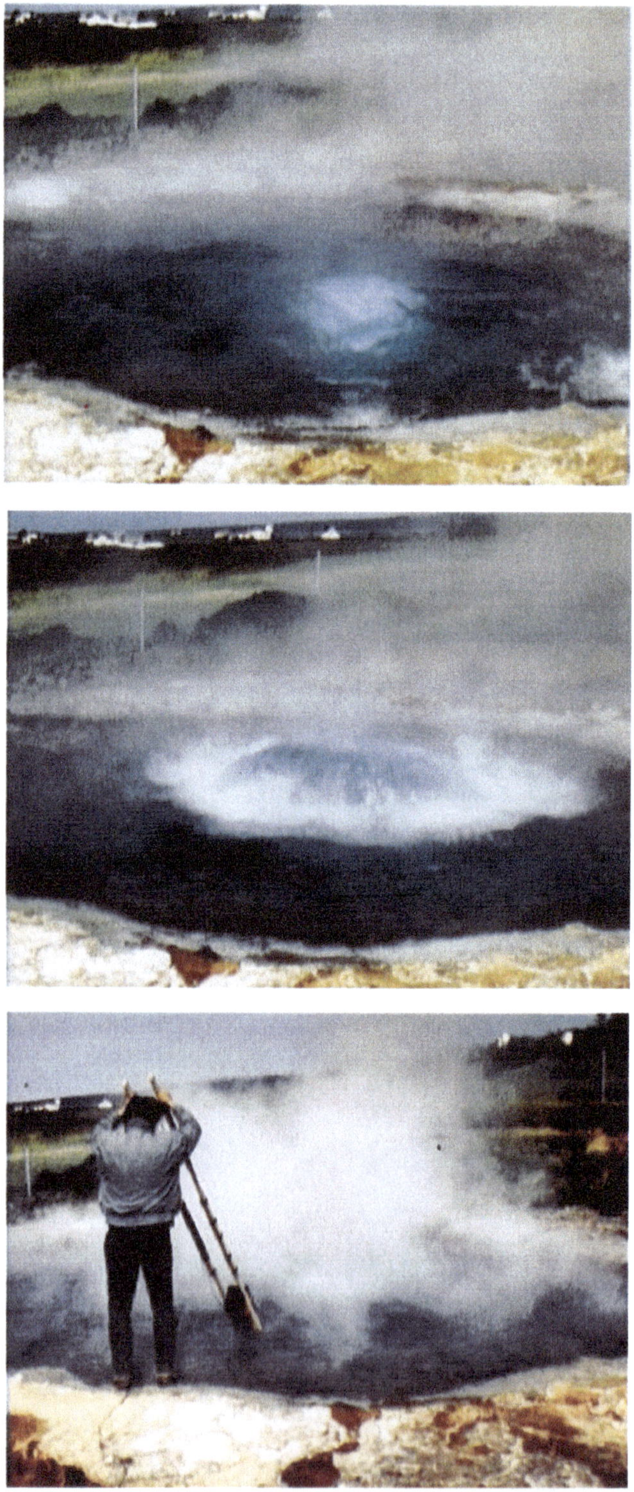

Roman Signer

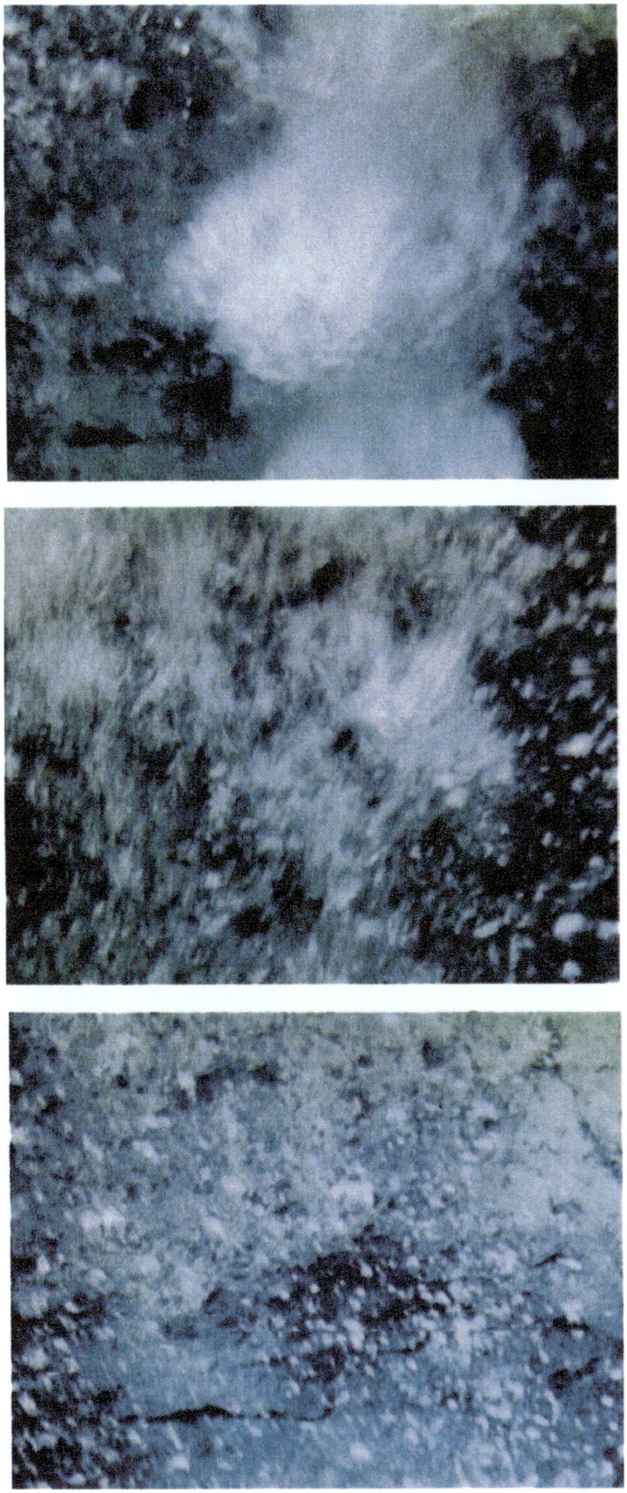

Roman Signer

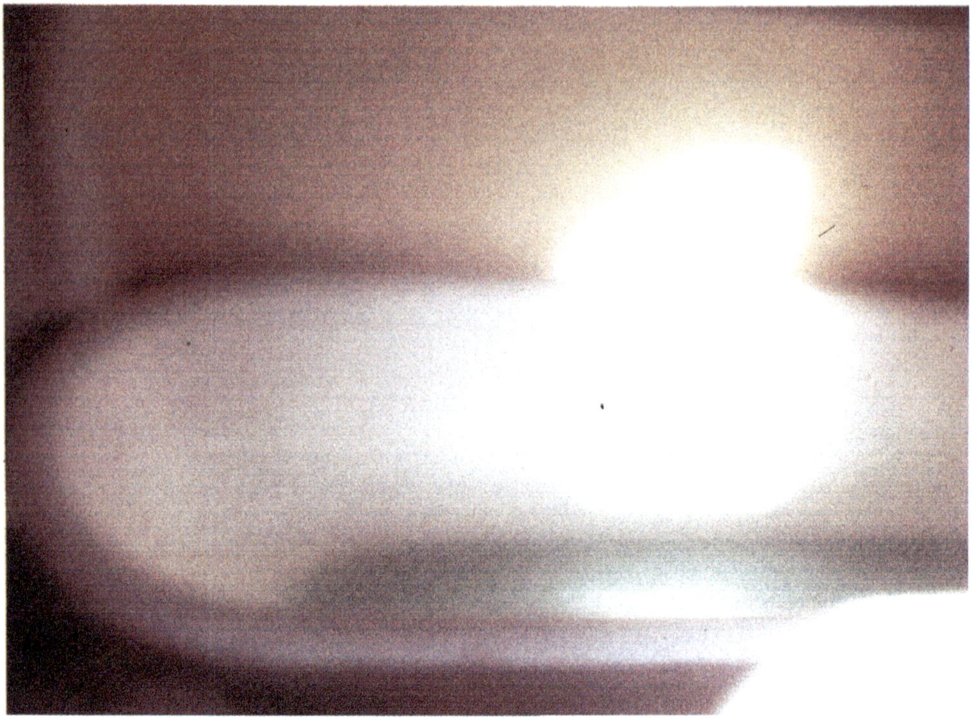

Andrea Ehrat

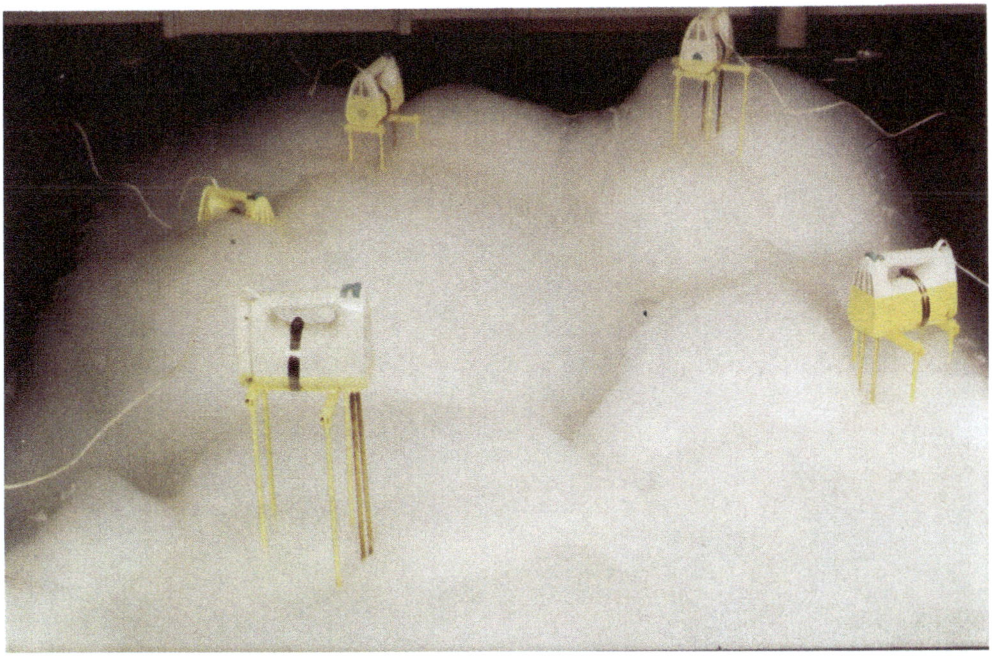

Brigitt Lademann

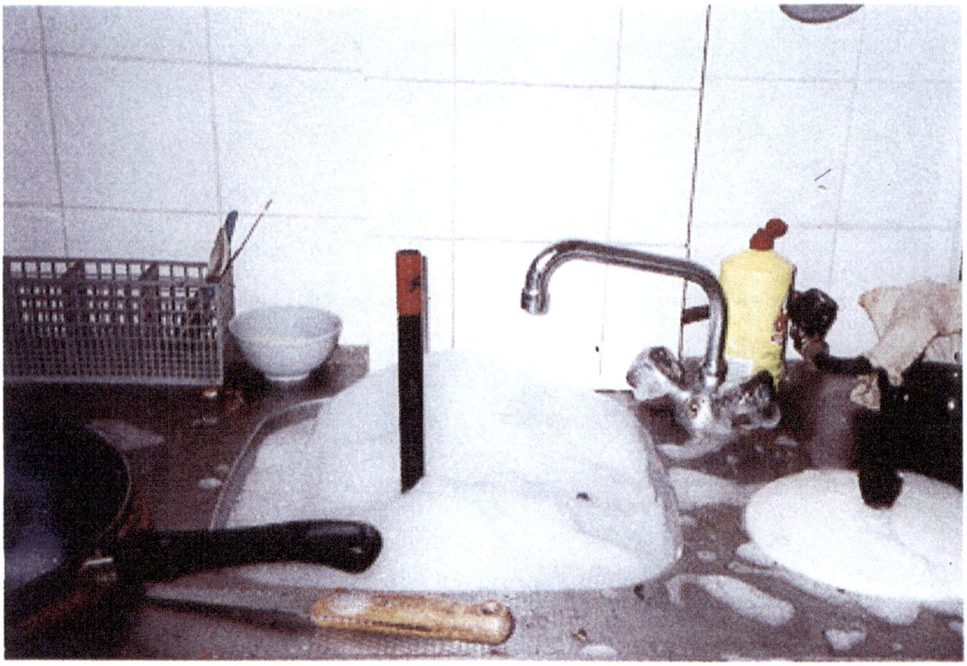

Nadia Gyr

GPSR Compliance

The European Union's (EU) General Product Safety Regulation (GPSR) is a set of rules that requires consumer products to be safe and our obligations to ensure this.

If you have any concerns about our products, you can contact us on ProductSafety@springernature.com

In case Publisher is established outside the EU, the EU authorized representative is:

Springer Nature Customer Service Center GmbH
Europaplatz 3
69115 Heidelberg, Germany

Batch number: 10755466

Printed by Printforce, the Netherlands